10 0652462 5
HAM
FROM THE LIBRARY

KU-548-383

Springer Optimization and Its Applications

VOLUME 54

Aims and Scope
Optimization has been expanding in all directions at an astonishing rate during the last few decades. New algorithmic and theoretical techniques have been developed, the diffusion into other disciplines has proceeded at a rapid pace, and our knowledge of all aspects of the field has grown even more profound. At the same time, one of the most striking trends in optimization is the constantly increasing emphasis on the interdisciplinary nature of the field. Optimization has been a basic tool in all areas of applied mathematics, engineering, medicine, economics, and other sciences.

The series *Springer Optimization and Its Applications* publishes undergraduate and graduate textbooks, monographs and state-of-the-art expository work that focus on algorithms for solving optimization problems and also study applications involving such problems. Some of the topics covered include nonlinear optimization (convex and nonconvex), network flow problems, stochastic optimization, optimal control, discrete optimization, multi-objective programming, description of software packages, approximation techniques and heuristic approaches.

For further volumes:
http://www.springer.com/series/7393

Pavel S. Knopov • Arnold S. Korkhin

Regression Analysis Under A Priori Parameter Restrictions

Springer

Pavel S. Knopov
Department of Mathematical Methods
of Operation Research
V.M. Glushkov Institute of Cybernetics
National Academy of Science of Ukraine
03187 Kiev
Ukraine
knopov1@yahoo.com

Arnold S. Korkhin
Department of Economical Cybernetics
and Information Technology
National Mining University
49005 Dnepropetrovsk
Ukraine
korkhina@nmu.org.ua

ISSN 1931-6828
ISBN 978-1-4614-0573-3 e-ISBN 978-1-4614-0574-0
DOI 10.1007/978-1-4614-0574-0
Springer New York Dordrecht Heidelberg London

Library of Congress Control Number: 2011935145

Printed on acid-free paper

Springer is part of Springer Science+Business Media (www.springer.com)

Preface

Regression analysis has quite a long history. It is conventional to think that it goes back to the works of Gauss on approximation of experimental data. Nowadays, regression analysis represents a separate scientific branch, which is based on optimization theory and mathematical statistics. Formally, there exist two branches of regression analysis: theoretical and applied.

Up to recent time, developments in regression analysis were based on the hypothesis that the domain of regression parameters has no restrictions. Divergence from that approach came later on when equality constraints were taken into account, which allowed use of some a priori information about the regression model. Methods of constructing the regression with equality constraints were first investigated in Rao (1965) and Bard (1974).

Usage of inequality constraints in a regression model gives much more possibilities to utilize available a priori information. Moreover, the representation of the admissible domain of parameters in the form of inequality constraints naturally includes the cases when constraints are given as equalities.

Properties of the regression with inequality constraints are investigated in many papers, in particular, in Zellner (1971), Liew (1976), Nagaraj and Fuller (1991) and Thomson and Schmidt (1982), where some particular cases are considered. Detailed qualitative analysis of the properties of estimates in case of linear regression with linear constraints is given in the monograph (Malinvaud 1969, Section 9.8).

Asymptotic properties of the estimates of regression parameters in regression with finite number of parameters under some known a priori information are studied in Dupacova and Wets (1986), Knopov (1997a–c), Korkhin (1985), Wang (1996), etc. We note that the results obtained in Korkhin (1985) and Wang (1996) under different initial assumptions, almost coincide. There are many results concerning practical implementation of regression models with inequality constraints, for example, Liew (1976), Rezk (1996) and McDonald (1999), Thomson (1982), Thomson and Schmidt (1982). This problem was also studied in Gross (2003, Subsection 3.3.2).

In this monograph, we present in full detail the results on estimation of unknown parameters in regression models under a priori information, described in the form of inequality constraints. The book covers the problem of estimation of regression parameters as well as the problem of accuracy of such estimation. Both problems are studied is cases of linear and nonlinear regressions. Moreover, we investigate the applicability of regression with constraints to problems of point and interval prediction.

The book is organized as follows.

In Chapter 1, we consider methods of calculation of parameter estimates in linear and nonlinear regression with constraints. In this chapter we describe methods of solving optimization problems which take into account the specification of regression analysis.

Chapter 2 is devoted to asymptotic properties of regression parameters estimates in linear and nonlinear regression. Both cases of equality and inequality constraints are considered.

In Chapter 3, we consider various generalizations of the estimation problem by the least squares method in nonlinear regression with inequality constraints on parameters. In particular, we discuss the results concerning robust Huber estimates and regressors which are continuous functions of time.

Chapter 4 is devoted to the problem of accuracy estimation in (linear and nonlinear) regression, when parameters are estimated by means of the least squares method.

In Chapter 5, we discuss/consider statistical properties of estimates of parameters in nonlinear regression, which are obtained on each iteration of the solution to the estimation problem. Here we use algorithms described in Chap. 1. Obtained results might be useful in practical implementation of regression analysis.

Chapter 6 is devoted to problems of prediction by linear regression with linear constraints.

Kiev, Ukraine
Dnepropetrovsk, Ukraine

Pavel S. Knopov
Arnold S. Korkhin

Acknowledgments

We are very grateful to the scientific editor of this book, Professor Panos Pardalos, senior publishing editor, Elizabeth Loew, and to the associate editor in mathematics, Nathan Brothers, for their helpful support and collaboration in preparation of the manuscript.

We thank our colleagues from V.M. Glushkov Institute of Cybernetics of National Academy of Science of Ukraine for many helpful discussions on the problems and results described and presented in this book.

We thank our colleagues L. Belyavina, L. Vovk, V. Knopova, Yu. Kolesnik, E. Odinzova, for invaluable help during the preparation of our book for publication.

Contents

Notation

m.s.e.	Mean square error
ECLS estimate	Estimate of the regression parameter my means of the least squares method with equality constraints
$\lvert I \rvert$	Cardinality of the set I
ICLS estimate	Estimate of the regression parameter my means of the least squares method with inequality constraints
$\mathbf{J}_n$	Unit matrix of order n
LS	Least squares method
LS estimate	Estimate of the regression parameter my means of the least squares method without restrictions
$\mathbf{M}'$	Transposition of a matrix (vector) $\mathbf{M}$
$\mathbf{O}_{mn}$	Zero $(m \times n)$ matrix
$\mathbf{O}_n$	Zero n-dimensional vector
$p \lim$	Means convergence in probability
$\mathbf{1}_n$	$n-$ dimensional vector with entries equal to 1
$\lvert\lvert \cdot \rvert\rvert$	Euclidean norm of a vector (matrix)
$\overset{p}{\Rightarrow}$	Convergence in distribution
$\boldsymbol{\varepsilon} \sim N(\mathbf{M}_1, \mathbf{M}_2)$	$\boldsymbol{\varepsilon}$ has a normal distribution with mean $\mathbf{M}_1$ and covariance $\mathbf{M}_2$

Chapter 1
Estimation of Regression Model Parameters with Specific Constraints

Consider the regression

$$y_t = \tilde{f}(\mathbf{x}_t, \boldsymbol{\alpha}^0) + \varepsilon_t, \quad t = 1, 2, \ldots, \tag{1.1}$$

where $y_t \in \Re^1$ is the dependent variable, $\mathbf{x}_t \in \Re^q$ is an argument (regressor), $\boldsymbol{\alpha}^0 \in \Re^n$ is a true regression parameter (unknown), $\tilde{f}(\mathbf{x}_t, \boldsymbol{\alpha})$ is some (nonlinear) function of $\boldsymbol{\alpha}$, ε_t is a noise, and t is an observation number.

In what follows the symbol "$'$" denotes the transposition.

We will use the function $\tilde{f}(\mathbf{x}_t, \boldsymbol{\alpha})$, where $\boldsymbol{\alpha} \in \Re^n$ is a dependent variable, for estimation of $\boldsymbol{\alpha}^0$ and for investigation of the obtained estimates.

For convenience we write

$$f_t(\boldsymbol{\alpha}) = \tilde{f}(\mathbf{x}_t, \boldsymbol{\alpha}), \quad t = 1, 2, \ldots \tag{1.2}$$

and call such a function the regression function.

Assume that a priori parameter constraints are known:

$$g_i(\boldsymbol{\alpha}^0) \le 0, \quad i = \overline{1, m}. \tag{1.3}$$

System of inequalities (1.3) involves equalities as a particular case due to the fact that any equality can be represented in the form of two inequalities:

$$g_i(\boldsymbol{\alpha}^0) \le 0 \quad \text{and} \quad - g_i(\boldsymbol{\alpha}^0) \le 0.$$

Suppose that for $t \in [1, T]$ the values of y_t and $\mathbf{x}_t \in \Re^n$ are known. In the present chapter the estimation of the parameter $\boldsymbol{\alpha}^0$ will be done by means of the least squares method, i.e.

$$S(\boldsymbol{\alpha}) = \frac{1}{2} \sum_{t=1}^{T} (y_t - f_t(\boldsymbol{\alpha}))^2 \to \min, \tag{1.4}$$

P.S. Knopov and A.S. Korkhin, *Regression Analysis Under A Priori Parameter Restrictions*, Springer Optimization and Its Applications 54,
DOI 10.1007/978-1-4614-0574-0_1,

under the constraints

$$g_i(\alpha) \leq 0, \quad i = \overline{1,m}, \tag{1.5}$$

where T is the length of the observed dynamic (time) series $\mathbf{x}_t$ and y_t.

Since the case of the linear regression and linear constraints on $\boldsymbol{\alpha}$ is extremely important and is used for nonlinear estimation algorithms, it will be discussed separately in Sect. 1.1.

Section 1.2 is dedicated to nonlinear estimation, i.e., to solving the problems (1.4) and (1.5) under rather general setting. Section 1.3 is dedicated to the perspective for economical applications in the case when the multivariate linear regression parameter with nonlinear equality constraints is analysed.

1.1 Estimation of the Parameters of a Linear Regression with Inequality Constraints

Assume that in (1.2) $f_t(\alpha) = \tilde{f}_t(\mathbf{x}_t, \alpha) = \mathbf{x}_t'\alpha, t = 1, 2, \ldots$ and take in (1.5) $g_i(\alpha) = \mathbf{g}_i'\alpha, i = \overline{1,m}$, where $\mathbf{g}_i \in \Re^m$, $i = \overline{1,m}$ are known vectors. Then the estimation problems (1.4) and (1.5) can be written in the following form:

$$S(\alpha) = \frac{1}{2}\sum_{t=1}^{T}(y_t - \mathbf{x}_t'\alpha)^2, \quad g_i(\alpha) = \mathbf{g}_i'\alpha - b_i \leq 0, \quad i = \overline{1,m} \tag{1.6}$$

or

$$\frac{1}{2}||\mathbf{Y} - \mathbf{X}\alpha||^2 \rightarrow \min, \quad \mathbf{G}\alpha \leq \mathbf{b}, \tag{1.7}$$

where $\mathbf{Y} = [y_1 \quad y_2 \ \ldots \ y_T]'$; $\mathbf{X}$ is some $(T \times n)$ matrix. The rows of this matrix are the vector rows $\mathbf{x}_t'$, $t = \overline{1,T}$; $\mathbf{G}$ is an $(m \times n)$ matrix with rows $\mathbf{g}_i'$, $i = \overline{1,m}$; $\mathbf{b} = [b_1 \quad b_2 \ \ldots \ b_m]'$.

We pose some additional assumptions on the regressor and the constraints, which will be used later on.

Assumption 1.1. Matrix $\mathbf{X}$ in (1.7) is of full rank.

Assumption 1.2. Matrix $\mathbf{G}$ in (1.7) is of full rank.

1.1.1 Method of Estimating the Solution to (1.7)

Taking into account the fact that the rank of $\mathbf{X}$ is equal to n (Assumption 1.1), we obtain its orthogonal expansion $\mathbf{X} = \mathbf{M}_1 \begin{bmatrix} \mathbf{M}_2 \\ \mathbf{O}_{T-n,n} \end{bmatrix} \mathbf{M}_3'$, $\mathbf{M}_1 = [\mathbf{M}_{11} \quad \mathbf{M}_{12}]$, where

$\mathbf{M}_1$ is an orthogonal $T \times T$ matrix, $T \times n$ is the dimension of the submatrix $\mathbf{M}_{11}$, $\mathbf{M}_2$ is a non-degenerate $(n \times n)$ matrix, $\mathbf{M}_3$ is an orthogonal $(n \times n)$ matrix.

Put $\mathbf{x} = \mathbf{M}_2\mathbf{M}_3^{-1}\boldsymbol{\alpha} - \mathbf{M}'_{11}\mathbf{Y}$. From the orthogonal decomposition of the matrix $\mathbf{X}$ and the properties of orthogonal matrixes mentioned above we obtain the following: for the cost function in (1.7),

$$\|\mathbf{Y} - \mathbf{X}\boldsymbol{\alpha}\|^2 = \left\| \mathbf{Y} - \mathbf{M}_1 \begin{bmatrix} \mathbf{J}_n \\ \mathbf{O}_{T-n,n} \end{bmatrix} (\mathbf{x} + \mathbf{M}'_{11}\mathbf{Y}) \right\|^2$$

$$= \mathbf{M}_1 \left\| \mathbf{M}'_1\mathbf{Y} - \begin{bmatrix} \mathbf{x} \\ \mathbf{O}_{T-n} \end{bmatrix} - \begin{bmatrix} \mathbf{M}'_{11}\mathbf{Y} \\ \mathbf{O}_{T-n} \end{bmatrix} \right\|^2$$

$$= \left\| \begin{bmatrix} \mathbf{O}_n \\ \mathbf{M}'_{12}\mathbf{Y} \end{bmatrix} - \begin{bmatrix} \mathbf{x} \\ \mathbf{O}_{T-n} \end{bmatrix} \right\|^2 = ||\mathbf{M}'_{12}\mathbf{Y}||^2 + ||\mathbf{x}||^2,$$

while for the constraints in (1.7)

$$\mathbf{N}_1\mathbf{x} \le \mathbf{N}_2$$

holds true, where $\mathbf{N}_1 = \mathbf{G}\mathbf{M}_3\mathbf{M}_2^{-1}$, $\mathbf{N}_2 = \mathbf{b} - \mathbf{G}\mathbf{M}_3\mathbf{M}_2^{-1}\mathbf{M}'_{11}\mathbf{y}$.

Getting rid of the term independent of $\mathbf{x}$, we obtain the transformed problem (1.7):

$$\frac{1}{2}||\mathbf{x}||^2 \to \min, \quad \mathbf{N}_1\mathbf{x} \le \mathbf{N}_2. \tag{1.8}$$

This problem has a solution (as well as problem (1.7)) if the constraints are consistent.

Consider the following minimization problem (Lawson and Hanson 1974, Chapter 23 §5),

$$P(\mathbf{U}) = \frac{1}{2}||\mathbf{N}\mathbf{U} - \boldsymbol{\Phi}||^2 \to \min, \quad \mathbf{U} \ge \mathbf{O}_m, \tag{1.9}$$

where $\mathbf{U} \in \Re^m, \mathbf{N} = [\mathbf{N}_1 \;\vdots\; \mathbf{N}_2]', \boldsymbol{\Phi}' = [\mathbf{O}'_n \;\vdots\; 1]$.

Unlike (1.8), (1.9) always has a solution. In order to establish the connection between the problems (1.9) and (1.8) we introduce the following notation: $\hat{\mathbf{U}}$ is the solution to (1.9), $\mathbf{r} = \mathbf{N}\hat{\mathbf{U}} - \boldsymbol{\Phi}$.

The necessary and sufficient conditions for the existence of the minimum in (1.9) are:

$$\mathbf{N}'(\mathbf{N}\hat{\mathbf{U}} - \boldsymbol{\Phi}) + \boldsymbol{\Lambda} = \mathbf{O}_m, \quad \boldsymbol{\Lambda} \ge \mathbf{O}_m, \hat{\mathbf{U}}'\boldsymbol{\Lambda} = 0. \tag{1.10}$$

Hence, we obtain

$$\mathbf{N}'(\mathbf{N}\hat{\mathbf{U}} - \boldsymbol{\Phi}) \le \mathbf{O}_m \tag{1.11}$$

and

$$\hat{\mathbf{U}}'\mathbf{N}'(\mathbf{N}\hat{\mathbf{U}} - \boldsymbol{\Phi}) = \mathbf{O}_m. \tag{1.12}$$

By arguments similar to those given in Lawson and Hanson (1974, Chapter 23 §4), we have $||\mathbf{r}||^2 = \mathbf{r}'\mathbf{r} = \hat{\mathbf{U}}'\mathbf{N}'(\mathbf{N}\hat{\mathbf{U}} - \boldsymbol{\Phi}) - r_{n+1}$, where r_{n+1} is $(n+1)$th component of $\mathbf{r}$. Using this equality and (1.12) we obtain $||\mathbf{r}||^2 = -r_{n+1} \geq 0$.

Suppose that $||\mathbf{r}|| > 0$, and assume that $\hat{\mathbf{x}} = -r_{n+1}\mathbf{N}_1'\hat{\mathbf{U}}$. Then

$$\mathbf{N}'(\mathbf{N}\hat{\mathbf{U}} - \boldsymbol{\Phi}) = \begin{bmatrix}\mathbf{N}_1 & \vdots & \mathbf{N}_2\end{bmatrix}\begin{bmatrix}\mathbf{N}_1'\hat{\mathbf{U}} \\ r_{n+1}\end{bmatrix} = \begin{bmatrix}\mathbf{N}_1 \vdots \mathbf{N}_2\end{bmatrix}\begin{bmatrix}\hat{\mathbf{x}} \\ -1\end{bmatrix}(-r_{n+1})$$
$$= (\mathbf{N}_1\hat{\mathbf{x}} - \mathbf{N}_2)||\mathbf{r}||^2 \leq \mathbf{O}_m, \tag{1.13}$$

which implies $\mathbf{N}_1\hat{\mathbf{x}} \leq \mathbf{N}_2$. We also would like to mention that if $||\mathbf{r}|| = 0$ the constraints in (1.8) are not consistent, see Lawson and Hanson (1974, Chapter 23 §4).

Now we can demonstrate that $\hat{\mathbf{x}}$ is the solution to (1.8).

Theorem 1.1. *If the constraints in (1.8) are consistent, then the solution is given by* $\mathbf{x} = \hat{\mathbf{x}} = ||\mathbf{r}||^{-2}\mathbf{N}_1'\hat{\mathbf{U}}$, *where* $\hat{\mathbf{U}}$ *is the solution to (1.9).*

Proof. The necessary and sufficient conditions for the existence of the minimum in (1.8) are:

$$\mathbf{x} + \mathbf{N}_1'\boldsymbol{\lambda} = \mathbf{O}_n, \quad \boldsymbol{\lambda} \geq O_m, \quad \boldsymbol{\lambda}_i'(\mathbf{N}_{1i}\mathbf{x} - \mathbf{N}_{2i}) = 0, \quad i = 1, \ldots, m, \tag{1.14}$$

where $\mathbf{N}_{1i}$ is the ith row of the matrix $\mathbf{N}_1$, $\mathbf{N}_{2i}$ is the ith component of the vector $\mathbf{N}_2$, $\boldsymbol{\lambda} \in \Re^m$ is the Lagrange multiplier, and $\boldsymbol{\lambda}_i$ denotes the ith component of $\boldsymbol{\lambda}$.

Substituting in (1.14) $\mathbf{x} = \hat{\mathbf{x}} = ||\mathbf{r}||^{-2}\mathbf{N}_1'\hat{\mathbf{U}}$, we obtain

$$\boldsymbol{\lambda} = ||\mathbf{r}||^{-2}\hat{\mathbf{U}} \geq \mathbf{O}_m. \tag{1.15}$$

Next we show that $\boldsymbol{\lambda}$ also satisfies the third condition in (1.14). From (1.10), (1.11), and (1.15) we derive

$$\hat{\mathbf{U}}'\boldsymbol{\Lambda} = 0 = \hat{\mathbf{U}}'\mathbf{N}'(\mathbf{N}\hat{\mathbf{U}} - \boldsymbol{\Phi}) = \hat{\mathbf{U}}'(\mathbf{N}_1\hat{\mathbf{x}} - \mathbf{N}_2)||\mathbf{r}||^2 = \boldsymbol{\lambda}'(\mathbf{N}_1\hat{\mathbf{x}} - \mathbf{N}_2).$$

Taking into account that $\boldsymbol{\lambda} \geq \mathbf{O}_m$, and according to (1.13) $\mathbf{N}_1\hat{\mathbf{x}} - \mathbf{N}_2 \leq \mathbf{O}_m$, we obtain from the latter equation the third condition in (1.14). Then the pair $(\hat{\mathbf{x}}, \boldsymbol{\lambda})$ satisfies the necessary and sufficient conditions for existence of the minimum in (1.8). Therefore, $\hat{\mathbf{x}}$ is the solution to (1.8). Theorem is proved. □

Thus, the solution to the problem (1.9) allows us to answer two questions: to determine the compatibility of the constraints in (1.8) (and, consequently, in (1.7)), and in case of compatibility to obtain the solution $\hat{\boldsymbol{\alpha}}$ by means of relatively easy transformation of the solution to (1.9). Namely,

$$\hat{\boldsymbol{\alpha}} = \mathbf{M}_3\mathbf{M}_2^{-1}(||\mathbf{N}\hat{\mathbf{U}} - \boldsymbol{\Phi}||^{-2}\mathbf{N}_1'\hat{\mathbf{U}} + \mathbf{M}_{11}'\mathbf{y}).$$

Corollary 1.1. *If Assumption 1.1 holds true and the problem (1.7) has a solution, then the related vector of Lagrange multipliers is given by* $\boldsymbol{\lambda} = \hat{\mathbf{U}}||\mathbf{N}\hat{\mathbf{U}} - \boldsymbol{\Phi}||^{-2}$.

Proof. The necessary and sufficient conditions for the existence of the minimum to (1.7) are:

$$\mathbf{X}'\mathbf{X}\boldsymbol{\alpha} - \mathbf{X}'\mathbf{Y} + \mathbf{G}\overline{\boldsymbol{\lambda}} = \mathbf{O}_n, \quad \overline{\boldsymbol{\lambda}}'(\mathbf{G}\boldsymbol{\alpha} - \mathbf{b}) = 0, \overline{\boldsymbol{\lambda}} \geq \mathbf{O}_m.$$

From above, using the orthogonal transformation $\mathbf{X}$, we obtain

$$\mathbf{x} + \mathbf{N}_1'\overline{\boldsymbol{\lambda}} = \mathbf{O}_n, \quad \overline{\boldsymbol{\lambda}} \geq O_m, \qquad \overline{\boldsymbol{\lambda}}_i'(\mathbf{N}_{1i}\mathbf{x} - \mathbf{N}_{2i}) = 0, \quad i = 1, \ldots, m,$$

where $\overline{\boldsymbol{\lambda}}_i$ is the ith component of $\overline{\boldsymbol{\lambda}}$.

We see that these relations are satisfied when $\mathbf{x} = \hat{\mathbf{x}}$, $\overline{\boldsymbol{\lambda}} = \boldsymbol{\lambda}$, compared with (1.14). However, the pair of vectors $(\hat{\mathbf{x}}, \overline{\boldsymbol{\lambda}})$ is unique due to uniqueness implied by Assumption 1.1. On the other hand, as it was shown in the proof of Theorem 1.1, $\boldsymbol{\lambda}$ is given by (1.15). Hence the corollary follows. □

1.1.2 *Algorithm of Finding the Solution to (1.9)*

Assume that Assumptions 1.1 and 1.2 are satisfied. According to Lawson and Hanson (1974, Chapter 23 §3), we can proceed as follows.

Step 1 Let $\mathrm{P} = \varnothing$, $\Im = \{1, 2, \ldots, m\}$, $\mathbf{U} := \mathbf{O}_m$.
Step 2 Calculate the vector $\mathbf{w} = \mathbf{N}'(\boldsymbol{\Phi} - \mathbf{N}\mathbf{U}) \in \Re^m$.
Step 3 If the set $\Im$ is empty or $w_j \leq 0$ for all $j \in \Im$, go to Step 12. Here w_j is the jth component of $\mathbf{w}$.
Step 4 Find the index $i \in \Im$ such that $w_i = \max(w_j, j \in \Im)$.
Step 5 Move the index i from the set $\Im$ to the set P.
Step 6 Denote by $\mathbf{N}_\mathrm{P}$ the $((n+1) \times m_\mathrm{P})$-matrix, whose jth column is jth column of matrix $\mathbf{N}$, if $j \in \mathrm{P}$, $j = \overline{1, m}$.
Here m_P is the number of columns in the matrix $\mathbf{N}_\mathrm{P}$.
If $n + 1 \geq m$, then calculate the vector $\mathbf{z}_\mathrm{P} = (\mathbf{N}_\mathrm{P}'\mathbf{N}_\mathrm{P})^{-1}\mathbf{N}_\mathrm{P}'\boldsymbol{\Phi} \in \Re^{m_\mathrm{P}}$.
If $n + 1 < m$, then find $\mathbf{z}_\mathrm{P}$ with a minimal norm: $\mathbf{z}_\mathrm{P} = \mathbf{N}_\mathrm{P}'(\mathbf{N}_\mathrm{P}\mathbf{N}_\mathrm{P}')^{-1}\boldsymbol{\Phi} \in \Re^{m_\mathrm{P}}$.
Denote the components $\mathbf{z}_\mathrm{P}$ by z_j, $j \in \mathrm{P}$.
Put $z_j = 0$, $j \in \Im$.
Form a vector $\mathbf{z} = [z_j]$, $j = \overline{1, m}$.
Note that if $n+1 < m$, then the first $m_\mathrm{P} - 1$ components of $\mathbf{z}_\mathrm{P}$ are zero, the m_P of the component is equal to the element $(m_\mathrm{P}, (n+1))$ of the matrix $\mathbf{N}_\mathrm{P}'(\mathbf{N}_\mathrm{P}\mathbf{N}_\mathrm{P}')^{-1}$ provided that $n + 1 < m$.
Step 7 If $z_j > 0$ for all $j \in \mathrm{P}$, then put $\mathbf{U} := \mathbf{z}$ and go to Step 2.
Step 8 Find the index $k \in \mathrm{P}$ such that

$$\frac{U_k}{U_k - z_k} = \min\left\{\frac{U_j}{U_j - z_j} : z_j \leq 0, j \in \mathrm{P}\right\}.$$

Step 9 Put $\gamma := U_k / U_k - z_k$.
Step 10 Put $\mathbf{U} := \mathbf{U} + \gamma(\mathbf{z} - \mathbf{U})$.
Step 11 Move all indices $j \in \mathrm{P}$ for which $U_j = 0$ from the set P to the set $\Im$. Then go to Step 6.
Step 12 Stop. The solution $\hat{\mathbf{U}} = \mathbf{U}$ is obtained.

1.1.3 Special Case of the Problem (1.7)

Usually a regression has a free term on which the constraints are often not imposed. Let us show that in the case when $\mathbf{x}_t = [1 \quad \tilde{\mathbf{x}}_t']'$, $\tilde{\mathbf{x}}_t \in \Re^{n-1}$, the solution of the estimation problem can be simplified by reducing the number of variables to one. The theorem presented below takes place.

Theorem 1.2. *If Assumption 1.1 holds true and no constraints are imposed on the free term* α_1, *then the solution to the problem (1.7) is of the form* $\hat{\boldsymbol{\alpha}} = [\hat{\alpha}_1 \quad \hat{\tilde{\boldsymbol{\alpha}}}']'$, *where* $\hat{\alpha}_1 = \overline{y} - \hat{\tilde{\boldsymbol{\alpha}}}'\overline{\tilde{\mathbf{x}}}$, $\hat{\tilde{\boldsymbol{\alpha}}} \in \Re^{n-1}$ *is the solution to*

$$\frac{1}{2}\tilde{\boldsymbol{\alpha}}'\mathbf{r}\tilde{\boldsymbol{\alpha}} - \tilde{\boldsymbol{\alpha}}'\mathbf{d} \rightarrow \min, \quad \mathbf{A}\tilde{\boldsymbol{\alpha}} \leq \mathbf{b}. \tag{1.16}$$

Here $\tilde{\boldsymbol{\alpha}} \in \Re^{n-1}$, $\mathbf{r} = \sum_{t=1}^{T} (\tilde{\mathbf{x}}_t - \overline{\tilde{\mathbf{x}}})(\tilde{\mathbf{x}}_t - \overline{\tilde{\mathbf{x}}})'$, $\mathbf{d} = \sum_{t=1}^{v} (\tilde{\mathbf{x}}_t - \overline{\tilde{\mathbf{x}}})(y_t - \overline{y})$, $\overline{\tilde{\mathbf{x}}} = \sum_{t=1}^{T} \tilde{\mathbf{x}}_t / T$, $\overline{y} = \sum_{t=1}^{T} y_t / T$, *and* $\mathbf{A}$ *is the* $m \times (n-1)$ *matrix composed of* $n - 1$ *last columns of the matrix* $\mathbf{G}$.

Proof. We write the Lagrange function for the minimization problem (1.7) in the form $L(\boldsymbol{\alpha}, \boldsymbol{\lambda}) = \frac{1}{2}\boldsymbol{\alpha}'\mathbf{R}\boldsymbol{\alpha} - \boldsymbol{\alpha}'\mathbf{X}'\mathbf{Y} + \boldsymbol{\lambda}'(\mathbf{G}\boldsymbol{\alpha} - \mathbf{b})$, where $\boldsymbol{\lambda}$ is the m-dimensional vector of Lagrange multipliers.

According to Assumption 1.1, the necessary and sufficient conditions for the existence of the minimum in (1.7) are of the form

$$\nabla_{\alpha} L(\boldsymbol{\alpha}, \boldsymbol{\lambda}) = \mathbf{R}\boldsymbol{\alpha} - \mathbf{X}'\mathbf{Y} + \mathbf{G}'\boldsymbol{\lambda} = \mathbf{O}_n, \tag{1.17}$$

$$\lambda_i (\mathbf{g}_i'\boldsymbol{\alpha} - b_i) = 0, \quad \lambda_i \geq 0, \ i = \overline{1, m}, \tag{1.18}$$

where $\nabla_{\alpha} L(\boldsymbol{\alpha}, \boldsymbol{\lambda})$ is the gradient of the Lagrange function along the vector $\boldsymbol{\alpha}$, and λ_i is ith component of $\boldsymbol{\lambda}$.

Since no constraints are imposed on the free term α_1, the matrix $\mathbf{G}$ and its ith row $\mathbf{g}'_i$ are of the form

$$\mathbf{G} = \begin{bmatrix} \mathbf{O}_m & \vdots & \mathbf{A} \end{bmatrix}, \quad \mathbf{g}'_i = [0 \ \mathbf{A}_i], \tag{1.19}$$

where $\mathbf{A}_i$ is the ith row of the matrix $\mathbf{A}$. Then we have

$$\mathbf{G}'\boldsymbol{\lambda} = \begin{bmatrix} \mathbf{O}'_m \\ \mathbf{A}'\boldsymbol{\lambda} \end{bmatrix}. \tag{1.20}$$

Let us consider the condition (1.17). We consider the first of these equations, which by (1.20) can be rewritten as

$$\frac{\partial L(\boldsymbol{\alpha}, \boldsymbol{\lambda})}{\partial \alpha_1} = \alpha_1 T + \alpha_2 \sum_{t=1}^{T} x_{t1} + \alpha_2 \sum_{t=1}^{T} x_{t2} + \cdots + \alpha_n \sum_{t=1}^{T} x_{t,n-1} - \sum_{t=1}^{T} y_t = 0.$$

Dividing both sides of the above equation by the number of observations T, we obtain

$$\alpha_1 + \alpha_2 \overline{x}_1 + \alpha_3 \overline{x}_2 + \cdots + \alpha_n \overline{x}_{n-1} - \overline{y} = 0, \tag{1.21}$$

where $\overline{\mathbf{x}}_i = \sum_{t=1}^{T} x_{ti}/T, i = \overline{1, n-1}$, is the ith component of $\overline{\tilde{\mathbf{x}}}$.

Equation (1.21) must be satisfied for the estimates of the parameters, i.e.,

$$\hat{\alpha}_1 = \overline{y} - \hat{\alpha}_2 \overline{x}_1 - \hat{\alpha}_3 \overline{x}_2 - \cdots - \hat{\alpha}_n \overline{x}_{n-1} = \overline{y} - \hat{\tilde{\alpha}}' \overline{\tilde{\mathbf{x}}}, \tag{1.22}$$

which proves the first statement of the theorem.

Consider the ith equation ($i = \overline{2, n}$) in the system of equations (1.17),

$$\begin{aligned} \frac{\partial L(\boldsymbol{\alpha}, \boldsymbol{\lambda})}{\partial \alpha_i} &= \alpha_1 \sum_{t=1}^{T} x_{ti} + \alpha_2 \sum_{t=1}^{T} x_{ti} x_{t1} \\ &\quad + \alpha_3 \sum_{t=1}^{T} x_{ti} x_{t2} + \cdots + \alpha_n \sum_{t=1}^{T} x_{ti} x_{t,n-1} \\ &\quad - \sum_{t=1}^{T} x_{ti} y_t + \mathbf{a}_i \boldsymbol{\lambda} = 0, \quad i = \overline{2, n}, \end{aligned}$$

where $\mathbf{a}_i$ is the ith row of the matrix $\mathbf{A}'$.

Substituting α_i from equality (1.21) in the above equation, we obtain

$$\sum_{j=2}^{n} \alpha_j \left[-\overline{x}_{j-1} \sum_{t=1}^{T} x_{ti} + \sum_{t=1}^{T} x_{ti} x_{t,j-1} \right] - \left[-\overline{y} \sum_{t=1}^{T} x_{ti} + \sum_{t=1}^{T} x_{ti} y_t \right] + a_i \lambda = 0. \tag{1.23}$$

After transformations, we get

$$\begin{aligned} -\overline{x}_{j-1} \sum_{t=1}^{T} x_{ti} + \sum_{t=1}^{T} x_{ti} x_{t,j-1} &= \sum_{t=1}^{T} (x_{ti} - \overline{x}_i)(x_{t,j-1} - \overline{x}_{j-1}) - \overline{y} \sum_{t=1}^{T} x_{ti} + \sum_{t=1}^{T} x_{ti} y_t \\ &= \sum_{t=1}^{T} (x_{ti} - \overline{x}_i)(y_t - \overline{y}). \end{aligned}$$

Substituting the last two expressions in equality (1.23), we find

$$\sum_{j=2}^{n} \alpha_j \sum_{t=1}^{T} (x_{ti} - \overline{x}_i)(x_{t,j-1} - \overline{x}_{j-1}) - \sum_{t=1}^{T} (x_{ti} - \overline{x}_i)(y_t - \overline{y}) + \mathbf{a}_i \boldsymbol{\lambda} = 0, \quad i = \overline{2, n}.$$

We express the obtained system of equations in the vector form using the notation for problem (1.16):

$$\mathbf{r}\tilde{\boldsymbol{\alpha}} - \mathbf{d} + \mathbf{A}'\boldsymbol{\lambda} = \mathbf{O}_n. \tag{1.24}$$

Consider the condition (1.18). Taking into account the structure of the matrix $\mathbf{G}$ (see representation (1.19)), we obtain

$$\boldsymbol{\lambda}_i(\mathbf{A}_i\tilde{\boldsymbol{\alpha}} - b_i) = 0, \quad \boldsymbol{\lambda}_i \geq 0, \; i = \overline{1, m}. \tag{1.25}$$

Since (1.17) and (1.18) have a unique solution, the (1.24) and (1.25) obtained from them possess the same property. Then these equations are the necessary and sufficient conditions of the existence of the minimum in problem (1.16), which holds true for the vector $\tilde{\boldsymbol{\alpha}} = \hat{\tilde{\boldsymbol{\alpha}}}$. Hence, the subvector $\hat{\tilde{\boldsymbol{\alpha}}}$ of the vector $\hat{\boldsymbol{\alpha}}$ is the solution to (1.16). Theorem is proved. □

Based on Theorem 1.2, one can reduce the estimation of the regression parameter to solving a quadratic programming problem in which the elements of the matrix of the objective function are by modules less than one. Put

$$\boldsymbol{\beta} = \mathbf{B}\tilde{\boldsymbol{\alpha}}, \tag{1.26}$$

where

$$\mathbf{B} = \sigma_y^{-1}\boldsymbol{\sigma}_{\mathbf{x}}, \boldsymbol{\sigma}_{\mathbf{x}} = \operatorname{diag}(\sigma_{xi}), \quad i = \overline{1, n},$$

$$\sigma_y = \sqrt{T^{-1}\sum_{t=1}^{T} (y_t - \overline{y})^2}, \quad \sigma_{xi} = \sqrt{T^{-1}\sum_{t=1}^{T} (x_{ti} - \overline{x}_i)^2}, \quad i = \overline{1, n}.$$

Denoting $\mathbf{r}_\beta = \boldsymbol{\sigma}_{\mathbf{x}}^{-1}\mathbf{r}\boldsymbol{\sigma}_{\mathbf{x}}^{-1}$, $\mathbf{d}_\beta = (\sigma_y\boldsymbol{\sigma}_{\mathbf{x}})^{-1}\mathbf{d}$, and $\mathbf{A}_\beta = \mathbf{A}\mathbf{B}^{-1}$, we obtain from problem (1.16) that

$$\frac{1}{2}\boldsymbol{\beta}'\mathbf{r}_\beta\boldsymbol{\beta} - \boldsymbol{\beta}'\mathbf{d}_\beta \rightarrow \min, \quad \mathbf{A}_\beta\boldsymbol{\beta} \leq \mathbf{b}. \tag{1.27}$$

The advantage of the solution $\hat{\boldsymbol{\beta}}$ of such estimation problem is that $\hat{\boldsymbol{\beta}}$ does not depend on the scale of measurement of variables. The elements of the matrix $\mathbf{r}_\beta$ and of the vector $\mathbf{d}_\beta$ vary in the same range: from -1 to 1, which allows to reduce the round-up errors.

Therefore, as is well known (Draper and Smith 1998, Sect. 2.1.3; Maindonald 1984, Sects. 1.8 and 1.10), the numerical solution to the problem (1.27 is more exact than that obtained as a numerical solution to (1.7) (if no constraints are imposed on both problems). Under constraints, the accuracy of the numerical solution to (1.27) will also be higher than that of problem (1.7) since the main error is due to the inversion of matrices involved in the objective functions of the problems above.

The components of $\hat{\boldsymbol{\beta}}$ are standardized estimates of $(n-1)$ last component of the vector $\hat{\boldsymbol{\alpha}}$. We call them "beta weights" in analogy with the term used in regression analysis without constraints. Such weights can be conveniently used for estimation and comparison of the influence force of independent variables on the dependent variable.

For the problem described in (1.27) we use the calculation scheme described in Sect. 1.1.1. For this we reduce problem (1.27) to the least squares estimation problem with constraints, see (1.7).

Let $\mathbf{X}^{-}$ be a $T \times (n-1)$ matrix, with tth row $\tilde{\mathbf{x}}_t' - \bar{\tilde{\mathbf{x}}}'$, and let $\mathbf{Y}^{-}$ be a vector whose tth element is $y_t - \overline{y}$.

Let us transform the objective function in (1.16) by adding the element $\frac{1}{2}(\mathbf{Y}^{-})'\mathbf{Y}^{-}$, which is a constant; thus, adding this element has no impact on the solution of our optimization problem. After transforming (1.26) we have

$$\begin{aligned}\frac{1}{2}\tilde{\boldsymbol{\alpha}}'\mathbf{r}\tilde{\boldsymbol{\alpha}} - \tilde{\boldsymbol{\alpha}}'\mathbf{d} + \frac{1}{2}(\mathbf{Y}^{-})'\mathbf{Y}^{-} &= \frac{1}{2}\tilde{\boldsymbol{\alpha}}'(\mathbf{X}^{-})'\mathbf{X}^{-}\tilde{\boldsymbol{\alpha}} - \tilde{\boldsymbol{\alpha}}'(\mathbf{X}^{-})'\mathbf{Y}^{-} + \frac{1}{2}(\mathbf{Y}^{-})'\mathbf{Y}^{-}\\ &= \frac{1}{2}||\mathbf{Y}^{-} - \mathbf{X}^{-}\tilde{\boldsymbol{\alpha}}||^2 = \frac{1}{2}||\mathbf{Y}^0 - \mathbf{X}^0\boldsymbol{\beta}||^2\sigma_y^2\\ &= \left(\frac{1}{2}\boldsymbol{\beta}'\mathbf{r}_{\boldsymbol{\beta}}\boldsymbol{\beta} - \boldsymbol{\beta}'\mathbf{d}_{\boldsymbol{\beta}} + \frac{1}{2}(\mathbf{Y}^0)\mathbf{Y}^0\right)\sigma_y^2,\end{aligned}$$

where $\mathbf{X}^0$ is the $T \times (n-1)$ matrix with (t,i)th component given by $x_{ti} - \overline{x}_i/\sigma_{xi}$; and tth element of $\mathbf{Y}^0 \in \Re^T$ is $y_t - \overline{y}/\sigma_y, t = \overline{1,T}$.

Therefore we have

$$\frac{1}{2}\boldsymbol{\beta}'\mathrm{r}_{\boldsymbol{\beta}}\boldsymbol{\beta} - \boldsymbol{\beta}'\mathbf{d}_{\boldsymbol{\beta}} = \frac{1}{2}||\mathbf{Y}^0 - \mathbf{X}^0\boldsymbol{\beta}||^2 - \frac{1}{2}(\mathbf{Y}^0)'(\mathbf{Y}^0). \tag{1.28}$$

It follows from (1.28) that solution to the problem

$$\frac{1}{2}||\mathbf{Y}^0 - \mathbf{X}^0\boldsymbol{\beta}||^2 \to \min, \quad \mathbf{A}_{\boldsymbol{\beta}}\boldsymbol{\beta} \le \mathbf{b} \tag{1.29}$$

coincides with solution to (1.27).

The problem (1.29) is similar to (1.7), and thus can be solved by orthogonal transformation of the matrix $\mathbf{X}^0$ described in Sect. 1.1.1. We only need to replace $\mathbf{Y}$ by $\mathbf{Y}^0$, $\mathbf{X}$ by $\mathbf{X}^0$, $\boldsymbol{\alpha}$ by $\boldsymbol{\beta}$, $\mathbf{G}$ by $\mathbf{A}_{\boldsymbol{\beta}}$ in every equation of Sect. 1.1.1.

Thus we obtained the solution $\hat{\boldsymbol{\beta}}$ to (1.29), and taking into account (1.26) we can find $\hat{\hat{\boldsymbol{\alpha}}} = \mathbf{B}^{-1}\hat{\boldsymbol{\beta}}$, since the matrix $\mathbf{B}$ is non-degenerate and diagonal.

To find $\hat{\alpha}_1$ we use (1.22). Finally, the solution to the original problem (1.7) is given by $\hat{\boldsymbol{\alpha}} = [\hat{\alpha}_1 \quad \hat{\hat{\boldsymbol{\alpha}}}']'$.

1.2 Estimation of Parameters of Nonlinear Regression with Nonlinear Inequality Constraints

1.2.1 Statement of the Problem and a Method of Its Solution

Consider a regression $y_t = f_t(\boldsymbol{\alpha}^0) + \varepsilon_t$, where $y_t \in \Re^1$ is the dependent variable, $\boldsymbol{\alpha}^0 \in \Re^n$ is the unknown parameter; ε_t is the noise; t is the index of the observation.

1.2.1.1 Estimation with Constraints

Let $y_t \in \Re^1$ and $\mathbf{x}_t \in \Re^n$ be known vectors, $t \in [1, T]$. The estimate $\hat{\boldsymbol{\alpha}}$ for the regression parameters can be found by solving the problem (1.4) and (1.5):

$$S(\boldsymbol{\alpha}) = \frac{1}{2}\sum_{t=1}^{T}(y_t - f_t(\boldsymbol{\alpha}))^2 \rightarrow \min,$$

$$g_i(\boldsymbol{\alpha}) \leq 0, \quad i \in I = \{1, 2, \ldots, m\}. \tag{1.30}$$

The solution to (1.30) can be obtained by iterations. Let us linearize at each iteration the components $\mathbf{f}(\boldsymbol{\alpha}) = [f_1(\boldsymbol{\alpha}) \ \ldots \ f_T(\boldsymbol{\alpha})]'$, and the functions $g_i(\boldsymbol{\alpha}), i = \overline{1, m}$, in neighborhood of the point determined at the previous iteration.

The auxiliary problem obtained after linearization has the following form at the current point $\boldsymbol{\alpha}$:

$$\begin{cases} \frac{1}{2}||\mathbf{Y} - \mathbf{f}(\boldsymbol{\alpha}) - \mathbf{D}(\boldsymbol{\alpha})\mathbf{X}||^2 + \frac{1}{2}v\mathbf{X}'A(\boldsymbol{\alpha})\mathbf{X} \rightarrow \min, \\ \mathbf{G}_\delta(\boldsymbol{\alpha})\mathbf{X} + \mathbf{g}_\delta(\boldsymbol{\alpha}) \leq \mathbf{O}_{m_\delta}, \end{cases} \tag{1.31}$$

where $\mathbf{Y}$ is the vector defined in (1.7), v is a positive number; $\mathbf{D}(\boldsymbol{\alpha})$ is a $T \times n$ matrix, $\mathbf{D}(\boldsymbol{\alpha}) = [\partial f_t(\boldsymbol{\alpha})/\partial \alpha_j]$, $t = \overline{1, T}$, $j = \overline{1, n}$, the matrix $\mathbf{G}_\delta(\boldsymbol{\alpha}) = [\partial g_i(\boldsymbol{\alpha})/\partial \alpha_j]$, $i \in I_\delta(\boldsymbol{\alpha}) \subseteq I$, $j = \overline{1, n}$, is of dimension $m_\delta \times n$, where m_δ is the number of elements in $I_\delta(\boldsymbol{\alpha})$; $\mathbf{g}_\delta(\boldsymbol{\alpha}) = [g_i(\boldsymbol{\alpha})]$, $i \in I_\delta(\boldsymbol{\alpha})$, $\mathbf{A}(\boldsymbol{\alpha})$ is a positive definite matrix with elements independent of $\mathbf{X}$.

Transformation of the objective function in the auxiliary problem yields

$$\frac{1}{2}\mathbf{X}'\tilde{\mathbf{R}}(\alpha)\mathbf{X} + \nabla S'(\alpha)\mathbf{X} \rightarrow \min, \quad \mathbf{G}_\delta(\alpha)\mathbf{X} + \mathbf{g}_\delta(\alpha) \leq \mathbf{O}_{m_\delta}, \tag{1.32}$$

where $\tilde{\mathbf{R}}(\alpha) = \mathbf{D}'(\alpha)D(\alpha) + v\mathbf{A}(\alpha)$, $\nabla S(\alpha) = -\mathbf{D}'(\alpha)(\mathbf{Y} - \mathbf{f}(\alpha))$.

In applied problems it might happen that the matrix $\mathbf{R}(\alpha) = \mathbf{D}'(\alpha)\mathbf{D}(\alpha)$ is degenerate (within the computation accuracy). In such cases the solution to (1.31) is Tikhonov regularized at $v > 0$ (Tikhonov and Arsenin 1979). It may be treated as a ridge estimate of the linear regression parameters obtained by taking into account the imposed constraints, where the vector $\mathbf{Y} - \mathbf{f}(\alpha)$ represents the values of the dependent variable, and the related columns of the matrix $\mathbf{D}(\alpha)$ represent the values of independent variables.

To construct an algorithm for calculation of regression parameters estimates which implements the iterative computing scheme described above, we pose some conditions on the regression function $f_t(\alpha)$. The Definition below is taken from Demidenko (1989, Chapter 1 §1).

Definition 1.1. The value $S_E = \lim_{r\rightarrow\infty} \inf_{||\alpha||\geq r} S(\alpha)$ is called the greatest lower bound of the function $S(\alpha)$ at infinity.

Methods of finding of S_E are described in Demidenko (1989).

Put

$$\Phi(\alpha) = \max_{1\leq i\leq m} \{0, g_i(\alpha)\}, \quad I_\delta(\alpha) = \{i = \overline{1,m} : g_i(\alpha) \geq \Phi(\alpha) - \delta\}. \tag{1.33}$$

Assumption 1.3. There exists the initial approximation α_0 and the constants $\Psi > 0, \delta > 0$, such that:

(a) $u(\alpha_0, \Psi) < S_E$
(b) For $\alpha \in \mathbf{K}_\Psi = \{\alpha : S(\alpha) \leq u(\alpha_0, \Psi)\}$ problem (1.31) has a solution, and its Lagrange multipliers $\lambda_i(\alpha), i \in I_\delta(\alpha)$ satisfy the condition $\sum_{i\in I_\delta(\alpha)} \lambda_i(\alpha) \leq \Psi$, $\alpha \in \mathbf{K}_\Psi$. Here $u(\alpha, \Psi) = S(\alpha) + \Psi\Phi(\alpha)$.

Assumption 1.4. The functions $f_t(\alpha)$ and $g_i(\alpha), i = \overline{1,m}$, are differentiable on $\Re^n$ and on any compact set $\mathbf{K}$ their gradients satisfy the Lipschitz condition with constants which might depend on $\mathbf{K}$ for e.g.

$$||\nabla f_t(\alpha_1) - \nabla f_t(\alpha_2)|| \leq l_{0t}||\alpha_1 - \alpha_2||, \quad l_{0t} > 0, \ t = \overline{1,T}, \tag{1.34}$$

$$||\nabla g_i(\alpha_1) - \nabla g_i(\alpha_2)|| \leq l_1||\alpha_1 - \alpha_2||, \quad l_1 > 0, \ i = \overline{1,m}. \tag{1.35}$$

Lemma 1.1. *Let $f_t(\alpha), t = \overline{1,T}$ be a family of functions continuous on $\Re^n$ and satisfying Assumption 1.3. Then the set $\mathbf{K}_\Psi$ is compact.*

The proof follows by continuity of $S(\alpha)$ on $\Re^n$, see Demidenko (1989, Theorem 1.1).

Algorithm A1.1

Step 1 Set the initial approximation $\boldsymbol{\alpha}_0$ and positive values v_0, Δ_1, Δ_2. Put $k = 0$.
Step 2 Find the solution $\mathbf{X}_k = \mathbf{X}(\boldsymbol{\alpha}_k)$ and the Lagrange multipliers corresponding to problem (1.31).
Step 3 If $\boldsymbol{\Phi}(\boldsymbol{\alpha}_k) \leq \Delta_1$ and $||\mathbf{X}(\boldsymbol{\alpha}_k)|| \leq \Delta_2$, then stop.
Step 4 Determine a step factor γ_k by halving a unity until the condition

$$u(\boldsymbol{\alpha}_k + \gamma_k \mathbf{X}_k, \Psi) - u(\boldsymbol{\alpha}_k, \Psi) \leq -\gamma_k^2 \mathbf{X}'_k \tilde{\mathbf{R}}(\boldsymbol{\alpha}_k)\mathbf{X}_k \tag{1.36}$$

is satisfied.
Step 5 Calculate $\boldsymbol{\alpha}_{k+1} = \boldsymbol{\alpha}_k + \gamma_k \mathbf{X}_k$.
Step 6 Calculate the regularization parameter

$$v_{k+1} = \begin{cases} v_k & \text{if} \quad \gamma_k = 1, \\ v_k & \text{if} \quad \gamma_k < 1, c v_k > V, \\ c v_k & \text{if} \quad \gamma_k < 1, c v_k \leq V. \end{cases} \tag{1.37}$$

The value V is set in order to avoid large values of the matrix $\tilde{\mathbf{R}}(\boldsymbol{\alpha}_k)$, which can make impossible (computationally) its inversion, especially for large n. In (1.37) c is a positive constant.
Step 7 Put $k = k + 1$ and proceed to Step 2.

Let us consider the convergence of this algorithm.

Lemma 1.2. *Under Assumption 1.4 the gradient of the function $S(\boldsymbol{\alpha})$ on a compact set K satisfies the Lipschitz condition:*

$$||\nabla S(\boldsymbol{\alpha}_1) - \nabla S(\boldsymbol{\alpha}_2)|| \leq l_2 ||\boldsymbol{\alpha}_1 - \boldsymbol{\alpha}_2||, \quad \boldsymbol{\alpha}_1, \boldsymbol{\alpha}_2 \in \boldsymbol{\Omega}. \tag{1.38}$$

Proof. After some transformations we obtain from (1.38)

$$||\nabla S(\boldsymbol{\alpha}_1) - \nabla S(\boldsymbol{\alpha}_2)|| \leq \sum_{t=1}^{T} [||\nabla f_t(\boldsymbol{\alpha}_1) - \nabla f_t(\boldsymbol{\alpha}_2)||(|y_t| + |f_t(\boldsymbol{\alpha}_2)| + ||\nabla f_t(\boldsymbol{\alpha}_1)|| \cdot |f_t(\boldsymbol{\alpha}_2) - f_t(\boldsymbol{\alpha}_1)|]. \tag{1.39}$$

Since the set $\mathbf{K}$ is compact the following inequalities hold true:

$$|f_t(\boldsymbol{\alpha})| \leq l_{1t}, \quad ||\nabla f_t(\boldsymbol{\alpha})|| \leq l_{2t}, \quad \boldsymbol{\alpha} \in \mathbf{K}, t = \overline{1, T}, \tag{1.40}$$

where $l_{1t} > 0, l_{2t} > 0$.

By differentiability of $f_t(\boldsymbol{\alpha})$ on $\mathbf{K}$ we have

$$|f_t(\boldsymbol{\alpha}_2) - f_t(\boldsymbol{\alpha}_1)| \leq l_{3t} ||\boldsymbol{\alpha}_1 - \boldsymbol{\alpha}_2||, \quad l_{3t} > 0, t = \overline{1, T}. \tag{1.41}$$

Taking into account (1.34), we obtain from (1.39), (1.41) the estimate (1.38) with $l_2 = \sum_{t=1}^{T} [l_{0t}(l_{1t} + l_4) + l_{2t}l_{3t}]_2$ since there always exists a number $l_4 > 0$ such that $|y_t| \leq l_4, t = \overline{1,T}$. Lemma is proved. □

Theorem 1.3. *Let Assumptions 1.3 and 1.4 be fulfilled. Then the process of computations by Algorithm A1.1 has the following properties:*

1. $\lim_{k\to\infty} \mathbf{X}_k = \mathbf{O}_n$
2. $\lim_{k\to\infty} \mathbf{\Phi}(\alpha_k) = 0$
3. *Any limit point* $\hat{\alpha}$ *of the sequence* α_k, $k = 0, 1, ...$, *satisfies the necessary conditions for the existence of the minimum in the (1.30)*
4. *If* V *is sufficiently large there exists an index* k_0, *such that for all* $k \geq k_0$ *we have* $\gamma_k = 1$, $v_k = \text{const}$.

Proof. 1. Expanding $S(\alpha)$ in Taylor series in the neighborhood of $\alpha = \alpha_k$, we obtain

$$S(\alpha_k + \gamma\mathbf{X}_k) = S(\alpha_k) + \gamma(\mathbf{X}_k, \nabla S(\alpha_k)) + \gamma(\mathbf{X}_k, \nabla S(\alpha_k + \theta_2\gamma\mathbf{X}_k) - \nabla S(\alpha_k)), \quad \theta_2 \in [0, 1]. \tag{1.42}$$

The necessary and sufficient conditions for the existence of the minimum in (1.32) are:

$$\nabla S(\alpha_k) + \tilde{\mathbf{R}}(\alpha_k)\mathbf{X} + \sum_{i\in I_\delta(\alpha_k)} \lambda_i(\alpha_k)\nabla g_i(\alpha_k) = \mathbf{O}_n,$$
$$\lambda_i(\alpha_k)[\nabla g_i'(\alpha_k)\mathbf{X} + g_i(\alpha_k)] = 0, \quad i \in I_\delta(\alpha_k). \tag{1.43}$$

From (1.32), the first equation in (1.43) and the Lipschitz property (1.38) we obtain

$$S(\alpha_k + \gamma\mathbf{X}_k) \leq S(\alpha_k) - \gamma\mathbf{X}'_k\tilde{\mathbf{R}}(\alpha_k)\mathbf{X}_k + \gamma\sum_{i\in I_\delta(\alpha_k)} \lambda_i(\alpha_k)g_i(\alpha_k) + \gamma^2 l_2||\mathbf{X}_k||^2, \tag{1.44}$$

where we used the equality

$$\mathbf{X}'_k \sum_{i\in I_\delta(\alpha_k)} \lambda_i(\alpha_k)\nabla g_i(\alpha_k) = -\sum_{i\in I_\delta(\alpha_k)} \lambda_i(\alpha_k)g_i(\alpha_k),$$

which follows from the second condition in (1.43).

Expanding $g_i(\alpha)$, $i \in I_\delta(\alpha_k)$ in Taylor series in the neighborhood of $\alpha = \alpha_k$, we obtain

$$g_i(\alpha_k + \gamma\mathbf{X}_k) \leq g_i(\alpha_k) + \gamma\mathbf{X}'_k\nabla g_i(\alpha_k) + \gamma\mathbf{X}'_k(\nabla g_i(\boldsymbol{\theta}_{ik}) - \nabla g_i(\alpha_k)),$$

where $\boldsymbol{\theta}_{ik} = \alpha_k + \xi_{ik}\gamma\mathbf{X}_k$, $0 \leq \xi_{ik} \leq 1$.

Taking into account that $\mathbf{X}_k$ is the solution to (1.31) we derive

$$\mathbf{X}'_k \nabla g_i(\boldsymbol{\alpha}_k) \leq -g_i(\boldsymbol{\alpha}_k).$$

According to (1.35) we have

$$\mathbf{X}'_k(\nabla g_i(\boldsymbol{\theta}_{ik}) - \nabla g_i(\boldsymbol{\alpha}_k)) \leq \gamma l_1 ||\mathbf{X}_k||^2.$$

From the last three inequalities and (1.33) we obtain

$$g_i(\boldsymbol{\alpha}_k + \gamma \mathbf{X}_k) \leq (1-\gamma)\boldsymbol{\Phi}(\boldsymbol{\alpha}_k) + \gamma^2 l_1 ||\mathbf{X}_k||, \quad i \in I_\delta(\boldsymbol{\alpha}_k). \tag{1.45}$$

Applying the mean value theorem to the constraint with the index $i \notin I_\delta(\boldsymbol{\alpha}_k)$, we obtain

$$g_i(\boldsymbol{\alpha}_k + \gamma \mathbf{X}_k) \leq g_i(\boldsymbol{\alpha}_k) + \gamma \mathbf{X}'_k \nabla g_i(\boldsymbol{\theta}_{ik}) \leq \boldsymbol{\Phi}(\boldsymbol{\alpha}_k) - \delta + \gamma\kappa ||\mathbf{X}_k||,$$

where $\boldsymbol{\kappa}$ is a constant bounding the norms of gradients of the objective function and constraints in $\mathrm{K}_{\boldsymbol{\Psi}}$: $\nabla S(\boldsymbol{\alpha}) \leq \boldsymbol{\kappa}$, $||g_i(\boldsymbol{\alpha})|| \leq \boldsymbol{\kappa}, i \in I, \boldsymbol{\alpha} \in \mathbf{K}_{\boldsymbol{\Psi}}$.

Put

$$\gamma \leq \gamma_{1k} = \min\left(1, \frac{\delta}{\boldsymbol{\Phi}(\boldsymbol{\alpha}_k) + \kappa ||\mathbf{X}_k||}\right). \tag{1.46}$$

From (1.46) we obtain

$$(1-\gamma)\boldsymbol{\Phi}(\boldsymbol{\alpha}_k) \geq \boldsymbol{\Phi}(\boldsymbol{\alpha}_k) - \delta + \gamma\kappa ||\mathbf{X}_k||.$$

From above and from inequality (1.45), we have

$$g_i(\boldsymbol{\alpha}_k + \gamma \mathbf{X}_k) \leq g_i(\boldsymbol{\alpha}_k) + \gamma \mathbf{X}'_k \nabla g_i(\boldsymbol{\theta}_{ik}) \leq (1-\gamma)\boldsymbol{\Phi}(\boldsymbol{\alpha}_k), \quad i \notin I_\delta(\boldsymbol{\alpha}_k).$$

From this inequality and (1.45) we obtain (see the notation in (1.33))

$$\boldsymbol{\Phi}(\boldsymbol{\alpha}_{k+1}) \leq (1-\gamma)\boldsymbol{\Phi}(\boldsymbol{\alpha}_k) + \gamma^2 l_1 ||\mathbf{X}_k||^2. \tag{1.47}$$

Let $\boldsymbol{\Psi}$ satisfy Assumption 1.3. Then taking to account Assumption 1.3(b) (the condition $\sum_{i \in I_\delta(\boldsymbol{\alpha})} \lambda_i(\boldsymbol{\alpha}) \leq \boldsymbol{\Psi}$) and formulas (1.33) and (1.47), we obtain from (1.44)

$$\begin{aligned} S(\boldsymbol{\alpha}_k + \gamma \mathbf{X}_k) + \boldsymbol{\Psi}\boldsymbol{\Phi}(\boldsymbol{\alpha}_k + \gamma \mathbf{X}_k) \leq S(\boldsymbol{\alpha}_k) + \boldsymbol{\Psi}\boldsymbol{\Phi}(\boldsymbol{\alpha}_k) - \gamma \mathbf{X}'_k \tilde{\mathbf{R}}(\boldsymbol{\alpha}_k)\mathbf{X}_k \\ + \gamma^2 l(\boldsymbol{\Psi} + 1)||\mathbf{X}_k||^2, \end{aligned} \tag{1.48}$$

where $l = \max(l_1, l_2)$.

Since $v_k > 0$, the matrix $\tilde{\mathbf{R}}(\boldsymbol{\alpha})$ is positive definite for $\boldsymbol{\alpha} \in \mathrm{K}_{\boldsymbol{\Psi}}$. Therefore

$$\mathbf{X}_k' \tilde{R}(\boldsymbol{\alpha})\mathbf{X}_k \geq v_k \mu_{\min}(\mathbf{A}(\boldsymbol{\alpha}_k))||\mathbf{X}_k||^2, \tag{1.49}$$

where $\mu_{\min}(\mathbf{A}) > 0$ is the minimal eigenvalue of the matrix $\mathbf{A}$. Taking into account (1.49), (1.48), putting $\gamma = \gamma_k$ and assuming that the inequality

$$1 - \frac{\gamma_k l(\boldsymbol{\Psi} + 1)}{v_k \mu_{\min}(\mathbf{A}(\boldsymbol{\alpha}_k))} \geq \gamma_k \tag{1.50}$$

holds true, we obtain the condition tested in Step 4 of the algorithm. From (1.50) we see that

$$\gamma_k \leq \gamma_{2k} = \frac{v_k \mu_{\min}(\mathbf{A}(\boldsymbol{\alpha}_k))}{l(\boldsymbol{\Psi} + 1) + v_k \mu_{\min}(\mathbf{A}(\boldsymbol{\alpha}_k))}. \tag{1.51}$$

It follows from (1.45) and (1.51) that the inequality which is tested in Step 4 will be satisfied for fixed v_k after finite number of subdivisions of unity, i.e.,

$$\frac{1}{2}\min(\gamma_{1k}, \gamma_{2k}) < \gamma_k \leq \min(\gamma_{1k}, \gamma_{2k}). \tag{1.52}$$

Since $\tilde{\mathbf{R}}(\boldsymbol{\alpha}_k)$ is positive definite, the function $u(\boldsymbol{\alpha}_k)$ is decreasing (see (1.36)) as k increases. Therefore all points $\boldsymbol{\alpha}_k$ belong to the set $\mathbf{K}_{\boldsymbol{\Psi}}$. Further, since $\mathrm{K}_{\boldsymbol{\Psi}}$ is compact and the function $u(\boldsymbol{\alpha})$ is continuous on $\Re^n$, it is bounded on K_{Ψ}. Thus,

$$\lim_{k\to\infty} \gamma_k^2 \mathbf{X}'_k \tilde{\mathbf{R}}(\boldsymbol{\alpha}_k)\mathbf{X}_k = 0. \tag{1.53}$$

Assume that $\mathbf{X}_k' \tilde{\mathbf{R}}(\boldsymbol{\alpha}_k)\mathbf{X}_k$ does not converge to 0. Then according to (1.53), $\gamma_k \to 0$ for some subsequence of indexes k. Then for large k it follows from (1.46), (1.51), and (1.52) that

$$\gamma_k \geq \frac{\delta}{2(\boldsymbol{\Phi}(\boldsymbol{\alpha}_k) + \kappa||\mathbf{X}_k||)}. \tag{1.54}$$

Since the function $\boldsymbol{\Phi}(\boldsymbol{\alpha})$ is continuous on the compact set $\mathbf{K}_{\boldsymbol{\Psi}}$, it has an upper bound. Therefore the right-hand side of (1.54) tends to 0 as $||\mathbf{X}_k|| \to \infty$. By positive definiteness of $\tilde{\mathbf{R}}(\boldsymbol{\alpha})$ we have from first equation in (1.43) and the condition $\sum_{i\in I_\delta(\alpha)} \lambda_i(\boldsymbol{\alpha}) \leq \boldsymbol{\Psi}$, $\boldsymbol{\alpha} \in \mathbf{K}_{\boldsymbol{\Psi}}$, that $||\mathbf{X}_k|| \leq \kappa(\Psi + 1)||\tilde{\mathbf{R}}^{-1}(\boldsymbol{\alpha}_k)|| < \infty$. Since this contradicts to our assumption, we arrive at $\mathbf{X}_k' \tilde{\mathbf{R}}(\boldsymbol{\alpha}_k)\mathbf{X}_k \to 0$. Since $v_k \geq v_0 > 0$, it follows from (1.49) that $\mathbf{X}_k' \tilde{\mathbf{R}}(\boldsymbol{\alpha}_k)\mathbf{X}_k \geq v_0 \mu_{\min}(\mathbf{A}(\boldsymbol{\alpha}_k))||\mathbf{X}_k||^2$. Hence, we obtain the statement (1).

2. From the constraints in (1.32) and the formula (1.33), we have

$$0 \leq \mathbf{\Phi}(\boldsymbol{\alpha}_k) = \max_{1 \leq i \leq m} g_i(\boldsymbol{\alpha}_k) \leq \kappa ||\mathbf{X}_k||.$$

The statement (2) follows from the last inequality.

3. Since $\boldsymbol{\alpha}_k \in \mathrm{K}_{\Psi}$ and K_{Ψ} is a bounded set, the limit point $\hat{\boldsymbol{\alpha}}$ of the sequence $\{\boldsymbol{\alpha}_k\}$, always exists. Since $\mathbf{X}_k$ is the solution of (1.32), we have $\lambda_i(\boldsymbol{\alpha}_k) = 0, i \notin I_\delta(\boldsymbol{\alpha})$. Further, from (1.43) we see that

$$\nabla S(\boldsymbol{\alpha}_k) + \tilde{\mathbf{R}}(\boldsymbol{\alpha}_k)\mathbf{X}_k + \sum_{i=1}^{m} \lambda_i(\boldsymbol{\alpha}_k) \nabla g_i(\boldsymbol{\alpha}_k) = \mathbf{O}_n. \quad (1.55)$$

According to Assumption 1.3(b), we have

$$\sum_{i=1}^{m} \lambda_i c(\boldsymbol{\alpha}_k) \leq \mathbf{\Psi}.$$

It follows from the last two expressions that there exists the limit of the sequence $\lambda_i(\boldsymbol{\alpha}_k)$. Suppose that $\lambda_i(\boldsymbol{\alpha}_k) \to \hat{\lambda}_i$ as $k \to \infty$. Then from the statement (1) and (1.55) it follows that $\nabla S(\hat{\boldsymbol{\alpha}}) + \sum_{i=1}^{m} \hat{\lambda}_i \nabla g_i(\hat{\boldsymbol{\alpha}}) = \mathbf{O}_n$. From second condition in (1.43) we have $\hat{\lambda}_i g_i(\hat{\boldsymbol{\alpha}}) = 0, i = \overline{1, m}$. Both last equalities hold by the necessary conditions for existence of the minimum of the problem (1.31), which proves statement (3).

4. Put $V_c = l(\mathbf{\Psi} + 1)/\mu_{\min}(\mathbf{A}(\boldsymbol{\alpha}_k))$. According to the assumptions of the theorem, this quantity is bounded. Suppose that $V_c \leq V$, and for some $k = k_1$, $v_{k_1} \geq V_c$. Then according to (1.51), $\gamma_{2k} \geq 1$. On the other hand, from (1.46) and $\lim_{k \to \infty} \mathbf{\Phi}(a_k) = 0$ we have $\gamma_{1k} = 1$ for $k \geq k_2$, which together with (1.52) and (1.37) implies that $v_k = \text{const}$, $\gamma_k = 1$, $k \geq k_0 = \max(k_1, k_2)$. This proves (4).

Theorem is proved. □

We can divide Algorithm A1.1 in two cases: only γ changes, $v = 0$ (Algorithm A1.2); only v changes, $\gamma = 1$ (Algorithm A1.3).

Consider the calculations performed by Algorithm A1.2, which can be used for estimation of the regression parameters provided that the condition below is fulfilled.

Assumption 1.5. The matrix $\mathbf{D}(\boldsymbol{\alpha})$ is of full rank (i.e., in any compact set the multicollinearity in regression models is absent).

In this case, the inequality $\mathbf{X}'\tilde{\mathbf{R}}(\boldsymbol{\alpha})\mathbf{X} \geq \mu_{\min}(\mathbf{R}(\boldsymbol{\alpha}))||\mathbf{X}||^2$ holds true, where $\tilde{\mathbf{R}}(\boldsymbol{\alpha}) = \mathbf{R}(\boldsymbol{\alpha}) = \mathbf{D}'(\boldsymbol{\alpha})\mathbf{D}(\boldsymbol{\alpha})$, $\mu_{\min}(\mathbf{R}(\boldsymbol{\alpha}))$ is the smallest eigenvalue of the matrix $\mathbf{R}(\boldsymbol{\alpha})$. Clearly, $\mu_{\min}(\mathbf{R}(\boldsymbol{\alpha})) \geq \mu > 0, \boldsymbol{\alpha} \in \mathbf{\Omega}_{\Psi}$. Put $1 - (\gamma l(\mathbf{\Psi} + 1)/\mu) \geq \eta > 0$. From here and (1.48) we obtain the condition for determination of the step factor γ:

$$u(\boldsymbol{\alpha}_k + \gamma_k \mathbf{X}_k, \mathbf{\Psi}_{k+1}) - u(\boldsymbol{\alpha}_k, \mathbf{\Psi}_{k+1}) \leq -\eta \gamma_k \mathbf{X}'_k \mathbf{R}(\boldsymbol{\alpha}_k)\mathbf{X}_k, \quad 0 < \eta < 1, \quad (1.56)$$

where γ_k is determined by halving the unity until (1.56) holds true. It is easy to verify that $\gamma_k > \frac{1}{2}\min(\gamma_{1k}, \mu((1-\eta)/l(\Psi+1)))$, where γ_{1k} is determined by expression (1.46).

Algorithm A1.2

Step 1 Set the initial approximation α_0 and positive quantities Δ_1, Δ_2, and $\eta \in (0,1)$. Put $v=0, k=0$.

Steps 2, 3 Coincide with the Steps 2, 3 of Algorithm A1.1.

Step 4 Determine the step factor γ_k from Condition (1.56).

Steps 5, 6 Coincide with the Steps 5, 7 of Algorithm A1.1.

Theorem 1.4. *Let Assumptions 1.3–1.5 be satisfied. Then for the iterative process of calculation by Algorithm A1.2 statements (1)–(3) of Theorem 1.3 hold true.*

This result obviously follows from Theorem 1.3. It was also obtained in Pshenichnyi (1983) who as applied to the general problem of nonlinear programming without indication of the method of calculation of $\mathbf{R}(\boldsymbol{\alpha})$.

Consider Algorithm A1.3 which can be obtained from Algorithm A1.1 for $v_k \geq v_0, k = 1,2,\ldots; \gamma_k = 1, k = 0,1,2,\ldots$ From (1.48) to (1.49) we derive for $\gamma = 1$ the condition for finding v_k:

$$u(\boldsymbol{\alpha}_k + \mathbf{X}_k, \boldsymbol{\Psi}_{k+1}) - u(\boldsymbol{\alpha}_k, \boldsymbol{\Psi}_{k+1}) \leq -\eta \mathbf{X}'_k \tilde{\mathbf{R}}(\boldsymbol{\alpha}_k)\mathbf{X}_k, \quad 0 < \eta < 1. \tag{1.57}$$

For this condition to be satisfied, it is sufficient that

$$v_k \geq \frac{l(\boldsymbol{\Psi}+1)}{\mu_{\min}(\mathbf{A}(\boldsymbol{\alpha}_k))(1-\eta)}.$$

Algorithm A1.3

Step 1 Set the initial approximation $\boldsymbol{\alpha}_0$ and positive quantities Δ_1, Δ_2, $v = 0$, and $\eta \in (0,1)$. Put $k=0$.

Steps 2, 3 Coincide with the Steps 2, 3 of Algorithm A1.1

Step 4 Determine the regularization parameter under the assumption that $v_k = v_0 \cdot 10^k$, where v_k is the minimal number of the sequence $k = 0,1,2,\ldots$, for which inequality (1.57) is satisfied.

Step 5 Put $\boldsymbol{\alpha}_{k+1} = \boldsymbol{\alpha}_k + \mathbf{X}_k$.

Step 6 Coincides with the Step 7 of Algorithm A1.2.

The statement below follows from Theorem 1.3.

Theorem 1.5. *Let Assumptions 1.3 and 1.4 be satisfied. Then for the iterative process of calculations by Algorithm A1.3, statements (1)–(3) of Theorem 1.3 hold true.*

1.2.1.2 Estimation Without Constraints

Consider the case when the constraints are absent ($\boldsymbol{\Psi} = 0$). Then the algorithms for the regression parameters estimation follow from the algorithms described above:

$$\boldsymbol{\alpha}_{k+1} = \boldsymbol{\alpha}_k + \gamma_k \mathbf{X}_k, \quad k = 0, 1, 2, \ldots,$$

where $\mathbf{X}_k = -\tilde{\mathbf{R}}(\boldsymbol{\alpha}_k)^{-1}\nabla S(\boldsymbol{\alpha}_k)$, $\tilde{\mathbf{R}}(\boldsymbol{\alpha}_k) = \mathbf{D}'(\boldsymbol{\alpha}_k)\mathbf{D}(\boldsymbol{\alpha}_k) + v_k\mathbf{A}(\boldsymbol{\alpha}_k)$.

We describe below algorithms for estimation of parameters of nonlinear regression, which naturally follow from Algorithms A1.1 to A1.3. We use the notation B (the first letter in the name of the algorithm) for the algorithms describing estimation without constraints.

Algorithm B1.1

Parameter γ_k is determined by sequential halving of a unity until the condition

$$S(\boldsymbol{\alpha}_k + \gamma_k \mathbf{X}_k) \leq S(\boldsymbol{\alpha}_k) - \gamma_k^2 \mathbf{X}'_k \tilde{\mathbf{R}}(\boldsymbol{\alpha}_k)\mathbf{X}_k$$

is satisfied. The parameter v_k is determined according to the expression (1.37).

Algorithm B1.2

Parameter γ_k is determined by sequential halving of a unity until the condition

$$S(\boldsymbol{\alpha}_k + \gamma_k \mathbf{X}_k) \leq S(\boldsymbol{\alpha}_k) - \eta\gamma_k \mathbf{X}'_k \mathbf{R}(\boldsymbol{\alpha}_k)\mathbf{X}_k, \quad 0 < \eta < 1 \tag{1.58}$$

is satisfied. In this case, $v_k = 0$ for all k.

Algorithm B1.3

Parameter $\gamma_k = 1$ for all k, and $v_k = v_0 \cdot c^{v_k}$, where $c > 0$, and v_k is the minimal number in the sequence $k = 0, 1, 2, \ldots$, for which condition

$$S(\boldsymbol{\alpha}_k + \gamma_k \mathbf{X}_k) \leq S(\boldsymbol{\alpha}_k) - \eta \mathbf{X}'_k \tilde{\mathbf{R}}(\boldsymbol{\alpha}_k)\mathbf{X}_{ck}, \quad 0 < \eta < 1$$

is satisfied.

This inequality is obtained from (1.57) when $\boldsymbol{\Psi} = 0$.

For $\mathbf{A}(\boldsymbol{\alpha}) = \text{diag}(r_{ii}((\boldsymbol{\alpha}))$, $i = \overline{1,n}$ (where $r_{ii}(\boldsymbol{\alpha})$ is an element of the matrix $\mathbf{R}(\boldsymbol{\alpha})$), the above algorithms are modifications of the well-known Levenberg-Marquardt algorithm which is the best algorithm in nonlinear optimization without constraints for estimation of nonlinear regression parameters by the least squares method.

An important part of these algorithms is the solution of the auxiliary problem. We consider this solution below in more general situation.

1.2.2 Solution to the Auxiliary Problem

Consider the solution to (1.31).

In this case, we assume that the rows of $\mathbf{G}(\boldsymbol{\alpha}_k)$ are sorted in decreasing order of $g_i(\boldsymbol{\alpha}_k)$. Putting $\mathbf{e}(\boldsymbol{\alpha}) = \mathbf{y} - \mathbf{f}(\boldsymbol{\alpha})$,

$$\mathbf{E}(\boldsymbol{\alpha}_k) = \begin{bmatrix} \mathbf{D}(\boldsymbol{\alpha}_k) \\ v_k^{1/2}\mathbf{A}^{1/2}(\boldsymbol{\alpha}_k) \end{bmatrix}, \quad \mathbf{F}(\boldsymbol{\alpha}_k) = \begin{bmatrix} \mathbf{e}(\boldsymbol{\alpha}_k) \\ \mathbf{O}_n \end{bmatrix},$$

we obtain the transformed problem (1.31):

$$\begin{cases} \frac{1}{2}||\mathbf{E}(\boldsymbol{\alpha}_k)\mathbf{X} - \mathbf{F}(\boldsymbol{\alpha}_k)||^2 \to \min, \\ \mathbf{G}_\delta(\boldsymbol{\alpha}_k)\mathbf{X} \leq -\mathbf{g}_\delta(\boldsymbol{\alpha}_k). \end{cases} \tag{1.59}$$

Up to notation (1.59) coincides with the problem (1.7). Comparing (1.59) with (1.7), we see that $\mathbf{F}(\boldsymbol{\alpha}_k) = \mathbf{y}$, $\mathbf{E}(\boldsymbol{\alpha}_k) = \mathbf{X}$, $\mathbf{X} = \boldsymbol{\alpha}$, $\mathbf{G}_\delta(\boldsymbol{\alpha}_k) = \mathbf{G}$, $-\mathbf{g}_\delta(\boldsymbol{\alpha}_k) = \mathbf{b}$ for a fixed $\boldsymbol{\alpha}_k$. Therefore, to solve the problem (1.59) one can use all the results obtained in Sect. 1.1.

Taking into account the fact that the rank of $\mathbf{E}(\boldsymbol{\alpha}_k)$ is equal to n, we obtain its orthogonal expansion $\mathbf{E}(\boldsymbol{\alpha}_k) = \mathbf{M}_1 \begin{bmatrix} \mathbf{M}_2 \\ \mathbf{O}_{Tn} \end{bmatrix} \mathbf{M}'_3$, $\mathbf{M}_1 = [\mathbf{M}_{11}\ \mathbf{M}_{12}]$, where $\mathbf{M}_1$ is an orthogonal $(T+n) \times (T+n)$ matrix, $(T+n) \times n$ is the dimension of its submatrices $\mathbf{M}_{11}$, $\mathbf{M}_{12}$ is a non-degenerate $(n \times n)$ matrix, $\mathbf{M}_3$ is an orthogonal $(n \times n)$ matrix. Hereafter, the argument $\boldsymbol{\alpha}_k$ in the transformation matrices is dropped for brevity.

According to Sect. 1.1.1, to solve (1.59) we need to solve

$$P(\mathbf{U}) = \frac{1}{2}||\mathbf{N}\mathbf{U} - \boldsymbol{\Phi}||^2 \to \min, \quad \mathbf{U} \geq \mathbf{O}_{m_\delta}, \tag{1.60}$$

where $\mathbf{U} \in \Re^{m_\delta}$, $\mathbf{N} = [\mathbf{N}_1 \vdots \mathbf{N}_2]'$, $\boldsymbol{\Phi}' = [\mathbf{O}'_{m_\delta} \vdots 1]$. Here $\mathbf{N}_1 = \mathbf{G}(\boldsymbol{\alpha}_k)\mathbf{M}_3\mathbf{M}_2^{-1}$, $\mathbf{N}_2 = -\mathbf{g}(\boldsymbol{\alpha}_k) - \mathbf{G}(\boldsymbol{\alpha}_k)\mathbf{M}_3\mathbf{M}_2^{-1}\mathbf{M}'_{11}\mathbf{F}(\boldsymbol{\alpha}_k)$.

The algorithm for solving (1.60) is described in Sect. 1.1.2.

Let $\hat{\mathbf{U}}$ be the solution to (1.60). According to Sect. 1.1.1, the solution to the auxiliary problem (1.31) is

$$\mathbf{X}_k = \mathbf{X}(\boldsymbol{\alpha}_k) = \mathbf{M}_3\mathbf{M}_2^{-1}(||\mathbf{N}\hat{\mathbf{U}} - \boldsymbol{\Phi}||^{-2}\mathbf{N}'_1\hat{\mathbf{U}} + \mathbf{M}'_{11}\mathbf{F}(\boldsymbol{\alpha}_k)).$$

1.2.3 Compatibility of Constraints in the Auxiliary Problem

An important aspect of solving problem (1.59) lies in finding compatible constraints. According to Sect. 1.1.1, constraints are incompatible at $p = P(\hat{\mathbf{U}}) = 0$, where $\hat{\mathbf{U}}$

is the solution to problem (1.60). In case of incompatibility of constraints, we have to pick up a set of indices of joint constraints included in the auxiliary problem. To reduce the amount of time used for calculations it is desirable not to solve the problem (1.60) several times in one iteration. This might be particularly important if the dimension of (1.60), or the number of constraints in (1.59), is large. Consider the problem more in detail.

There might be two possibilities:

1. The number of constraints is $m_\delta \leq n + 1$. We assume that $\mathbf{G}(\boldsymbol{\alpha}_k)$ is of full rank. Indeed, if the rank $\mathbf{G}(\boldsymbol{\alpha}_k)$ is incomplete, then due to round-off errors this matrix is regarded as having a completed rank. Thus the rank of $\mathbf{N}_1$ is $\min(n, m_\delta)$. Then the rank of N is equal to m_δ. Denote by $\mathbf{U}^*$ the solution to the problem

$$P(\mathbf{U}) \to \min . \tag{1.61}$$

Then we find the rule by which we can regard the constraints as compatible without solving (1.60). The constraints are compatible if $P(\mathbf{U}^*) > 0$ regardless of the sign of $\mathbf{U}^*$, because $P(\hat{\mathbf{U}}) \geq P(\mathbf{U}^*)$. If $P(\mathbf{U}^*) = 0$ and the constraints in (1.60) are not fulfilled for $\mathbf{U} = \mathbf{U}^*$, then the constraints of the auxiliary problem are also not compatible. Let us calculate $P(\mathbf{U}^*)$.

The matrix $\mathbf{N}$ can be transformed as $\mathbf{QN} = \mathbf{M}$, where $\mathbf{M}$ is an upper triangular matrix of dimension $((n+1) \times m_\delta)$, $\mathbf{Q}$ is an orthogonal $(n+1) \times (n+1)$ matrix. According to Sect. 1.1.1, $P(\mathbf{U}) = \frac{1}{2}||\mathbf{r}||^2$, where $\mathbf{r}$ is the $(n+1) - m_\delta$ vector, whose components are equal to the last $(n+1) - m_\delta$ components of the vector $\mathbf{Q\Phi}$, respectively. Thus, we obtain $P(\mathbf{U}^*) = 0$ at $m_\delta = n + 1$ and

$$P(\mathbf{U}^*) = \frac{1}{2} \sum_{i=m+1}^{n+1} Q^2_{n+1,i}, \quad m_\delta < n + 1, \tag{1.62}$$

where Q_{ni} is the (n, i)th element of the matrix $\mathbf{Q}$. Using (1.62), we recursively determine compatibility of constraints. Introduce the matrix $\mathbf{N}_l = [\mathbf{z}_1\ \mathbf{z}_2\ \ldots\ \mathbf{z}_l]$, $i = \overline{1, m_\delta}$, where $\mathbf{z}_i$ is the ith column of the matrix $\mathbf{N}(\mathbf{N}_{m_\delta} = \mathbf{N})$. The orthogonal transformation which reduces $\mathbf{N}_l$ to an upper triangular matrix is determined by the $(n+1) \times (n+1)$ matrix $\mathbf{Q}_l$. Here, according to Lawson and Hanson 1974, Chapter 24), $\mathbf{Q}_{l+1} = \mathbf{q}_{l+1}\mathbf{Q}_l$, where $\mathbf{q}_{l+1}$ is the Hauscholder transformation, which is selected so that components of the vector $\mathbf{q}_{l+1}\mathbf{Q}_l\mathbf{z}_{l+1}$ with indexes $l + 2, \ldots, n + 1$ are equal to zero. According to (1.62),

$$P(\mathbf{U}_l^*) = \frac{1}{2}||\mathbf{N}_l\mathbf{U}_l^* - \mathbf{\Phi}||^2 = \frac{1}{2} \sum_{i=l+1}^{n+1} (Q^{(l)}_{n+1,i})^2,$$

where $Q^{(l)}_{n+1,i}$ is the (n, i)th element of the matrix $\mathbf{Q}_l$, $\mathbf{U}_l^*$ is a solution to (1.61) for $\mathbf{N} = \mathbf{N}_l$.

We have

$$P(\mathbf{U}_1^*) = \frac{1}{2}\left(1 - \left(\frac{z_{1,n+1}}{||\mathbf{z}_1||}\right)^2\right).$$

For $l > 1$, taking into account the fact that $\mathbf{Q}_l$ is the orthogonal matrix, we can speed up the computation of $P(\mathbf{U}_l^*)$ by using the following relation:

$$P(\mathbf{U}_l^*) = \begin{cases} \frac{1}{2}\left(1 - \sum_{i=1}^{l} (Q_{n+1,i}^{(l)})^2\right) & \text{if } l < n+1-l, \\ \frac{1}{2} \sum_{i=l+1}^{n+1} (Q_{n+1,i}^{(l)})^2 & \text{if } l \geq n+1-l. \end{cases}$$

Thus, by recursion relation for calculation of the matrix $\mathbf{Q}_l$ and the formula for $P(\mathbf{U}_l^*)$, the constraints with indexes $2, 3, \ldots, m_\delta$, are sequentially tested for compatibility (it is assumed that the constraints are arranged in the increasing order of the sequence $\{g_i(\boldsymbol{\alpha}_k)\}$).

2. The number of constraints is $m_\delta \geq n+1$. Similar to the first case, we assume that $\mathbf{G}$ is of full rank. Then for $l > n+1$ the rank of $\mathbf{N}_l$ is equal to $n+1$. In this case, $P(\mathbf{U}_l^*) = 0$, $l > n+1$ and the solution to (1.61) is not unique. It has the form

$$\mathbf{U}_l^* = \mathbf{Q}_l \begin{bmatrix} \bar{\mathbf{y}} \\ \mathbf{y} \end{bmatrix} = \mathbf{Q}_{1l}\bar{\mathbf{y}} + \mathbf{Q}_{2l}\mathbf{y} = \mathbf{U}_l^0 + \mathbf{Q}_{2l}\mathbf{y},$$

where $\mathbf{Q}_{1l}$ and $\mathbf{Q}_{2l}$ are submatrices of the orthogonal matrix $\mathbf{Q}_l$ of dimensions $l \times (n+1)$ and $l \times h$, respectively, $h = l - n - 1$, $\bar{\mathbf{y}}$ and $\mathbf{y}$ are the vectors of dimensions $n+1$ and h, respectively. The vector $\bar{\mathbf{y}}$ is a solution of the system of equations $\overline{\mathbf{M}}\bar{\mathbf{y}} = \boldsymbol{\Phi}$, where the lower triangular matrix $\overline{\mathbf{M}}$ is determined from the relation $\mathbf{N}_l\mathbf{Q}_l = [\overline{\mathbf{M}}\mathbf{O}_{n+1,h}]$. If the solution to (1.61) with minimal norm $\mathbf{U}_l^* = \mathbf{U}_l^0 = \mathbf{Q}_{1l}\bar{\mathbf{y}} \geq \mathbf{O}_{n+1}$, then the first l constraints of the auxiliary problem are compatible. Otherwise, it is necessary to establish whether there is at least one solution $\mathbf{U}_l^*$ with the non-minimal norm, which satisfies the constraints in (1.60) (then the constraints in (1.59) are incompatible). For this purpose, we solve the problem

$$F(\mathbf{y}) = \frac{1}{2} \sum_{i \in J_l} [\max(0, \mathbf{a}'_i\mathbf{y} - u_{il}^0)]^2 \to \min, \tag{1.63}$$

where J_l is the set of indices of the first l constraints that are included in the set $I_\delta(\boldsymbol{\alpha})$ (the constraints are arranged in increasing order of sequence $\{g_i(\boldsymbol{\alpha}_k)v\}$); $\mathbf{a}_i$ is the ith line of $\mathbf{Q}_{2l}$ taken with the inverse sign; u_{il}^0 is the ith component of $\mathbf{U}_l^0$.

We solve (1.63) iteratively:

$$\mathbf{y}_{j+1} = \mathbf{y}_j + \nu_j\mathbf{d}_j, \quad j = 0, 1, 2, \ldots, \mathbf{y}_0 = \mathbf{O}_h. \tag{1.64}$$

Let us determine $\mathbf{d}_j$. Denote $J_l(y_j) = \{i : \mathbf{a}'_i\mathbf{y} - u^0_{il} > 0,\ i \in J_l\}$, $\mathbf{A}_j$ is a $(\mu_j \times h)$ matrix composed of the rows $\mathbf{a}'_i$, $i \in J_l(\mathbf{y}_j)$; $\mathbf{B}_j$ is a vector of dimension μ_j with the components u^0_{il}, $i \in J_l(\mathbf{y}_j)$. Then

$$\mathbf{d}_j = -\mathbf{y}_j + \mathbf{A}^+_j\mathbf{B}_j, \tag{1.65}$$

where

$$\mathbf{A}^+_j = \begin{cases} (\mathbf{A}'_j\mathbf{A}_j)^{-1}\mathbf{A}'_j\mathbf{B}_j & \text{if } \mu_j > l - n - 1, \\ \mathbf{A}'_j(\mathbf{A}_j\mathbf{A}'_j)\mathbf{B}_j & \text{if } \mu_j < l - n - 1. \end{cases}$$

We find the step factor ν_j from the condition

$$\min_{\nu \geq 0} F(\mathbf{y}_j + \nu\mathbf{d}_j). \tag{1.66}$$

From here we obtain the condition for determination of ν_j:

$$\sum_{i \in J_l} [\max(0, c_0(i) - c_1(i)\nu)]c_1(i)\nu = 0, \tag{1.67}$$

where $c_0(i) = \mathbf{a}'_i\mathbf{y}_j - u^0_{il}$, $c_1(i) = -\mathbf{a}'_i\mathbf{d}_j$.

We solve (1.67) by taking into account the fact that the left-hand side in (1.67) is a piecewise linear function. Denote by $\mathrm{K} = \{i : c_0(i)c_1(i) > 0, i \in J_l\}$ the set of indices of the constraints, which change their sign when ν changes. Let $\overline{\nu}(i) = c_0(i)/c_1(i)$, $i \in \mathrm{K}$. Arranging $\overline{\nu}(i)$ in increasing order we obtain the sequence $\{\overline{\nu}(i_q)\} = 0$, $q = 0, 1, \ldots$, where $\overline{\nu}(i_1) = 0$. For $\nu \in [\overline{\nu}(i_{q-1}), \overline{\nu}(i_q)]$ the indices of the constraints for which $a'_i(y_j + \nu d_j) - u^0_{il} > 0$ do not depend on ν. Moreover, there is an index L such that $\nu_j \in [\overline{\nu}(i_{L-1}), \overline{\nu}(i_L)]$. From here and from (1.33) we obtain the algorithm for finding ν_j.

Algorithm A1.4

1. Put $l = 1$. Calculate

$$w_1 = \sum_{i \in J_l(y_j)} c_1^2(i) + \sum_{i \in \overline{J}_l(y_j)} c_1^2(i), \quad \nu_1 = \sum_{i \in J_l(y_j)} c_0(i)c_1(i),$$

   where $\overline{J}_l = \{i := 0, c_1(i) < 0, J_l\}$.
2. Calculate $\nu^*(l) = \nu_l/w_l$.
3. If $\overline{\nu}(i_{l-1}) \leq \nu^*(l) \leq \overline{\nu}(i_l)$, then $\nu_j = \nu^*(l)$. Otherwise, put $l = l + 1$ and calculate

$$w_l = w_{l-1} + \theta(i_{l-1})c_l^2(i_{l-1}), \quad \nu_l = \nu_{l-1} + \theta(i_{l-1}) + c_0(i_{l-1})c_1(i_{l-1}),\ i_{l-1} \in \mathrm{K},$$

   where $\theta(i_l) = -\operatorname{sign} c_1(i_l)$; the indices i_l increase with the increase of the indices in the sequence $\{\overline{\nu}(i_l)\}$. Go over to the Step 2.

Theorem 1.6. *Suppose that the matrix* $\mathbf{G}(\boldsymbol{\alpha}_k)$ *is of full rank. Then the iterative process (1.64)–(1.66) converges to the solution of (1.63) in finite number of steps* $H \leq 2^l$.

Proof. Denote the set of the minimal points by $M^* = \arg\min_{\mathbf{y}\in\Re^h} F(\mathbf{y})$, where $h = l - n - 1$; $J_l^* = \{i : \mathbf{a}'_i\mathbf{y} - u_{il}^0 > 0, i \in J_l, y \in M^*\}$.

The necessary and sufficient conditions for the existence of the minimum of $F(\mathbf{y})$ are:

$$\nabla F(\mathbf{y}) = \sum_{i\in J_l} [\max(0, \mathbf{a}'_i\mathbf{y} - u_{il}^0)]^2 \mathbf{a}_i = \mathbf{O}_h. \tag{1.68}$$

If $\mathbf{a}'_i\mathbf{y} - u_{il}^0 \leq 0$, $i \in J_l$, then $\nabla F(\mathbf{y}) = \mathbf{O}_h$, $\mathbf{y} \in M^*$, $J_l^* = \varnothing$.

Let $\mathbf{a}'_i\mathbf{y} - u_{il}^0 > 0$, $i \in J_l^* \neq \varnothing$, $\mathbf{y} \in M^*$. From (1.68), we have $\mathbf{A}'_*\mathbf{A}_*\mathbf{y} - \mathbf{A}'_*\mathbf{B}_* = \mathbf{O}_h$, $\mathbf{y} \in M^*$, where $\mathbf{A}_*$ is a $\mu_* \times h$-matrix composed of the rows $\mathbf{a}'_i$, $i \in J_l^*$; u_{il}^0, $i \in J_l^*$ are the components of the vector $\mathbf{B}_*$. If $\mu_* \leq h$, then $\mathbf{A}'_*\mathbf{y} - \mathbf{B}_* = \mathbf{O}_{\mu^*}$ for $\mathbf{y} \in M^*$, which contradicts to the condition $J_l^* = \varnothing$. Therefore, if $J_l^* \neq \varnothing$ the number of elements in the set J_l^* is $\mu^* > h$, and M^* contains one element.

It follows from above that

$$\mathbf{y}_j \in M^* \quad \text{if } J_l(\mathbf{y}_j) = J_l^*,\ j = 0, 1, \ldots. \tag{1.69}$$

We show that if $\mathbf{y}_j \notin M^*$ then $J_l(\mathbf{y}_{j+1}) \neq J_l(\mathbf{y}_j)$. Assume the converse:

$$J_l(\mathbf{y}_{j+1}) = J_l(\mathbf{y}_j), \quad \mathbf{y}_j \notin M^*. \tag{1.70}$$

From (1.65) to (1.67), we obtain

$$\mathbf{d}'_j\mathbf{A}'_j\mathbf{A}_j\mathbf{d}_j(\nu - 1) = 0. \tag{1.71}$$

Let $\mu_j \geq h$. Since $\mathbf{y}_j \notin M^*$, we have $\nabla F(\mathbf{y}_j) = \mathbf{A}'_j\mathbf{A}_j\mathbf{y}_j - \mathbf{A}'_j\mathbf{B}_j \neq \mathbf{O}_h$. It follows from (1.65) that $(\mathbf{A}'_j\mathbf{A}_j)\mathbf{d}_j = -(\mathbf{A}'_j\mathbf{A}_j)\mathbf{y}_j + \mathbf{A}'_j\mathbf{B}_j \neq \mathbf{O}_h$, hence $\mathbf{d}_j \neq \mathbf{O}_h$. By positive definiteness of the matrix $\mathbf{A}'_j\mathbf{A}_j$, we have $\mathbf{d}'_j\mathbf{A}'_j\mathbf{A}_j\mathbf{d}_j > 0$. From above and from (1.71) we obtain $\nu_j = 1$. Then, according to (1.64) and (1.65), $\mathbf{y}_{j+1} = (\mathbf{A}'_j\mathbf{A}_j)^{-1}\mathbf{A}'_j\mathbf{B}_j$. However, $J_l(\mathbf{y}_{j+1}) = J_l(\mathbf{y}_j)$; therefore, $\nabla F(\mathbf{y}_{j+1}) = \mathbf{A}'_j\mathbf{A}_j\mathbf{y}_{j+1} - \mathbf{A}'_j\mathbf{B}_j = \mathbf{O}_h$, i.e., $\mathbf{y}_{j+1} \in M^*$, and thus $J_l(\mathbf{y}_{j+1}) = J_l^*$. Therefore, by our assumption $J_l(\mathbf{y}_j) = J_l^*$. However, according to (1.69), $\mathbf{y}_j \in M^*$, which contradicts (1.70).

Let $\mu_j < h$. Since $\mathbf{A}_j\mathbf{y}_j - \mathbf{B}_j > \mathbf{O}_{\mu_j}$, then $\mathbf{d}'_j\mathbf{A}'_j\mathbf{A}_j\mathbf{d}_j = ||\mathbf{A}_j\mathbf{y}_j - \mathbf{B}_j||^2 > 0$. Then, by (1.71) we have $\mu_j = 1$. Repeating the arguments for the case where $\mu_j \geq h$, we conclude that for any μ_j and h, the sets $J_l(\mathbf{y}_j)$ will be different at different iterations of the algorithm. Taking into account the fact that the objective function is bounded from below and decreases at each iteration, we obtain the statement of the theorem. □

The results presented for $m_\delta \leq n+1$ allow us to solve (1.63) and to determine the compatibility of constraints in (1.59): the constraints are incompatible if $\min_{\mathbf{y}\in\Re^h} F(\mathbf{y}) = 0$, and are compatible otherwise. The compatibility of the constraints arranged in increasing order of $g_i(\boldsymbol{\alpha}_k)$ is sequentially determined for $l = 2, 3, \ldots$.

1.2.4 Calculation of the Constants $\boldsymbol{\Psi}$ and δ

In Sect. 1.2.1 we assumed that Ψ and δ are known, which is not always the case. If $\boldsymbol{\Psi}$ and δ are unknown, they can be estimated using the results of Sect. 1.2.3. In order to include their estimates in our general estimation problem, we need to modify Algorithm A1.1.

Algorithm A1.1.1

1. Set $\boldsymbol{\alpha}_0, \boldsymbol{\Psi}_0 > 0, \delta_0 = \infty, \nu_0, \Delta_1, \Delta_2$. Put $k = 0$
2. Determine the set $I_{\delta_k}(\boldsymbol{\alpha}_k)$. Arrange the constraints, whose indices belong to this set, in the decreasing order according to $g_i(\boldsymbol{\alpha}_k)$. Determine according to Sect. 1.2.3 the set of compatible constraints $I_{l_k}(\boldsymbol{\alpha}_k) \subseteq I_{\delta_k}(\boldsymbol{\alpha}_k)$, where l_k is the number compatible constraints in $I_{\delta_k}(\boldsymbol{\alpha}_k)$. Put
$$\delta_{k+1} = \boldsymbol{\Phi}(\boldsymbol{\alpha}_k) - \min_{i\in I_{l_k}(\boldsymbol{\alpha}_k)} g_i(\boldsymbol{\alpha}_k). \tag{1.72}$$
3. Put $I_{\delta_k}(\boldsymbol{\alpha}_k) \subseteq I_{l_k}(\boldsymbol{\alpha}_k)$. By (1.15), find the Lagrange multipliers $\boldsymbol{\lambda}(\boldsymbol{\alpha}_k) = \boldsymbol{\lambda} = \hat{\mathbf{U}}||\mathbf{N}\hat{\mathbf{U}}-\boldsymbol{\Phi}||^{-2}$, $i \in I_{\delta_k}(\boldsymbol{\alpha}_k)$ of problem (1.31) and its solution $\mathbf{X}_k = \mathbf{X}(\boldsymbol{\alpha}_k)$. Here $\hat{\mathbf{U}}$ is the solution to (1.60).
4. If $\boldsymbol{\Phi}(\boldsymbol{\alpha}_k) \leq \Delta_1$ and $||\mathbf{X}(\boldsymbol{\alpha}_k)|| \leq \Delta_2$, then stop.
5. If the condition $\boldsymbol{\Psi}_k^0 = \sum_{i\in I_{\delta_k}(\boldsymbol{\alpha}_k)} \lambda_i(\boldsymbol{\alpha}_k) \leq \boldsymbol{\Psi}_k$ is fulfilled, then $\boldsymbol{\Psi}_{k+1} = \boldsymbol{\Psi}_k$. Otherwise $\boldsymbol{\Psi}_{k+1} = 2\boldsymbol{\Psi}_k^0$.
6. Determine $\boldsymbol{\alpha}_{k+1}$ according to Steps 4–6 of Algorithm 1.2.1.1 (see Sect. 1.2.1). Put $k = k+1$ and go over to Step 2.

We have $\min_{i\in I_{l_k}(\boldsymbol{\alpha}_k)} g_i(\boldsymbol{\alpha}_k) \geq \Phi(\boldsymbol{\alpha}_k) - \delta_k$. Then using (1.72) we obtain $\delta_{k+1} \leq \delta_k$. According to Algorithm A1.1.1, $\boldsymbol{\Psi}_{k+1} \geq \boldsymbol{\Psi}_k$. Since δ_k is decreasing and $\boldsymbol{\Psi}_k$ is increasing, the algorithm will stop when some values δ and $\boldsymbol{\Psi}$ are achieved, see Assumption 1.3.

In Pshenichnyi (1983, Chapter 3 §9), compatible constraints are sequentially determined assuming that $\delta_k := \delta_k/2$ until the compatible constraints in (1.31) are selected.

According to Sect. 1.1, to determine compatible constraints, it is not necessary to solve (1.31). If $l_k \leq n+1$, then the compatible constraints can be determined rather easily. When $l_k > n+1$ and $n+1 \leq l \leq l_k$, the unconstrained-minimization problem is solved with $l-(n+1)$ variables instead of l. This method allows us to reduce significantly the time spent for determination of the sets $I_\delta(\boldsymbol{\alpha}_k)$ in (1.31).

1.3 Estimation of Multivariate Linear Regression Parameters with Nonlinear Equality Constraints

In Sect. 1.2 we described algorithms for estimation of parameters under some constraints which can be given in the form of equalities or inequalities. However if we have only equality constraints, the calculations are much simpler, since the auxiliary problem (1.31) can be solved in one step. Taking into account that the problem of estimation with equality constraints is rather specific, consider its solution in more detail. Such a problem appears, for example, in econometrics, when one needs to construct multidimensional regression models. Therefore, we consider the case of multidimensional regression, which generalizes the case treated in (1.30).

Multivariate regression which generalizes (1.1) is of the form:

$$y_{pt} = f_{pt}(\boldsymbol{\alpha}^0) + \varepsilon_{pt}, \quad p = \overline{1,P},\ t = \overline{1,T}, \tag{1.73}$$

where y_{pt} is the dependent variable, $f_{pt}(\cdot)$ is the regression function, ε_{pt} is the noise, $\boldsymbol{\alpha}^0 \in \Re^n$ is a vector of required parameters, and P is the number of regression equations.

Parameters of the regression (1.73) are subjected to a priori constraints

$$g_i(\boldsymbol{\alpha}^0) = 0,\ i \in I = \{1, 2, \ldots, m\},\ m < n.$$

Let $\boldsymbol{\varepsilon}_t = [\varepsilon_{1t}, \ldots, \varepsilon_{pt}]'$. We assume that ε_t satisfies the assumption below.

Assumption 1.6. Assume that the expectation of the vector $\boldsymbol{\varepsilon}_t$ is 0, and there exists the covariance matrix $\sigma^2\mathbf{s}^0$. For $t \neq \tau$ the vectors $\boldsymbol{\varepsilon}_t$ and $\boldsymbol{\varepsilon}_\tau$ are independent. The $(P \times P)$-matrix $\mathbf{s}^0$ is known and is positive definite. If the matrix $\mathbf{s}^0$ is unknown, then its estimate $\hat{\mathbf{s}}$ is known, which is also positive definite. Suppose that σ^2 is unknown.

Define $\boldsymbol{\Sigma}^0 = \mathbf{s}^0 \otimes \mathbf{J}_T$, $\hat{\boldsymbol{\Sigma}} = \hat{\mathbf{s}} \otimes \mathbf{J}_T$. We estimate $\boldsymbol{\alpha}^0$ under the condition

$$S(\boldsymbol{\alpha}) = \frac{1}{2}(\mathbf{Y} - \mathbf{f}(\boldsymbol{\alpha}))'\boldsymbol{\Sigma}^{-1}(\mathbf{Y} - \mathbf{f}(\boldsymbol{\alpha})) \to \min \tag{1.74}$$

and the constraints

$$g_i(\boldsymbol{\alpha}) = 0,\ i \in I = \{1,\ 2, \ldots, m\}, \quad m < n. \tag{1.75}$$

In (1.74) the matrix $\boldsymbol{\Sigma}$ can be equal to $\boldsymbol{\Sigma}^0$ or to $\hat{\boldsymbol{\Sigma}}$, $\mathbf{Y}$ and $\mathbf{f}(\boldsymbol{\alpha})$ are N-dimensional vectors, $N = PT$. Their $(p-1)T+t$th components $(p = \overline{1,P}, t = \overline{1,T})$ are equal, respectively, to y_{pt} and $f_{pt}(\boldsymbol{\alpha})$. According to Bard (1974, §4.3), the criterion (1.74) allows to get the estimate of $\boldsymbol{\alpha}^0$ close to the optimal, provided that T is finite, and the Assumption 1.6 holds true.

The solution to (1.74) and (1.75) can be found by means of Algorithms A1.2 with the following modifications. The function which satisfies (1.75) is of the form $\Phi(\boldsymbol{\alpha}) = \max_{1 \leq i \leq m} |g_i(\boldsymbol{\alpha})|$. The auxiliary problem at the current point $\boldsymbol{\alpha}$ becomes

$$\begin{cases} \frac{1}{2}(\mathbf{y}-\mathbf{f}(\boldsymbol{\alpha}) - \mathbf{D}(\boldsymbol{\alpha})\mathbf{X})\Sigma^{-1}(\mathbf{y} - \mathbf{f}(\boldsymbol{\alpha})-\mathbf{D}(\boldsymbol{\alpha})\mathbf{X})+\frac{1}{2}\nu\mathbf{X}'\mathbf{A}(\boldsymbol{\alpha})\mathbf{X} \to \min, \\ \mathbf{G}_{\delta}(\boldsymbol{\alpha})\mathbf{X} + \mathbf{g}_{\delta}(\boldsymbol{\alpha}) \leq \mathbf{O}_{m_{\delta}}, \end{cases} \tag{1.76}$$

where $\nu > 0$; $\mathbf{D}(\boldsymbol{\alpha}) = \text{diag}(\mathbf{D}_1(\boldsymbol{\alpha}), \ldots, \mathbf{D}_P(\boldsymbol{\alpha}))$ is the matrix of the size $N \times n$, $n = \sum_{l=1}^{P} n_l$, $\partial f_{lt}(\boldsymbol{\alpha})/\partial \alpha_j$, $j = \overline{1, n_l}$ are the elements of the $T \times n_l$ submatrix $\mathbf{D}_l(\boldsymbol{\alpha})$. Matrix $\mathbf{G}_{\delta}(\boldsymbol{\alpha}) = [\partial g_i(\boldsymbol{\alpha})/\partial \alpha_j]$, $j = \overline{1, n}$, $i \in I_{\delta}(\boldsymbol{\alpha}) = \{i : |g_i(\boldsymbol{\alpha})| \geq \Phi(\boldsymbol{\alpha}) - \delta, i = \overline{1, m}\}$, $\delta > 0$, is $m_{\delta} \times n$ dimensional, where m_{δ} is the number of elements in $I_{\delta}(\boldsymbol{\alpha})$, $\mathbf{g}_{\delta}(\boldsymbol{\alpha}) = [g_i(\boldsymbol{\alpha})]$, $i \in I_{\delta}(\boldsymbol{\alpha})$, and $\mathbf{A}(\boldsymbol{\alpha})$ is a positive definite matrix.

If δ is such that the constraints in (1.76) are consistent, then the solution to (1.76) is

$$\mathbf{X}(\boldsymbol{\alpha}) = (\mathbf{J}_n - \mathbf{E}'(\boldsymbol{\alpha})\mathbf{G}_{\delta}(\boldsymbol{\alpha}))\tilde{\mathbf{R}}^{-1}(\boldsymbol{\alpha})\mathbf{D}'(\boldsymbol{\alpha})\boldsymbol{\Sigma}^{-1}(\mathbf{y} - \mathbf{f}(\boldsymbol{\alpha})) - \mathbf{E}'(\boldsymbol{\alpha})\mathbf{g}_{\delta}(\boldsymbol{\alpha}), \tag{1.77}$$

where

$$\tilde{\mathbf{R}}(\boldsymbol{\alpha}) = \mathbf{D}'(\boldsymbol{\alpha})\boldsymbol{\Sigma}^{-1}\mathbf{D}(\boldsymbol{\alpha}) + \nu\mathbf{A}(\boldsymbol{\alpha}),$$
$$\mathbf{E}(\boldsymbol{\alpha}) = (\mathbf{G}_{\delta}(\boldsymbol{\alpha})\tilde{\mathbf{R}}^{-1}(\boldsymbol{\alpha})\mathbf{G}'_{\delta}(\boldsymbol{\alpha}))^{-1}\mathbf{G}_{\delta}(\boldsymbol{\alpha})\tilde{\mathbf{R}}^{-1}(\boldsymbol{\alpha}).$$

Passing to the limit in (1.76) as $\nu \to \infty$, we get

$$\begin{aligned} ||\mathbf{X}(\boldsymbol{\alpha})|| &\to \mathbf{g}'_{\delta}(\boldsymbol{\alpha})(\mathbf{G}_{\delta}(\boldsymbol{\alpha})\mathbf{A}^{-1}(\boldsymbol{\alpha})\mathbf{G}'_{\delta}(\boldsymbol{\alpha}))^{-1}\mathbf{G}_{\delta}(\boldsymbol{\alpha})\mathbf{A}^{-2}(\boldsymbol{\alpha}) \\ &\times\mathbf{G}'_{\delta}(\boldsymbol{\alpha})(\mathbf{G}_{\delta}(\boldsymbol{\alpha})\mathbf{A}^{-1}(\boldsymbol{\alpha})\mathbf{G}'_{\delta}(\boldsymbol{\alpha}))^{-1}\mathbf{g}_{\delta}(\boldsymbol{\alpha}). \end{aligned} \tag{1.78}$$

Under the convergence of Algorithm A1.1 it is necessary to make some remarks: Assumptions 1.3 and 1.4 should be replaced by Assumptions 1.3′ and 1.4′, see below.

Assumption 1.3′ There exists an initial approximation $\boldsymbol{\alpha}_0$ and constants $\Psi > 0$, $\delta > 0$, such that

(a) $u(\boldsymbol{\alpha}_0, \Psi) < S_E$;
(b) For $\boldsymbol{\alpha} \in \mathrm{K}_{\Psi} = \{\boldsymbol{\alpha} : S(\boldsymbol{\alpha}) \leq u(\boldsymbol{\alpha}_0, \Psi)\}$ the problem (1.76) has a solution, and its Lagrange multipliers $\lambda_i(\boldsymbol{\alpha}), i \in I_{\delta}(\boldsymbol{\alpha})$ satisfy the condition $\sum_{i \in I_{\delta}(\boldsymbol{\alpha})} |\lambda_i(\boldsymbol{\alpha})| \leq \Psi$, $\boldsymbol{\alpha} \in \mathrm{K}_{\Psi}$.

Assumption 1.4′ The functions $f_{pt}(\boldsymbol{\alpha})$ and $g_i(\boldsymbol{\alpha})$, $i = \overline{1, m}$ are differentiable on $\Re^n$ and their gradients satisfy in any compact set K the Lipschitz condition with constants that may depend on K:

$$\begin{aligned} &||\nabla f_{pt}(\boldsymbol{\alpha}_1) - \nabla f_{pt}(\boldsymbol{\alpha}_2)|| \leq l_{pt}||\boldsymbol{\alpha}_1 - \boldsymbol{\alpha}_2||, l_{pt} > 0, \quad p = \overline{1, P}, t = \overline{1, T}, \\ &||\nabla g_i(\boldsymbol{\alpha}_1) - \nabla g_i(\boldsymbol{\alpha}_2)|| \leq l_1||\boldsymbol{\alpha}_1 - \boldsymbol{\alpha}_2||, \quad l_1 > 0, i = \overline{1, m}. \end{aligned} \tag{1.79}$$

Similarly to Lemma 1.2, we have

Lemma 1.3. *Suppose that the matrix $\mathbf{\Sigma}$ is positive definite and (1.79) is valid for any compact set. Then the gradient of the function $S(\boldsymbol{\alpha})$ defined in (1.74) satisfies in this set the Lipschitz condition (1.38).*

Proof. If the matrix $\mathbf{\Sigma}$ is positive definite, then $\mathbf{\Sigma} = (\mathbf{\Gamma}'\mathbf{\Gamma})^{-1}$, where $\mathbf{\Gamma}$ is a non-degenerate matrix with bounded elements. Let $\mathbf{F}(\boldsymbol{\alpha}) = \mathbf{\Gamma}\mathbf{f}(\boldsymbol{\alpha})$, $\mathbf{Y}^* = \mathbf{\Gamma}\mathbf{Y}$. Then $S(\boldsymbol{\alpha}) = ||\mathbf{Y}^* - \mathbf{F}(\boldsymbol{\alpha})||^2$.

By (1.79), $||\nabla F_q(\boldsymbol{\alpha}_1) - \nabla F_q(\boldsymbol{\alpha}_2)|| \leq L_{0q}||\boldsymbol{\alpha}_1 - \boldsymbol{\alpha}_2||$, where $F_q(\boldsymbol{\alpha})$ is the component of $\mathbf{F}(\boldsymbol{\alpha})$, and $L_{0q} = \sum_{i=1}^{N} |\Gamma_{qj}|\omega_j > 0$. Here Γ_{qj} is the matrix element $\mathbf{\Gamma}$, and $\omega_j = l_{pt}$ when $j = (p-1)T + t$, $p = \overline{1,P}$, $t = \overline{1,T}$.

The proof follows by the same arguments as the proof of Lemma 1.2, if we substitute $\nabla f_t(\boldsymbol{\alpha})$ and $f_t(\boldsymbol{\alpha})$ by $\nabla F_q(\boldsymbol{\alpha})$ and $F_q(\boldsymbol{\alpha})$, respectively. □

By Lemma 1.3 one can show that the convergence results given above remain true for the estimation of regression parameters which are subject to inequality constraints.

Now we discuss a special case of the regression (1.73), where regression function and restrictions are linear in $\boldsymbol{\alpha}^0$, i.e. $f_{pt}(\boldsymbol{\alpha}^0) = \mathbf{x}'_{pt}\boldsymbol{\alpha}^0$, $p = \overline{1,P}$, $t = \overline{1,T}$, where $\mathbf{x}_{pt} \in \Re^n$ is a regressor in pth regression and $\boldsymbol{\alpha}^0$ satisfies $\mathbf{G}\boldsymbol{\alpha}^0 \leq \mathbf{b}$, where $\mathbf{G}$ is the $(m \times n)$ dimensional matrix, and $\mathbf{b}$ is the m dimensional vector.

In such a way (1.74) can be written as $\mathbf{f}(\boldsymbol{\alpha}) = \mathbf{D}\boldsymbol{\alpha}$, where $\mathbf{D} = \text{diag}(\mathbf{D}_1, \ldots, \mathbf{D}_P)$ is the $N \times n$ matrix, $n = \sum_{l=1}^{P} n_l$, $\mathbf{D}_l$ is the $T \times n_l$ submatrix, $l = \overline{1,P}$, whose tth row is equal to $\mathbf{x}'_{pt}$. Obviously $\mathbf{D}$ is a special case of $\mathbf{D}(\boldsymbol{\alpha})$ from (1.76).

Therefore the estimation problems (1.74) and (1.75) can be rewritten as

$$\tfrac{1}{2}(\mathbf{Y} - \mathbf{D}\boldsymbol{\alpha})'\Sigma^{-1}(\mathbf{Y} - \mathbf{D}\boldsymbol{\alpha}) \to \min, \tag{1.80}$$

$$\mathbf{G}\boldsymbol{\alpha} \leq \mathbf{b}. \tag{1.81}$$

The problems (1.80) and (1.81) has a solution under weaker conditions (compared with the general nonlinear case). If Assumptions 1.2 and 1.6 hold true and matrixes $\mathbf{D}_l$, $l = \overline{1,P}$ have full rank, then this solution is unique. It can be obtained by comparing (1.76), (1.80), and (1.81). Therefore, putting in (1.77) $v = 0$, $\mathbf{G}_\delta(\boldsymbol{\alpha}) = \mathbf{G}$, $\mathbf{D}(\boldsymbol{\alpha}) = \mathbf{D}$, $\mathbf{g}_\delta(\boldsymbol{\alpha}) = -\mathbf{b}$ and replacing $\mathbf{Y} - \mathbf{f}(\boldsymbol{\alpha})$ by $\mathbf{Y}$, we derive

$$\hat{\boldsymbol{\alpha}} = (\mathbf{J}_n - \mathbf{E}'\mathbf{G})\hat{\boldsymbol{\alpha}}^* + \mathbf{E}'b, \tag{1.82}$$

where $\hat{\boldsymbol{\alpha}}^* = (\mathbf{D}'\mathbf{\Sigma}^{-1}\mathbf{D})^{-1}\mathbf{D}'\mathbf{\Sigma}^{-1}\mathbf{Y}$ is the solution to (1.80),

$$\mathbf{E} = (\mathbf{G}(\mathbf{D}'\mathbf{\Sigma}^{-1}\mathbf{D})^{-1}\mathbf{G}')^{-1}\mathbf{G}(\mathbf{D}'\mathbf{\Sigma}^{-1}\mathbf{D})^{-1}. \tag{1.83}$$

In case of only one regression equation ($P = 1$), which corresponds to the regression problem

$$\frac{1}{2}||\mathbf{Y} - \mathbf{X}\alpha||^2 \rightarrow \min, \quad \mathbf{G}\alpha = \boldsymbol{b},$$

(in notations of (1.7)), estimation of the regression parameter is given by (1.82) and (1.83) with $\mathbf{D} = \mathbf{X}$, $\mathbf{\Sigma} = \mathbf{J}_T$. Therefore

$$\mathbf{E} = (\mathbf{G}(\mathbf{X}'X)^{-1}\mathbf{G}')^{-1}\mathbf{G}(\mathbf{X}'X)^{-1}. \tag{1.84}$$

The proposed method allows to obtain easily well-known results for special cases of (1.74) and (1.75); (1.82)–(1.84) were proposed earlier, for example, by Johnston (1963, Section 5.6).

Chapter 2
Asymptotic Properties of Parameters in Nonlinear Regression Models

In this chapter, we investigate some regression models with unknown coefficients. We assume that the parametric set of unknown parameters is closed and, generally speaking, unbounded. The case of open sets is easier to study, because in most of the cases the asymptotic distribution of estimates is normal. This is not always true when the constraints are compact sets. Everywhere in the text we consider discrete time observations. It is known that the observation errors taken at different times can be dependent. We do not consider here the continuous time version, although in that case many statements listed below also take place.

In Sect. 2.1 we give some general statements which are useful for proving consistency of estimates. Using them, we prove the results on consistency for types of estimates that are best known in practice, including least squares and least modules estimates. In Sect. 2.2 we investigate the asymptotic distribution of the least squares estimates. In Sects. 2.3 and 2.4 we investigate the asymptotic properties of the least squares estimates method in the context of problems where the set of restrictions on the parameter is convex or non-convex. In Sect. 2.5 a linear regression model with non-stationary variables is studied. As a rule, by $||\cdot||$ we denote the norm in $\Re^l$ or in some functional space, if it would not lead to misunderstanding.

2.1 Consistency of Estimates in Nonlinear Regression Models

In this section, we focus on getting the consistency conditions for nonlinear regression models of rather general form. From these conditions one can derive the consistency conditions for many known estimates, for example, for the least squares estimates, which we is discussed later.

P.S. Knopov and A.S. Korkhin, *Regression Analysis Under A Priori Parameter Restrictions*, Springer Optimization and Its Applications 54,
DOI 10.1007/978-1-4614-0574-0_2, © Springer Science+Business Media, LLC 2012

We give some examples of regression models, which are widely known to specialists in the field of theoretical and applied statistics.

1.
$$y_t = \sum_{i=1}^{p} x_{it}\alpha_i^0 + \varepsilon(t). \tag{2.1}$$

Here $\varepsilon(t)$ are independent or stationary dependent random variables, $\mathbf{x}_t = \{x_{it}, i = \overline{1, p}\}$, $t = 1, \ldots, n$ are independent identically distributed random vectors with $Ex_t = a_t \neq 0$, independent of $\varepsilon(t)$ (or observed without errors), such that

$$\frac{1}{T}\sum_{t=1}^{T} x_{it}^2 \to b_i^2 \quad \text{as } T \to \infty. \tag{2.2}$$

The vector $\boldsymbol{\alpha}^0 = (\alpha_1^0, \ldots, \alpha_p^0)$ is unknown and is to be estimated.

2.
$$y_t = f(t, \boldsymbol{\alpha}^0) + \varepsilon(t), \quad t = 1, \ldots, T. \tag{2.3}$$

This model is nonlinear in the vector of the unknown parameters $\boldsymbol{\alpha}^0$; the errors $\varepsilon(t)$ are the same as in model 1.

3.
$$y_t = f(\mathbf{x}(t), \boldsymbol{\alpha}^0) + \varepsilon(t), \quad \mathbf{x}(t) \in R^p, \tag{2.4}$$

where the p-dimensional vector $\mathbf{x}(t)$ and $\varepsilon(t)$ are mutually independent, and each of the sequences $\{\mathbf{x}(t)\}$ and $\{\varepsilon(t)\}$, $t = 1, \ldots, T$, is the sequence of independent or stationary random vectors or variables.

4.
$$y_{tT} = \alpha^0\left(\frac{t}{T}\right) + \varepsilon_{tT}, \quad t = 1, \ldots, T, \ T \geq 1, \tag{2.5}$$

where ε_{tT} is a sequence of series of independent random variables. This is a nonparametric regression model, because we have to estimate the function $\alpha^0(t) \in K$, where K is a compact set from some Banach space.

We restrict ourselves with these examples, and only note that each of the models is of independent interest. Of course, the examples above do not exhaust all regression models.

In the models above we assume that the parameter vector $\boldsymbol{\alpha}^0$ or the function α^0 is unknown and is a priori contained in some set $\tilde{K}$. The specific form of the set $\tilde{K}$ is discussed later.

Let us consider some cost functions characterizing the accuracy of the estimate.

1. *Least squares method.* For our model the cost functions are, respectively, of the form

$$Q_T(\boldsymbol{\alpha}) = \frac{1}{T}\sum_{t=1}^{T}\left[y_t - \sum_{i=1}^{p} x_{it}\alpha_i\right]^2, \tag{2.6}$$

$$Q_T(\boldsymbol{\alpha}) = \frac{1}{T}\sum_{t=1}^{T}[y_t - f(t,\boldsymbol{\alpha})]^2, \tag{2.7}$$

$$Q_T(\boldsymbol{\alpha}) = \frac{1}{T}\sum_{t=1}^{T}[y_t - f(\mathbf{x}(t),\boldsymbol{\alpha})]^2, \tag{2.8}$$

$$Q_T(\alpha) = \frac{1}{T}\sum_{t=1}^{T}\left[y_t - \alpha\left(\frac{t}{T}\right)\right]^2. \tag{2.9}$$

2. *Least absolute values method.* In formulas (2.6)–(2.9), the square of the difference is replaced by the absolute value of the difference. Least squares estimates and least absolute values estimates are particular cases of L_p-estimates, which use a cost function with the absolute value of the difference raised to the power p.
3. *Maximum likelihood method.* The distribution of observed random variables is assumed to be known, and the cost function for model (2.1), for instance, has the form

$$Q_T(\boldsymbol{\alpha}) = \frac{1}{T}\prod_{t=1}^{T} f\left(y_t - \sum_{i=1}^{p} x_{it}\alpha_i\right). \tag{2.10}$$

Similar criteria are used for maximum likelihood estimations in other models.

Here we give only the best known classical estimation criteria of unknown parameters in various observation models. Of course, these criteria do not exhaust all possible estimates. In particular, we consider estimates of functions of the form

$$Q_T(\boldsymbol{\alpha}) = \frac{1}{T}\sum_{t=1}^{T}\rho(\mathbf{x}_t,\boldsymbol{\alpha}), \tag{2.11}$$

where $\rho(\mathbf{x},\boldsymbol{\alpha})$ is known and depends on observations and an unknown parameter $\boldsymbol{\alpha}$.

The function $\rho(\mathbf{x},\boldsymbol{\alpha})$ may be quite general and is not necessarily everywhere differentials (Ermoliev and Norkin 2003). Estimates of the unknown parameter $\boldsymbol{\alpha}$ are also classified as Huber's M-estimates (Huber 1981; Hampel et al. 1986; Vapnik 1982, 1996). Various non-classical examples of M-estimates are given in Ermol'eva and Knopov (1986), Ermoliev and Knopov (2006), Knopov (1997a), Liese and Vajda (1994), and Van de Geer (1995). Many researches focus on consistency of estimates of linear regression in the case of independent observations. Main results are presented in Demidenko (1989), Dorogovtsev (1982), Ivanov (1984a, b, 1997), Kukush (1989), Le Cam (1953), and Malinvaud (1969), and we will not dwell on them in detail. For the case of dependent observations in discrete time we mention the works (Ajvazyan and Rozanov 1963; Holevo 1971; Dorogovtsev 1982).

Apparently, the analysis of consistency conditions for nonlinear regression models originated from classical works of Jennrich (1969) and Malinvaud (1969), further significant progress in this area was obtained in Dorogovtsev (1982) and

Ivanov (1997), and in many others. We discuss some of these results below. Let us focus on some essential points that arise in proving the consistency of estimates.

Some statements from which the properties of consistency or strong consistency of the above assessments follow, are given in Dorogovtsev (1982). We present theorems that give some common properties for all these estimates.

Theorem 2.1. *1. Let* $(\mathbf{\Omega}, F, P)$ *be a probability space,* $\Im_n \subset \Im$ *a sequence of* σ*-algebras such that* $\Im_n \subset \Im_{n+1}$*, and* K *a compact set from some Banach space with norm* $||\cdot||$.

2. Let $\{\mathbf{\Phi}_n(s) = \mathbf{\Phi}_n(s,\omega), (s,\omega) \in K \times \mathbf{\Omega}, n \geq 1\}$ *be a sequence of a separable real functions continuous everywhere on sets of the form* $||s - s_0|| \geq \delta$ *for any* $\delta > 0$*, where* $s_0 \in K$*, semi-continuous at the point* s_0 *for a fixed* $\omega \in \mathbf{\Omega}$*, and* $\Im_n$*-measurable for fixed* $n \geq 1$ *and* $s \in K$.

3. $P\{\lim_{n\to\infty} \mathbf{\Phi}_n(s,\omega) = \mathbf{\Phi}(s)\} = 1$*, where* $\mathbf{\Phi}(s)$ *is a separable real function continuous everywhere on sets of the form* $||s - s_0|| > \delta$ *for any* $\delta > 0$*, semi-continuous at the point* s_0*, and such that* $\mathbf{\Phi}(s) > \mathbf{\Phi}(s_0), s \in K, s \neq s_0$.

4. For any $\delta > 0$ *there exist* $\gamma_0 > 0$ *and function* $c(\gamma)$*,* $c(\gamma) \to 0$ *for* $\gamma \to 0$*, such that for any* $s' \in K$ *and any* $0 \leq \gamma < \gamma_0$ *we have*

$$P\left\{\lim_{n\to\infty} \sup_{\{||s-s'||\leq\gamma, ||s-s_0||\geq\delta\}} |Q_n(s) - Q_n(s')| < c(\gamma)\right\} = 1.$$

Assume that $s_n = s_n(\omega) \in K$ *is defined by* $Q_n(s_n) = \min_{s\in K} Q_n(s)$*. Then the element* s_n *can be chosen* $\Im_n$*-measurable and satisfying*

$$P\{\lim_{n\to\infty} ||s_n - s_0|| = 0\} = 1.$$

Remark 2.1. Theorem 2.1 for continuous $Q_n(s)$ was proved in Dorogovtsev (1982), but with slight changes in the proof the theorem remains valid also when $Q_n(s)$ are only lower semicontinuous at the point s_0.

Remark 2.2. If conditions 3 and 4 of Theorem 2.1 hold true in the sense of convergence in probability, then the assertion also takes place in the sense of convergence in probability.

In some cases the following modification of Theorem 2.1, given in Nemirovskii et al. (1984) is useful.

Let $(\Omega, \Im, P)$ be a probability space, $(F, \mathfrak{A})$ is a measurable space, $(E, ||\cdot||)$ is a Banach space containing F, $\{J_n\}$ is a sequence of functionals, which map $\mathbf{\Omega} \times F$ in $[-\infty, \infty]$. The functional $R : \Omega \times F \to [-\infty, \infty]$ is called the *normal integrand* on $\Omega \times F$, if it is measurable on $\Im \times \mathfrak{A}$, and lower semi-continuous in f (i.e. the sets $\{f : R(\omega, f) \leq c\}$ are closed) for almost every $\omega \in \Omega$. Suppose that the minimal contrast estimate, i.e. the measurable mapping $f_n : \Omega \to F$, satisfying the condition $f_n(\omega) = \arg\min_{f\in F} J_n(\omega, f), \omega \in \Omega$, is defined.

Let us give the results on the convergence of the sequence of minimal contrast estimates to some fixed $f^* \in F$. These results are simple modifications of the results proved in Dorogovtsev (1982), Huber (1981), Jennrich (1969), and Pfanzagl (1969), and therefore we formulate them without a proof. For brevity we drop the dependence on ω in J_n. Let $\Im_n(f) = J_n(f) - J_n(f^*)$. Fix $\varepsilon > 0$ and a natural number n_0.

Let W, U be some subsets from the class F. The assertion "condition (W, U) holds almost sure" means that

$$A_1. \quad \inf_{f \in W} \underline{\lim}_{n \to \infty} \Im_n(f) > 0 \quad \text{(a.s.)},$$

$$A_2. \quad \lim_{\delta \to 0} \overline{\lim}_{n \to \infty} \sup_{g:||g-f|| \leq \delta} |J_n(g) - J_n(f)| = 0 \quad \text{(a.s.)} \; \forall f \in U.$$

The statement "the condition (W, U) holds in probability" means that

$$A_1'. \; \exists d_w > 0 : \lim_{n \to \infty} P\{\Im_n(f) \geq d_w\} = 1, \quad \forall f \in W,$$

$$A_2'. \; \lim_{\delta \to 0} \overline{\lim}_{n \to \infty} \mathrm{P} \left\{ \sup_{g:||g-f|| \leq 0} |J_n(g) - J_n(f)| \geq \eta \right\} = 0, \quad \forall \eta > 0, \; \forall f \in U.$$

Remark 2.3. The supremum in A_2 and A_2' may be not σ-measurable. Then we need to assume that A_2 and A_2' are satisfied for its σ-measurable majorant.

Theorem 2.2. *Suppose that the functionals $J_n : \mathbf{\Omega} \times \mathrm{F} \to R^1$ are normal integrands on $\mathbf{\Omega} \times F$ for all $n > n_0$. Then $||f_n - f^*|| \to 0, n \to \infty$, a.s. (or in probability) in each of the following cases:*

1. *The set F is completely bounded in E and the condition $(S_\delta \cup (F \backslash B_\delta), F)$ is satisfied a.s. (in probability) for all $0 < \delta < \varepsilon$.*
2. *The set F is convex and locally completely bounded in, E the functionals J_n are convex in f, and the condition $(S_\delta, B_\varepsilon)$ is satisfied a.s. (in probability) for all $0 < \delta < \varepsilon$.*

(We say that F is a locally completely bounded set if all the balls in F are completely bounded.)

Remark 2.4. If we replace in Theorem 2.2 the requirement that U is completely bounded with the requirement that U is compact, then conditions A_1 and A_1' can be reduced, respectively, to

$$A_3. \quad \underline{\lim}_{n \to \infty} \Im_n(f) > 0, \quad \forall f \in \xi, \; f \neq f^*,$$

$$A_4. \quad \lim_{n \to \infty} P\{\Im_n(f) > 0\} = 1, \quad \forall f \in \xi, \; f \neq f^*.$$

Theorems 2.1 and 2.2 generalize known results of Le Cam (1953), Jennrich (1969), Pfanzagl (1969), Pfanzagl and Wefelmeyer (1985) and others, and are quite useful tools for proving consistency of estimates in different models. A wide range of regression models, for which the statements of consistency of estimates are proved by using Theorem 1, are given in Dorogovtsev (1982).

Traditionally, from the early works in the field of regression analysis, the estimation problem have been studied for the case when there are no restrictions on the range of admissible values of unknown regression parameters. There are many rather general conditions for consistency of some estimates (for example, least squares estimates, least modules estimates, maximum likelihood estimates and some others). The proof of consistency is more difficult and uses a priori restrictions on unknown parameters. Sometimes these difficulties are hardly possible to overcome. Therefore the statement concerning the relation between the consistency in case of a priori constraints on parameters and the consistency in case of estimation in the whole space might be useful. This statement is formulated in the lemma below.

Lemma 2.1. *Suppose that S is some set in a Banach space, $(\mathbf{\Omega}, \Im, P)$ is a probability space, $\{\Im_n, n \geq 1\}$ is a sequence of $\boldsymbol{\sigma}$-algebras such that $\Im_n \subset \Im_{n+1}$, $\Im_n \subset \Im$, $n \geq 1$. Assume that the following conditions hold true:*

1. *$\{Q_n(s) = Q_n(s,\omega), (s,\omega) \in S \times \mathbf{\Omega}, n \geq 1\}$ is a sequence of real functions, which are $\Im_n$-measurable for fixed s, and are uniformly continuous in s for fixed n and ω.*
2. *There is a function $Q(s)$ and a unique element $s_0 \in K \subset S$ such that $P\{\lim_{n\to\infty} Q_n(s) = Q(s)\} = 1$ and $Q(s) > Q(s_0)$, $s \neq s_0$.*
3. *Let $s_n = \arg\min_{s\in S} Q_n(s)$, $\tilde{s}_n = \arg\min_{s\in K} Q_n(s)$, and $P\{\lim_{n\to\infty} ||s_n - s_0|| = 0\} = 1$. Then $P\{\lim_{n\to\infty} ||\tilde{s}_n - s_0|| = 0\} = 1$.*

Proof. It is obvious that $Q_n(s_0) \to Q(s_0)$ and $\lim_{n\to\infty} Q_n(s) = Q(s) > Q(s_0)$ with probability 1. The convergence is uniform on any set $\mathbf{\Phi}_\delta = \{\omega : ||s - s_0|| \geq \delta\}$ with $\delta > 0$.

We prove that $||\tilde{s}_n - s_0|| \to 0$ as $n \to \infty$ with probability 1. Indeed, if such a convergence does not take place, then there exists a subsequence $n_k \to \infty$ such that $s_{n_k} \to s' \neq s_0$ as $k \to \infty$. We have $Q_{n_k}(\tilde{s}_{n_k}) \to Q(s') > Q(s_0)$ as $k \to \infty$ because the convergence with probability 1 is uniform on $\mathbf{\Phi}_\delta$ with $\delta < ||s' - s_0||$. On the other hand, according to the definition of $\tilde{s}_{n_k}$, $Q_{n_k}(\tilde{s}_{n_k}) \leq Q_{n_k}(s_0)$, and $Q_{n_k}(s_0) \to Q(s_0)$ with probability 1. This contradiction proves the lemma. □

Before proceeding to the problem of finding the consistency conditions in some regression models we would like to discuss briefly the problem of the measurability of considered estimates. These problems were first investigated in the fundamental paper (Pfanzagl 1969). Further, in Dorogovtsev (1982) and Knopov and Kasitskaya (2002) some questions concerning the measurability of studied estimates are investigated. We present some of the results below.

Theorem 2.3. *Let X be an arbitrary subset of some separable metric space with a metric ρ, (Y, Q) be a measurable space, $f = f(x, y) : X \times Y \to R$ be a function, continuous in the first argument for each y and measurable in the second argument for each x. Then the mappings $g(y) = \inf_{x\in X} f(x, y)$, $h(y) = \sup_{x\in X} f(x, y)$, $y \in Y$ are Q-measurable.*

Proof. Let X' be a discrete everywhere dense subset of X. The properties of measurable functions imply that the mapping $g_1(y) = \inf_{x\in X'} f(x, y)$, $y \in Y$, is Q-measurable. Fix an arbitrary element $y \in Y$. It will be shown that $g(y) = g_1(y)$. It is sufficient to prove that $f(x, y) \geq g_1(y)$, $x \in X$.

Fix $x \in X$. There exists a sequence $\{x_n\}$ of elements from X', converging to x as $n \to \infty$. Since f is continuous in the first argument, $f(x_n, y) \to f(x, y)$ as $n \to \infty$. Then $f(x_n, y) \geq g_1(y)$, $n \in N$. Hence $f(x, y) \geq g_1(y)$. Then the function $g(y) = g_1(y)$, $y \in Y$ is Q-measurable.

We can write $h(y) = -\inf_{x\in X}(-f(x, y))$, $y \in Y$. The same arguments can be applied to the function $-f$. Hence the mapping $h(y)$, $y \in Y$ is Q-measurable. □

Theorem 2.4. *(Schmetterer 1974) Let S be an arbitrary closed or open subset of R^l, $l \geq 1$, $(X, \aleph)$ is some measurable space. Suppose that $f : S \times X \to [-\infty, \infty]$ is a function satisfying the following conditions:*

1. *$f(\mathbf{s}, x), \mathbf{s} \in S$, is continuous for all $x \in X$*
2. *$f(\mathbf{s}, x), x \in X$ is $\aleph$-measurable for each $\mathbf{s} \in S$*
3. *For any $x \in X$ there exists $\mathbf{s} \in S$ with $\tilde{f}(x) = \inf_{s\in S} f(\mathbf{s}, x)$.*

Then there exists a measurable mapping $\phi : X \to T$ such that

$$f(\phi(x), x) = \inf_{s\in S} f(\mathbf{s}, x), \quad x \in X.$$

To illustrate Theorem 2.1 we consider below a model in which the function is a nonlinear regression of unknown parameter $\boldsymbol{\alpha}^0$, time-dependent with parameter t. Consider the model (2.3) with criterion (2.7). For this purpose we formulate the problem more clearly.

1. *We have the observation model $y_t = f(t, \boldsymbol{\alpha}^0) + \varepsilon(t)$, where $\varepsilon(t)$ is a Gaussian stationary process with discrete time, $E\varepsilon(t) = 0$, $r(t) = E\varepsilon(t)\varepsilon(0)$ and $|r(t)| \leq c/|t|^{1+\delta}$, $\delta > 0$, $E|\varepsilon(0)|^4 \leq c$.*
2. *The function $f(t, \alpha)$ is continuous in the second argument and satisfies the following conditions:*
   a. $\lim_{T\to\infty}(1/T)\sum_{t=1}^{T}[f(t, \boldsymbol{\alpha}) - f(t, \boldsymbol{\alpha}^0)]^2 = \boldsymbol{\Phi}(\boldsymbol{\alpha}) > 0$ for $\boldsymbol{\alpha} \neq \boldsymbol{\alpha}^0$
   b. $|f(t, \boldsymbol{\alpha}) - f(t, \bar{\boldsymbol{\alpha}})| \leq c||\boldsymbol{\alpha} - \bar{\boldsymbol{\alpha}}||^{\beta}, 0 < \beta \leq 2$, *wherethe constant* $\mathbf{c}$ *does not depend on* $t, \boldsymbol{\alpha}, \bar{\boldsymbol{\alpha}}$;
   c. $f(t, \boldsymbol{\alpha}) \to \infty$, $||\boldsymbol{\alpha}|| \to \infty$

Let $Q_T(\boldsymbol{\alpha}) = (1/T)\sum_{t=1}^{T}[y_t - f(t, \boldsymbol{\alpha})]^2$ and $\boldsymbol{\alpha}_T \in \arg\min_{\boldsymbol{\alpha}\in J} Q_T(\boldsymbol{\alpha})$, where J is a closed subset in R^p. The following assertion holds true.

Theorem 2.5. *Let conditions 1 and 2 be satisfied. Then*

$$P\left\{\lim_{T\to\infty} ||\boldsymbol{\alpha}_T - \boldsymbol{\alpha}^0|| = 0\right\} = 1.$$

Proof. We first prove that there exists $c > 0$ such that with probability 1 only a finite number of elements in the sequence $\boldsymbol{\alpha}_T$ lies outside the sphere $K = \{\boldsymbol{\alpha} : ||\boldsymbol{\alpha}|| \le c\}$. Let $\varphi(y_t, \boldsymbol{\alpha}) = [y_t - f(t, \boldsymbol{\alpha})]^2$. Note that $\psi_t(c) = \inf_{||\boldsymbol{\alpha}|| \ge c} \varphi(y_t, \boldsymbol{\alpha}) \to \infty$ as $c \to \infty$ with probability 1. Then $E\psi_t(c) \to \infty$, $c \to \infty$. Consequently, there is $c > 0$ such that $E\psi_t(c) > E\varphi(y_t, \boldsymbol{\alpha}^0)$, and by ergodic theorem we have

$$\frac{1}{T}\sum_{t=1}^{T} \psi_t(c) > \frac{1}{T}\sum_{t=1}^{T} \varphi(y_t, \boldsymbol{\alpha}^0) = F_T(\boldsymbol{\alpha}^0)$$

with probability 1 for sufficiently large T.

Thus with probability 1 all $\boldsymbol{\alpha}_T$ belong to K for some T.

Using this fact, we check the conditions of Theorem 2.1.

Let

$$\boldsymbol{\Phi}_T(\boldsymbol{\alpha}) = \frac{1}{T}\sum_{t=1}^{T} [y_t - f(t, \boldsymbol{\alpha})]^2 - \frac{1}{T}\sum_{t=1}^{T} \varepsilon^2(t),$$

$$E\boldsymbol{\Phi}_T(\boldsymbol{\alpha}) = \frac{1}{T}\sum_{t=1}^{T} [f(t, \boldsymbol{\alpha}^0) - f(t, \boldsymbol{\alpha})]^2.$$

Let $\eta_T = \boldsymbol{\Phi}_T(\boldsymbol{\alpha}) - E\boldsymbol{\Phi}_T(\boldsymbol{\alpha})$. Then

$$E\eta_T^4 = E\left\{\frac{2}{T}\sum_{t=1}^{T} [f(t, \boldsymbol{\alpha}^0) - f(t, \boldsymbol{\alpha})]\varepsilon(t)\right\}^4$$
$$= \frac{16}{T^4}\sum_{t_1=1}^{T}\sum_{t_2=1}^{T}\sum_{t_3=1}^{T}\sum_{t_4=1}^{T} [f(t_1, \boldsymbol{\alpha}^0) - f(t_1, \boldsymbol{\alpha})][f(t_2, \boldsymbol{\alpha}^0) - f(t_2, \boldsymbol{\alpha})]$$
$$\times [f(t_3, \boldsymbol{\alpha}^0) - f(t_3, \boldsymbol{\alpha})][f(t_4, \boldsymbol{\alpha}^0) - f(t_4, \boldsymbol{\alpha})]E\varepsilon(t_1)\varepsilon(t_2)\varepsilon(t_3)\varepsilon(t_4) \le \frac{c}{T^2}.$$

Therefore, by Borel-Cantelli lemma $P\{\lim_{T\to\infty} \eta_T = 0\} = 1$ implying that

$$P\{\lim_{T\to\infty} \boldsymbol{\Phi}_T(\boldsymbol{\alpha}) = \boldsymbol{\Phi}(\boldsymbol{\alpha})\} = 1.$$

Thus, condition 3 of Theorem 2.1 is satisfied. Let us verify condition 4 of this theorem. For fixed $\boldsymbol{\alpha}$ we have

$$\sup_{||\boldsymbol{\alpha} - \bar{\boldsymbol{\alpha}}|| < \gamma} |\boldsymbol{\Phi}_n(\boldsymbol{\alpha}) - \boldsymbol{\Phi}_n(\bar{\boldsymbol{\alpha}})|$$
$$\le \sup_{||\boldsymbol{\alpha} - \bar{\boldsymbol{\alpha}}|| < \gamma} \frac{1}{T}\left|\sum_{t=1}^{T} [y_t - f(t, \boldsymbol{\alpha})]^2 - \sum_{t=1}^{T} [y_t - f(t, \bar{\boldsymbol{\alpha}})]^2\right|$$

$$\leq \sup_{\|\alpha-\bar{\alpha}\|<\gamma} \frac{1}{T} \left| \sum_{t=1}^{T} \{[f(t,\alpha^0) - f(t,\alpha)]^2 - [f(t,\alpha^0) - f(t,\bar{\alpha})]^2 \right.$$

$$\left. +2\,[f(t,\bar{\alpha}) - f(t,\alpha)]\varepsilon(t)\} \right|$$

$$\leq \sup_{\|\alpha-\bar{\alpha}\|<\gamma} \frac{1}{T} \sum_{t=1}^{T} |f(t,\bar{\alpha}) - f(t,\alpha)| \cdot |2f(t,\alpha^0) - f(t,\alpha) - f(t,\bar{\alpha})|$$

$$+ \sup_{\|\alpha-\bar{\alpha}\|<\gamma} \left| \frac{2}{T} \sum_{t=1}^{T} [f(t,\alpha) - f(t,\bar{\alpha})]\varepsilon(t) \right|$$

$$\leq c\gamma + 2 \sup_{\|\alpha-\bar{\alpha}\|<\gamma} \left\{ \frac{1}{T} \sum_{t=1}^{T} [f(t,\alpha) - f(t,\bar{\alpha})]^2 \right\}^{1/2} \cdot \left\{ \frac{1}{T} \sum_{t=1}^{T} \varepsilon^2(t) \right\}^{1/2}$$

$$\leq c\gamma + c_1 \left\{ \frac{1}{T} \sum_{t=1}^{T} \varepsilon^2(t) \right\}^{1/2} \gamma^{\beta} \to c\gamma + c_1\gamma^{\beta} r(0) \quad \text{as } T \to \infty.$$

Let $c(\gamma) = c\gamma + c_1\gamma^{\beta} r(0)$. Then condition 4 of Theorem 2.1 holds true, which finishes the proof of the theorem. □

In order to prove the consistency let us consider another example of application of Theorem 2.1.

Consider the regression model (2.4) with the least modules cost function (2.7).

Theorem 2.6. *Let the following conditions be satisfied:*

1. $E|\varepsilon(t)| < \infty$
2. $E\{\max_{\{\alpha\in J, \|\alpha\|\leq c\}} |f(\mathrm{x}(t),\alpha)|\} < \infty$

where $f(\alpha,\beta)$ is continuous with respect to the first argument and measurable in the second;

3. *For any sequence $\{\mathrm{u}_j\}$ such that $\|\mathrm{u}_j\| \to \infty$ as $j \to \infty$, we have $f(\mathrm{x}(t),\mathrm{u}_j) \to \infty$ as $j \to \infty$ with probability 1*
4. $P\{\varepsilon(t) < 0\} = \frac{1}{2}$ for $t = 1, 2, \ldots$
5. *α_T is the solution to the minimization problem*

$$Q_T(\alpha_T) = \min_{\alpha\in J} \frac{1}{T} \sum_{t=0}^{T} |y_t - f(\mathrm{x}(t),\alpha_t)|.$$

Then $P\{\lim_{T\to\infty} \|\alpha_T - \alpha^0\| = 0\} = 1$.

In the proof we essentially use the following lemma proved in Knopov and Kasitskaya (1995).

Lemma 2.2. *Let ξ be a random variable on the probability space $(\Omega, \Im, P)$, $E|\xi| < \infty$, and let $\beta > 0$ be some real number. Assume that*

1. $P\{\xi < 0\} = \frac{1}{2}$.

2. $P\{\xi \in A_\beta\} > 0, \quad$ *where* $A_\beta = \begin{cases} [0, \beta), & \beta > 0, \\ [\beta, 0), & \beta < 0. \end{cases}$

If the condition (1) is satisfied, then $E|\xi - \beta| \geq E|\xi|$. *If conditions (1) and (2) are satisfied, then* $E|\xi - \beta| > E|\xi|$.

The proof of Theorem 2.6 follows by the same scheme as the proof of Theorem 2.5. Using the strong law of large numbers we get

$$P\left\{\lim_{T\to\infty} Q_T(\boldsymbol{\alpha}) = E|y_t - f(\boldsymbol{x}(t), \boldsymbol{\alpha})|\right\} = 1,$$

where $Q_T(\boldsymbol{\alpha}) = (1/T)\sum_{t=0}^{T}|y_t - f(\boldsymbol{x}(t), \boldsymbol{\alpha})|$.

Applying Lemma 2.2 we see that the point $\boldsymbol{\alpha}^0$ *is the only minimum point of the function* $E|y_t - f(\boldsymbol{x}(t), \boldsymbol{\alpha})|$. *Using this fact, we can get the assertion of Theorem 2.6 in the same way as that of Theorem 2.5. The complete proof of this theorem is given inKnopov (1997a–c).*

Obviously, if $\boldsymbol{x}_t = \{x_{it}, i = \overline{1, p}\}$ *are independent or stationary dependent vectors, the model (2.3) includes the model (2.1). If* $\boldsymbol{x}_t$ *is a non-random vector satisfying (2.2), then one can show that if the criterion (2.6) holds true, the conditions of Theorem 2.1 are be fulfilled with the functions* $\boldsymbol{\Phi}(\boldsymbol{\alpha})$ *and* $\boldsymbol{\Phi}_T(\boldsymbol{\alpha})$ *of the following form:*

$$\boldsymbol{\Phi}(\boldsymbol{\alpha}) = \sum_{i=1}^{p} b_i^2(\alpha_i - \alpha_i^0)^2,$$

$$\boldsymbol{\Phi}_T(\boldsymbol{\alpha}) = \frac{1}{T}\sum_{t=1}^{T}\left[y_t - \sum_{i=1}^{p} x_{it}\alpha_i\right]^2 - \frac{1}{T}\sum_{t=1}^{T}[\varepsilon(t)]^2,$$

provided that the values of $\varepsilon(t)$ *are independent and* $E|\varepsilon(t)|^4 < \infty$. *Similar conclusion, but subject to the conditions of Lemma 2.2, can be made for model (2.1) in the case of the least modules criterion.*

2.2 Asymptotic Properties of Nonlinear Regression Parameters Estimates Obtained by the Least Squares Method Under a Priory Inequality Constraints (Convex Case)

2.2.1 *Introduction*

In Sect. 1.2 we formulated the problem of estimation of parameters of nonlinear regression. For studying the statistical properties of regression parameters estimates it is more convenient to rewrite the estimation problem (1.16) as below:

$$S_T(\boldsymbol{\alpha}) \to \min, \quad g_i(\boldsymbol{\alpha}) \leq \mathbf{0}, \mathbf{i} \in \boldsymbol{I} = \{1, 2, \ldots, m\}, \tag{2.12}$$

where

$$S_T(\boldsymbol{\alpha}) = (2T)^{-1}\sum_{t=1}^{T}(y_t - f_t(\boldsymbol{\alpha}))^2. \quad (2.13)$$

In what follows we use the index T to emphasize the dependence of the criterion on T. Further, in this section we use the notation $\boldsymbol{\alpha}_T$ for the solution to (2.12). The true value $\boldsymbol{\alpha}^0$ of the regression parameter may satisfy both inequality and equality constraint:

$$g_i(\boldsymbol{\alpha}^0) = 0, \quad i \in I_1^0,\ g_i(\boldsymbol{\alpha}^0) < 0,\ i \in I_2^0,\ I_1^0 \cup I_2^0 = I = \{1,\dots,m\}. \quad (2.14)$$

We assume that the number m_1 of elements in the set I_1^0 is less than n.

Set

$$\mathbf{G}(\boldsymbol{\alpha}) = [\nabla g_1(\boldsymbol{\alpha}),\dots,\nabla g_m(\boldsymbol{\alpha})]', \quad \mathbf{R}_T(\boldsymbol{\alpha}) = T^{-1}\mathbf{D}_T'(\boldsymbol{\alpha})\mathbf{D}_T(\boldsymbol{\alpha}),$$

where matrix $\mathbf{D}_T(\boldsymbol{\alpha}) = [\partial f_t(\boldsymbol{\alpha})/\partial\alpha_j]$, $t = \overline{1,T}$, $j = \overline{1,n}$ is of dimension $T \times n$. Thus, the elements of the matrix $\mathbf{R}_T(\boldsymbol{\alpha})$ are

$$r_{kl}^T(\boldsymbol{\alpha}) = T^{-1}\sum_{t=1}^{T}\frac{\partial f_t(\boldsymbol{\alpha})}{\partial\alpha_k}\frac{\partial f_t(\boldsymbol{\alpha})}{\partial\alpha_l}, \quad k,l = \overline{1,n}. \quad (2.15)$$

In the sequel we assume that the following assumptions hold true.

Assumption 2.1. Random variables $\boldsymbol{\varepsilon}_t$ are independent and identically distributed with zero mathematical expectation and variance σ^2.

Assumption 2.2. A. Functions $f_t(\boldsymbol{\alpha})$, $t=\overline{1,T}$ and $g_i(\boldsymbol{\alpha})$, $i\in I$, are twice continuously differentiable on $\Re^n$.

B. For all $\boldsymbol{\alpha}\in\mathbf{M}=\{\boldsymbol{\alpha} : \mathbf{g}(\boldsymbol{\alpha}) \le \mathbf{O}_n, \boldsymbol{\alpha} \in \Re^n\}$ and all possible values of independent variables x_t there exist constants c_1 and c_2 such that

$$\left|\frac{\partial f_t(\boldsymbol{\alpha})}{\partial\alpha_i}\right| \le c_1, \quad \left|\frac{\partial^2 f_t(\boldsymbol{\alpha})}{\partial\alpha_i\partial\alpha_j}\right| \le c_2, \quad t = \overline{1,T},\ i,j = \overline{1,n}.$$

C. Functions $g_i(\boldsymbol{\alpha})$, $i \in I$, and their derivatives up to and including the second order are bounded in the neighborhood of $\boldsymbol{\alpha} = \boldsymbol{\alpha}^0$.

Assumption 2.3. The estimate $\boldsymbol{\alpha}_T$ is consistent and converges in probability to $\boldsymbol{\alpha}^0$.

The problem of estimating the consistency of $\boldsymbol{\alpha}_T$ was discussed in the preceding Sect. 2.1.

Assumption 2.4. Gradients $\nabla g_i(\boldsymbol{\alpha}^0)$, $i \in I_1^0$ are linearly independent.

Assumption 2.5. The matrix $\mathbf{R}_T(\boldsymbol{\alpha}^0)$ converges as $T \to \infty$ to a positive definite matrix $\mathbf{R}(\boldsymbol{\alpha}^0)$.

Assumption 2.6. Functions $g_i(\boldsymbol{\alpha}), i \in I$ are convex.

Assumption 2.7. There exists $\boldsymbol{\alpha}^-$ such that $\mathbf{g}(\boldsymbol{\alpha}^-) < \mathbf{O}_m$.

This assumption means that there exists at least one interior point in some suitable set $\mathbf{M}$ (the Slater condition).

From the last two assumptions it follows that the gradients $\nabla g_i(\boldsymbol{\alpha}_T), i \in I_a$, $I_a = \{i : g_i(\boldsymbol{\alpha}_T) = 0, i \in I\}$, are linearly independent.

2.2.2 Auxiliary Results

We prove two lemmas on properties of random matrices. Here, as well as in the sequel, we use some theorems on convergence of random variables from Rao (1965). The precise references will be given later in the text.

Lemma 2.3. *Let $\mathbf{A}_T$ be a random symmetric matrix of dimension $n \times n$, converging in probability as $T \to \infty$ to a positive definite matrix $\mathbf{A}$. Then:*

1. *The eigenvalues $a_{kT}, k = \overline{1,n}$ of $\mathbf{A}_T$ converge in probability to the eigenvalues $a_k, k = \overline{1,n}$ of the matrix $\mathbf{A}$.*
2. *The model matrix $\mathbf{C}_T$ of $\mathbf{A}_T$ converges in probability to the model matrix $\mathbf{C}$ of $\mathbf{A}$.*
3. *$\lim_{T\to\infty} p_T = 1$, where p_T is the probability that $\mathbf{A}_T$ has positive eigenvalues.*

Proof. We arrange the eigenvalues of $\mathbf{A}_T$ in non-decreasing order; these eigenvalues are random, see Girko (1980, Chap. 3). Since eigenvalues of a matrix are continuous functions of its elements, we have

$$p \lim_{T\to\infty} a_{kT} = a_k, \quad k = \overline{1,n}. \tag{2.16}$$

For the proof of the second statement one requires that

$$||\mathbf{x}_{kT}|| = ||\mathbf{x}_k|| = 1, \quad k = \overline{1,n}, \tag{2.17}$$

where $\mathbf{x}_{kT}$ and $\mathbf{x}_k$ are, respectively, the eigenvectors of $\mathbf{A}_T$ and $\mathbf{A}$. Assume that the eigenvectors of $\mathbf{A}_T$ are distinct and random. If there are multiple eigenvalues (among the variables a_{kT}, $k = \overline{1,n}$, for some T there are identical ones), then, in addition, fix some components of $\mathbf{x}_{kT}$. Thus, with probability 1, the vectors are determined uniquely from (2.16) and from equations $\mathbf{A}_T\mathbf{x}_{kT} = a_{kT}\mathbf{x}_{kT}$, $k = \overline{1,n}$. Taking into account that $\mathbf{A}x_k = a_k\mathbf{A}_k$, we have

$$(\mathbf{A} - \mathbf{a}_k\mathbf{J}_n)(\mathbf{x}_k - \mathbf{x}_{kT}) = \mathbf{b}_{kT} - \mathbf{c}_{kT} = -\mathbf{B}_{kT}, \quad k = \overline{1,n}, \tag{2.18}$$

where

$$\mathbf{b}_{kT} = (a_k - a_{kT})\mathbf{x}_k - (\mathbf{A} - \mathbf{A}_T)\mathbf{x}_k,$$

$$\mathbf{c}_{kT} = [(\mathbf{A}_T - \mathbf{A}) - (a_{kT} - a)\mathbf{J}_n](\mathbf{x}_k - \mathbf{x}_{kT}).$$

Taking into account (2.17), we have

$$||\mathbf{c}_{kT}|| \leq (||\mathbf{A}_T - \mathbf{A}|| + |a_k - a_{kT}|)\, ||\mathbf{x}_k - \mathbf{x}_{kT}|| \leq 2\,(||\mathbf{A}_T - \mathbf{A}|| + |a_k - a_{kT}|)\,.$$

According to the conditions of the lemma and (2.16) we obtain $p\lim_{T\to\infty} \mathbf{b}_{kT} = p\lim_{T\to\infty} \mathbf{c}_{kT} = \mathbf{O}_n$, $k = \overline{1,n}$. From this equation and (2.18) we derive $p\lim_{T\to\infty} \mathbf{B}_{kT} = \mathbf{O}_n$, $k = \overline{1,n}$. Taking into account that $(\mathbf{A} - a_k\mathbf{J}_n)x_k = \mathbf{O}_n$, we get from (2.18) the equality $(\mathbf{A} - a_k\mathbf{J}_n)\mathbf{x}_{kT} = \mathbf{B}_{kT}$. The general solution of this inhomogeneous equation is the sum of any of its partial solutions $\bar{\mathbf{x}}_{kT}$ and the general solution of the homogeneous equation $(\mathbf{A} - a_k\mathbf{J}_n)\mathbf{y}=\mathbf{O}_n$, which is equal to $q\mathbf{x}_k$, where q is random scalar, and $\mathbf{x}_k$ belongs to the linear subspace of dimension r, generated by r linearly independent eigenvectors $\mathbf{A}$ corresponding to a_k, $1 \leq r \leq n-1$. Thus, we have $\mathbf{x}_{kT} = \bar{\mathbf{x}}_{kT} + q_T\mathbf{x}_k$, where q_T have be chosen such that $\mathbf{x}_{kT}$ is the solution to (2.17). We arrive at the equation for determining q_T:

$$||\mathbf{x}_{kT}||^2 = \sum_{i=1}^{n} \left[\bar{x}_{kT}^{(i)} + q_T x_k^{(i)}\right]^2 = 1,$$

where $\bar{x}_{kT}^{(i)}, x_k^{(i)}$ are components of $\bar{\mathbf{x}}_k$ and $\mathbf{x}_k$, respectively. By (2.17), the solutions to this equation are

$$q_T^{1,2} = -\sum_{i=1}^{n} \bar{x}_{kT}^{(i)} x_k^{(i)} \pm \sqrt{\left[\sum_{i=1}^{n} \bar{x}_{kT}^{(i)} x_k^{(i)}\right]^2 - ||\bar{\mathbf{x}}_{kT}||^2 + 1}.$$

Without loss of generality we may assume that $\bar{x}_{kT}^{(i)} = 0, i = n-r+1, \ldots, n$. Then

$$\bar{\bar{\mathbf{x}}}_{kT} = \begin{bmatrix} x_{kT}^{(1)} \\ \vdots \\ x_{kT}^{(n-r)} \end{bmatrix} = \mathbf{A}_{n-r}^{-1} \begin{bmatrix} B_{kT}^{(1)} \\ \vdots \\ B_{kT}^{(n-r)} \end{bmatrix},$$

where $\mathbf{A}_{n-r}^{-1}$ is a submatrix of the matrix $(\mathbf{A} - a_k\mathbf{J}_n)$, obtained by deleting the last r columns and rows, and $B_{kT}^{(i)}$ is ith component of vector $\mathbf{B}_{kT}$. Since the rank of $(\mathbf{A} - a_k\mathbf{J}_n)$ is $n-r$, the matrix $\mathbf{A}_{n-r}$ is non-degenerate. We have $p\lim_{T\to\infty} \bar{\bar{\mathbf{x}}}_{kT} = \mathbf{O}_{n-r}$, whence $p\lim_{T\to\infty} \bar{\mathbf{x}}_{kT} = \mathbf{O}_n$, $k = \overline{1,n}$. Thus $p\lim_{T\to\infty} q_T = 0$, and we obtain

$$p\lim_{T\to\infty} \mathbf{x}_{kT} = \mathbf{x}_k, \quad k = \overline{1,n}. \tag{2.19}$$

The columns of the matrix $\mathbf{C}_T$ are the vectors $\mathbf{x}_{kT}$, $k = \overline{1,n}$. From here and (2.19) we get $p\lim_{T\to\infty} \mathbf{C}_T = \mathbf{C}$, which proves the second statement of the lemma.

Let us prove the third assertion. According to (2.16), for arbitrary $\varepsilon>0$ and $\delta > 0$ there exists $T_{0k} > 0$, such that $P\{|a_{kT} - a_k| < \varepsilon\} \geq 1 - \delta$, $T > T_{0k}$, $k = \overline{1,n}$. We select ε such that it is less than the minimal eigenvalue of $\mathbf{A}$. Then

$$p_T \geq P\{|a_{kT} - a_k| < \varepsilon, k = \overline{1,n}\} \geq P\{|a_{1T} - a_1| < \varepsilon\}$$
$$- \sum_{i=2}^{n} P\{|a_{kT} - a_k| \geq \varepsilon\} \geq 1 - n\delta, \quad T > \max_{k=\overline{1,n}} T_{0k}.$$

Thus, the third statement of the lemma is proved. □

Lemma 2.4. *Assume that the random matrix $\mathbf{A}_T$ of dimension $n \times n$ converges in probability to the positive definite matrix A. Then $p\lim_{T\to\infty} \tilde{\mathbf{A}}_T = \mathbf{A}^{-1}$, where $\tilde{\mathbf{A}}_T = (\mathbf{A}_T + \gamma_T(|a_{1T}| + c)\mathbf{J}_n)^{-1}$. Here $\gamma_T = 1$ if the matrix $\mathbf{A}_T$ is degenerate and $\gamma_T = 0$ otherwise, a_{1T} is the least eigenvalue of $\mathbf{A}_T$, and c is some positive number.*

Proof. First we prove that γ_T converges in probability to 0. The determinant $\det \mathbf{A}_T$ of $\mathbf{A}_T$ is a continuous function of elements of $\mathbf{A}_T$; therefore, it converges in probability to $\det \mathbf{A} > 0$. We select a number $\varepsilon > 0$ such that $\varepsilon < \det \mathbf{A}$. Then for arbitrary $\varepsilon_1 > 0$ and $\delta_0 > 0$, we have

$$P\{|\gamma_T| < \varepsilon_1\} \geq P\{\gamma_T = 0\} = P\{\det \mathbf{A}_T \neq 0\}$$
$$\geq P\{|\det \mathbf{A}_T - \det \mathbf{A}| < \varepsilon\} \geq 1 - \delta_0, \quad T > T_0.$$

Thus, $p\lim_{T\to\infty} \gamma_T = 0$. By convergence in probability of $\mathbf{A}_T$ to $\mathbf{A}$ and of a_{1T} to the minimal eigenvalue a_1 of $\mathbf{A}$ (according to the first statement of Lemma 2.3), we derive the statement of the lemma. □

We establish some property of the solution to the quadratic programming problem, required in the sequel.

Lemma 2.5. *The quadratic programming problem*

$$\frac{1}{2}\boldsymbol{\alpha}'\mathbf{R}\boldsymbol{\alpha} - \mathbf{Q}'\boldsymbol{\alpha} \to \min, \quad \mathbf{G}\boldsymbol{\alpha} \leq \mathbf{b} \tag{2.20}$$

has a solution $\boldsymbol{\alpha}(\mathbf{G},\mathbf{Q})$, continuous in $\mathbf{G}$ and $\mathbf{Q}$. Here $\boldsymbol{\alpha}, \mathbf{Q} \in \Re^n$, $\mathbf{R}$ is the positive definite matrix of dimension $n \times n$, $\mathbf{G}$ is the matrix of dimension $m \times n$, $\mathbf{b} \in \Re^m$.

Proof. For a nonsingular matrix H there always exists a positive definite matrix R such that $\mathbf{R} = \mathbf{H}'\mathbf{H}$. Put $\boldsymbol{\beta} = \mathbf{H}\boldsymbol{\alpha}$, $\mathbf{P} = (\mathbf{H}^{-1})'\mathbf{Q}$, $\mathbf{S} = \mathbf{G}\mathbf{H}^{-1}$. Under such rearrangements the quadratic programming problem (2.20) is transformed to

$$\min\left\{\frac{1}{2}\boldsymbol{\beta}'\boldsymbol{\beta} - \mathbf{P}'\boldsymbol{\beta} \,\middle|\, \mathbf{S}\boldsymbol{\beta} \leq \mathbf{b}\right\}. \tag{2.21}$$

Denote its solution by $\boldsymbol{\beta}(\mathbf{S},\mathbf{P})$. For arbitrary $\mathbf{Q}_1$ and $\mathbf{Q}_2 = \mathbf{Q}_1 + \Delta\mathbf{Q}$, $\mathbf{G}_1$ and $\mathbf{G}_2 = \mathbf{G}_1 + \Delta\mathbf{G}$, we have $\mathbf{P}_2 = \mathbf{P}_1 + \Delta\mathbf{P}$, $\mathbf{S}_2 = \mathbf{S}_1 + \Delta\mathbf{S}$, where $\mathbf{P}_i = (\mathbf{H}^{-1})\mathbf{Q}_i$, $\mathbf{S}_i = \mathbf{H}^{-1}\mathbf{G}_i$, $i = 1, 2$. Then $\mathbf{P}_2 = \mathbf{P}_1 + \Delta\mathbf{P}$, $\mathbf{S}_2 = \mathbf{S}_1 + \Delta\mathbf{S}$ for random $\mathbf{Q}_1$ and $\mathbf{Q}_2 = \mathbf{Q}_1 + \Delta\mathbf{Q}$, $\mathbf{G}_1$ and $\mathbf{G}_2 = \mathbf{G}_1 + \Delta\mathbf{G}$; here $\mathbf{P}_i = (\mathbf{H}^{-1})\mathbf{Q}_i$. Hence

$$\mathbf{\Delta Q} = \mathbf{H}'\mathbf{\Delta P}, \quad \mathbf{\Delta G} = \mathbf{H\Delta S}'. \tag{2.22}$$

For $\boldsymbol{\alpha}(\mathbf{G}_i, \mathbf{Q}_i) = \mathbf{H}^{-1}\boldsymbol{\beta}(\mathbf{S}_i, \mathbf{P}_i)$, $i = 1, 2$, we have

$$\boldsymbol{\Delta\alpha} = \mathbf{H}^{-1}\boldsymbol{\Delta\beta}, \tag{2.23}$$

where $\boldsymbol{\Delta\alpha} = \boldsymbol{\alpha}(\mathbf{G}_2, \mathbf{Q}_2) - \boldsymbol{\alpha}(\mathbf{G}_1, \mathbf{Q}_1)$, $\boldsymbol{\Delta\beta} = \boldsymbol{\beta}(\mathbf{S}_2, \mathbf{P}_2) - \boldsymbol{\beta}(\mathbf{S}_1, \mathbf{P}_1)$.

Let $\mathbf{\Delta Q} \to \mathbf{O}_n$, $\mathbf{\Delta G} \to \mathbf{O}_{mn}$; then by non-degeneracy of $\mathbf{H}$ it follows from (2.22) that $\mathbf{\Delta S} \to \mathbf{O}_{mn}$ and $\mathbf{\Delta P} \to \mathbf{O}_n$, which leads to $\boldsymbol{\Delta\beta} \to \mathbf{O}_n$ (Khenkin and Volinsky 1976), which together with (2.23) gives $\boldsymbol{\Delta\alpha} \to \mathbf{O}_n$. Lemma is proved. □

We introduce now a series of relations based on the necessary conditions for the existence of the extremum in problem (2.12), satisfied by α_T (see Assumptions 2.6 and 2.7):

$$\nabla S_T(\boldsymbol{\alpha}_T) + \sum_{i=1}^{m} \lambda_{iT} \nabla g_i(\boldsymbol{\alpha}_T) = \mathbf{O}_n, \tag{2.24}$$

$$\lambda_{iT} g_i(\boldsymbol{\alpha}_T) = 0, \quad \lambda_{iT} \geq 0, \, i \in I, \tag{2.25}$$

where λ_{iT} are the Lagrange multipliers.

Set $F_T(\boldsymbol{\alpha}) = \nabla S_T(\boldsymbol{\alpha})$. Applying the mean value theorem to the kth component of $\mathbf{F}_T(\boldsymbol{\alpha})$, $k = \overline{1, n}$, we obtain

$$\mathbf{F}_T(\boldsymbol{\alpha}_T) = \mathbf{F}_T(\boldsymbol{\alpha}^0) + \boldsymbol{\Phi}_T(\boldsymbol{\alpha}_T - \boldsymbol{\alpha}^0), \tag{2.26}$$

where $\boldsymbol{\Phi}_T$ is the $n \times n$ matrix, whose (k, l)th elements are of the form

$$\boldsymbol{\Phi}_{kl}^T = r_{kl}^T(\boldsymbol{\xi}_{1T}) + T^{-1} \sum_{t=1}^{T} (y_t - f_t(\boldsymbol{\xi}_{1T})) \left. \frac{\partial^2 f_t(\boldsymbol{\alpha})}{\partial \alpha_k \, \partial \alpha_l} \right|_{\boldsymbol{\alpha} = \boldsymbol{\xi}_{1T}}. \tag{2.27}$$

Here $\boldsymbol{\xi}_{1T}$ is a random variable, satisfying $||\boldsymbol{\xi}_{1T} - \boldsymbol{\alpha}^0|| \leq ||\boldsymbol{\alpha}_T - \boldsymbol{\alpha}^0||$. It follows from (2.15) and (2.27) that $\boldsymbol{\Phi}_T$ is symmetric. Let us find its limit in probability. First we find the limit of $r_{kl}^T(\boldsymbol{\xi}_{1T})$. For arbitrary $\varepsilon > 0$ we have

$$\begin{aligned} P\left\{\left|r_{kl}^T(\boldsymbol{\xi}_{1T}) - r_{kl}(\boldsymbol{\alpha}^0)\right| \geq \varepsilon\right\} &\leq P\left\{\left|r_{kl}^T(\boldsymbol{\xi}_{1T}) - r_{kl}^T(\boldsymbol{\alpha}^0)\right| \geq \frac{\varepsilon}{2}\right\} \\ &+ P\left\{\left|r_{kl}^T(\boldsymbol{\alpha}^0) - r_{kl}(\boldsymbol{\alpha}^0)\right| \geq \frac{\varepsilon}{2}\right\}. \end{aligned} \tag{2.28}$$

To estimate the first term in the right-hand side of (2.28) we perform some calculations. By the mean value theorem

$$r_{kl}^T(\boldsymbol{\xi}_{1T}) = r_{kl}^T(\boldsymbol{\alpha}^0) + (\nabla r_{kl}((\boldsymbol{\xi}_{2T}))'(\boldsymbol{\xi}_{1T} - \boldsymbol{\alpha}^0),$$

where ξ_{2T} is a random variable, satisfying

$$||\boldsymbol{\xi}_{2T} - \boldsymbol{\alpha}^0|| \leq ||\boldsymbol{\xi}_{1T} - \boldsymbol{\alpha}^0||.$$

Obviously, $\boldsymbol{\xi}_{1T}, \boldsymbol{\xi}_{2T} \in \mathrm{M}$ by convexity of the set M (Assumption 2.6); therefore, if Assumption 2.2B holds true, then one can easily show, making use of (2.15), that $||\nabla r_{kl}^T(\boldsymbol{\xi}_{2T})|| \leq 2c_1c_2\sqrt{n}$. From the last two relations and the convergence in probability of $\boldsymbol{\xi}_{1T}$ to $\boldsymbol{\alpha}^0$, we have for arbitrary $\delta > 0$

$$P\left\{\left|r_{kl}^T(\boldsymbol{\xi}_{1T}) - r_{kl}^T(\boldsymbol{\alpha}^0)\right| \geq \frac{\varepsilon}{2}\right\} \leq P\left\{||\boldsymbol{\xi}_{1T} - \boldsymbol{\alpha}^0|| \geq \frac{\varepsilon}{4c_1c_2\sqrt{n}}\right\} < \delta, \quad T > T_1.$$

According to Assumption 2.5, the second term in the right-hand side of (2.28) is equal to 0 for all sufficiently large T. From the estimates in the right-hand side of (2.28) it follows that

$$p \lim_{T\to\infty} r_{kl}^T(\boldsymbol{\xi}_{1T}) = r_{kl}(\boldsymbol{\alpha}^0). \tag{2.29}$$

Consider the second term in (2.27), which we denote by B_T. Using again the mean value theorem, we get

$$f_t(\boldsymbol{\xi}_{1T}) = f_t(\boldsymbol{\alpha}^0) + (\nabla f_t(\boldsymbol{\xi}_{3T}))'(\boldsymbol{\xi}_{1T} - \boldsymbol{\alpha}^0),$$

where $\boldsymbol{\xi}_{3T}$ is a random variable, satisfying the condition

$$||\boldsymbol{\xi}_{3T} - \boldsymbol{\alpha}^0|| \leq ||\boldsymbol{\xi}_{1T} - \boldsymbol{\alpha}^0||.$$

Then

$$\begin{aligned} B_T &= \frac{1}{T}\sum_{t=1}^{T} \varepsilon_t \left.\frac{\partial^2 f_t(\boldsymbol{\alpha})}{\partial\alpha_k\,\partial\alpha_l}\right|_{\boldsymbol{\alpha}=\boldsymbol{\xi}_{1t}} - \frac{1}{T}\sum_{t=1}^{T}\left[(\nabla f_t(\boldsymbol{\xi}_{3t}))'(\boldsymbol{\xi}_{1t} - \boldsymbol{\alpha}^0)\left.\frac{\partial^2 f_t(\boldsymbol{\alpha})}{\partial\alpha_k\,\partial\alpha_l}\right|_{\boldsymbol{\alpha}=\boldsymbol{\xi}_{1t}}\right] \\ &= B_{1T} - B_{2T}. \end{aligned}$$

Consider the random variable $B_{3T} = T^{-1}\sum_{t=1}^{T}\varepsilon_t(\partial^2 f_t(\boldsymbol{\alpha})/\partial\alpha_k\,\partial\alpha_l)$. By Assumption 2.2B and the law of large numbers, $\boldsymbol{B}_{3T}$ converges in probability to 0 uniformly with respect to $\boldsymbol{\alpha} \in \mathbf{M}$. Then, since $\boldsymbol{\xi}_{1T}$ converges in probability to $\boldsymbol{\alpha}^0$ and $\boldsymbol{\alpha}^0 \in \mathbf{M}$ according to Wilks (1962, Sect. 4.3.8), we have $\boldsymbol{p}\lim_{T\to\infty}\boldsymbol{B}_{1T} = 0$.

By Assumption 2.2B, the fact that $\boldsymbol{\xi}_{3T} \in \mathbf{M}$ and M is convex, we obtain $||\nabla f_t(\boldsymbol{\xi}_{3T})||^2 \leq nc_1^2$. Then, since $\boldsymbol{\xi}_{1T}$ converges in probability to $\boldsymbol{\alpha}^0$ and the second derivative of $f_t(\boldsymbol{\alpha})$ at $\boldsymbol{\alpha} = \boldsymbol{\xi}_{1T} \in \mathrm{M}$ is bounded, we obtain for arbitrary $\boldsymbol{\varepsilon} > \mathbf{0}$ and $\boldsymbol{\delta} > \mathbf{0}$

$$\mathbf{P}\{|B_{2T}| \geq \varepsilon\} \leq P\left\{||\boldsymbol{\xi}_{1T} - \boldsymbol{\alpha}^0|| \geq \frac{\boldsymbol{\varepsilon}}{c_1 c_2 \sqrt{n}}\right\} < \boldsymbol{\delta}, \quad T > T_2.$$

Consequently, the limit in probability of B_{2T}, as well as that of B_{1T}, is equal to 0, implying $p\lim_{T\to\infty} B_T = 0$. From here, (2.27), and (2.29), we obtain under Assumptions 2.2B, 2.3, 2.5, and 2.6,

$$p\lim_{T\to\infty} \boldsymbol{\Phi}_T = \mathbf{R}(\boldsymbol{\alpha}^0). \tag{2.30}$$

According to (2.8) we have

$$\left.\frac{\partial S_T(\boldsymbol{\alpha})}{\partial \alpha_k}\right|_{\alpha=\alpha^0} = -T^{-1}\sum_{t=1}^{T} \varepsilon_t \left.\frac{\partial f_t(\boldsymbol{\alpha})}{\partial \alpha_k}\right|_{\alpha=\alpha^0}. \tag{2.31}$$

From (2.31), Assumption 2.2B and the law of large numbers, we get $p\lim_{T\to\infty} \nabla \mathbf{S}_T(v) = \mathbf{O}_n$. Then, using (2.26), (2.30), and Assumption 2.3 and taking into account that the elements of $\mathbf{R}(\boldsymbol{\alpha}^0)$ are finite, we obtain

$$p\lim_{T\to\infty} \nabla S_T(\boldsymbol{\alpha}_T) = \mathbf{O}_n. \tag{2.32}$$

From (2.25) and Assumption 2.2C we derive

$$p\lim_{T\to\infty} \lambda_{iT} = 0, \quad i \in I_2^0, \tag{2.33}$$

since by continuity of $g_i(\boldsymbol{\alpha})$ with respect to $\boldsymbol{\alpha}$ we have

$$p\lim_{T\to\infty} g_i(\boldsymbol{\alpha}_T) = g_i(\boldsymbol{\alpha}^0), \quad i \in I. \tag{2.34}$$

Similarly,

$$p\lim_{T\to\infty} \nabla g_i(\boldsymbol{\alpha}_T) = \nabla g_i(\boldsymbol{\alpha}^0), \quad i \in I. \tag{2.35}$$

According to Assumptions 2.6 and 2.7, gradients $\nabla g_i(\boldsymbol{\alpha}_T)$, $i \in I_a$ are linearly independent. Thus, from the system of (2.24) we can determine uniquely λ_{iT}, $i \in I_1^0 \cap I_a$, by using (2.14), and putting $\lambda_{iT} = 0$ if $i \in I_1^0 \backslash I_1^0 \cap I_a$. According to (2.32), (2.33), (2.35), and Assumption 2.2C, the right-hand side of this system converges to 0 in probability. Then according to Assumption 2.4 and the condition $m_1 < n$, we have $p\lim_{T\to\infty} \lambda_{iT} = 0, i \in I_1^0$. Thus,

$$p\lim_{T\to\infty} \lambda_{iT} = \lambda_i^0 = 0, \quad i \in I.$$

We have proved the following lemma.

Lemma 2.6. *Assume that Assumptions 2.1–2.7 hold true and the number of elements in the set I_1^0 is less than n. Then the Lagrange multipliers λ_{iT}, $i \in I$, in problem (2.12) converge in probability to 0 as $T \to \infty$.*

According to Lemma 2.6, set $\mathbf{V}_T = \sqrt{T}(\boldsymbol{\lambda}_T - \boldsymbol{\lambda}^0) = \sqrt{T}\boldsymbol{\lambda}_T$, $\mathbf{U}_T = \sqrt{T}(\boldsymbol{\alpha}_T - \boldsymbol{\alpha}^0)$, where $\boldsymbol{\lambda}_T = [\lambda_{1T} \quad \ldots \quad \lambda_{mT}]'$, $\boldsymbol{\lambda}^0 = [\lambda_1^0 \quad \ldots \quad \lambda_m^0]'$. From (2.24)–(2.26) and (2.31) we have

$$-\mathbf{D}'_T(\boldsymbol{\alpha}^0)\mathbf{E}_T + \boldsymbol{\Phi}_T\mathbf{U}_T + \mathbf{G}'(\boldsymbol{\alpha}_T)\mathbf{V}_T = \mathbf{O}_n, \tag{2.36}$$

$$v_{iT}g_i(\boldsymbol{\alpha}_T) = 0, \quad v_{iT} \geq 0,\ i \in I, \tag{2.37}$$

where $\tilde{\mathbf{D}}'_T(\boldsymbol{\alpha}^0) = (\sqrt{T})^{-1}\mathbf{D}_T(\boldsymbol{\alpha}^0)$, $\mathbf{E}_T = [\varepsilon_1 \quad \ldots \quad \varepsilon_T]'$, $\mathbf{G}(\boldsymbol{\alpha})$ is an $m \times n$ matrix, whose ith row is $\nabla g_i'(\boldsymbol{\alpha})$, $i \in I$, $\mathbf{V}_T = [v_{1T}, \quad \ldots, \quad v_{mT}]'$.

Set

$$\mathbf{Q}_T = \tilde{\mathbf{D}}'_T(\boldsymbol{\alpha}^0)\mathbf{E}_T. \tag{2.38}$$

To find the asymptotic behavior of the distribution function of $\mathbf{Q}_T$ we use Corollary 2.6.1 of Theorem 2.6.1 from Anderson (1971), which includes this particular case. The conditions of the corollary are satisfied if Assumptions 2.1 and 2.5 hold true and elements of $\mathbf{D}_T(\boldsymbol{\alpha}^0)$ are uniformly bounded (i.e., Assumption 2.2B holds true). Then $\mathbf{Q}_T$ converges in distribution to a normally distributed random variable Q with zero expectation and covariance matrix $\sigma^2\mathbf{R}(\boldsymbol{\alpha}^0)$.

Set $\mathbf{V}_T = [\mathbf{V}'_{1T} \;\vdots\; \mathbf{V}'_{2T}]'$, $\mathbf{G}(\boldsymbol{\alpha}) = [\mathbf{G}'_1(\boldsymbol{\alpha}) \;\vdots\; \mathbf{G}'_2(\boldsymbol{\alpha})]'$, where $\mathbf{V}_{1T}$ and $\mathbf{V}_{2T}$ are vectors with components v_{iT}, $i \in I_1^0$ and v_{iT}, $i \in I_2^0$, respectively; $\mathbf{G}_1(\boldsymbol{\alpha})$ is $m_1 \times n$ matrix whose ith row is $\nabla g_i'(\boldsymbol{\alpha})$, $i \in I_1^0$; $\mathbf{G}_2(\boldsymbol{\alpha})$ is $(m - m_1) \times n$ matrix, $\nabla g_i'(\boldsymbol{\alpha})$, $i \in I_2^0$ is its ith row. In terms of the introduced notation we obtain from (2.36) to (2.38)

$$-\mathbf{Q}_T + \boldsymbol{\Phi}_T\mathbf{U}_T + \mathbf{G}'(\boldsymbol{\alpha}_T)\mathbf{V}_{1T} + \mathbf{G}'_2(\boldsymbol{\alpha}_T)\mathbf{V}_{2T} = \mathbf{O}_n. \tag{2.39}$$

According to (2.7), $g_i(\boldsymbol{\alpha}_T) + z_{iT} = 0$, $i \in I$, where $z_{iT} \geq 0$, $i \in I$. From here and (2.25) we derive

$$\lambda_{iT}z_{iT} = 0, \quad i \in I$$

or

$$v_{iT}w_{iT} = 0, \quad i \in I, \tag{2.40}$$

where $w_{iT} = \sqrt{T}z_{iT}$. Moreover, according to (2.25),

$$v_{iT} \geq 0, \quad i \in I. \tag{2.41}$$

By definition, $w_{iT} \geq 0$, $i \in I$. Expanding $g_i(\boldsymbol{\alpha})$ in Taylor series in the neighborhood of $\boldsymbol{\alpha} = \boldsymbol{\alpha}^0$, we obtain

$$\psi_i(\boldsymbol{\alpha}_T)(\boldsymbol{\alpha}_T - \boldsymbol{\alpha}^0) + z_{iT} = 0, \quad i \in I_1^0, \tag{2.42}$$

where the row vector equals to

$$\psi_i(\boldsymbol{\alpha}_T) = \nabla g_i'(\boldsymbol{\alpha}^0) + \frac{1}{2}(\boldsymbol{\alpha}_T - \boldsymbol{\alpha}^0)' g_{i2}(\boldsymbol{\alpha}^0 + \boldsymbol{\theta}_{T1}\boldsymbol{\Delta\alpha}_T). \tag{2.43}$$

Here $\theta_1 \in [0,1]$; $\Delta\boldsymbol{\alpha}_T = \boldsymbol{\alpha}_T - \boldsymbol{\alpha}^0$; $g_{i2}(\boldsymbol{\alpha})$ *is the Hessian matrix of dimension* $n \times n$, *with elements* $\partial^2 g_i(\boldsymbol{\alpha})/\partial\alpha_j\partial\alpha_k$, $j,k = \overline{1,n}$. *Multiplying both sides of (2.42) by* $\sqrt{T}$, *we write this expression in the matrix form:*

$$\boldsymbol{\Psi}_1(\boldsymbol{\alpha}_T)\mathbf{U}_T + \mathbf{W}_{1T} = \mathbf{O}_{m_1}, \tag{2.44}$$

where $\boldsymbol{\Psi}_1(\boldsymbol{\alpha}_T)$ *is an* $m_1 \times n$ *matrix whose* i*th row is* $\psi_i(\boldsymbol{\alpha}_T)$, $i \in I_1^0$; $\mathbf{W}_{1T}$ *is* m_1*-dimensional vector with components* w_{iT}, $i \in I_1^0$.

Based on the results obtained above, we prove the following three lemmas.

Lemma 2.7. *Assume that Assumptions 2.1–2.7 hold true. Then*

$$p\lim_{T\to\infty} \mathbf{V}_{2T} = \mathbf{O}_{m_2}. \tag{2.45}$$

This expression follows easily from (2.34) to (2.37) since $g_i(\boldsymbol{\alpha}^0) < 0$, $i \in I_2^0$.

Lemma 2.8. *Assume that Assumptions 2.1–2.7 hold true. Then for any* $\delta > 0$ *there exist* $\varepsilon > 0$ *and* $T_0 > 0$ *such that*

$$P\{||\mathbf{V}_{1T}|| \geq \varepsilon\} < \delta, \quad T > T_0. \tag{2.46}$$

Proof. Denote by $\tilde{\boldsymbol{\Phi}}_T = (\boldsymbol{\Phi}_T + \gamma_T(|\Lambda_{1T}| + c)\mathbf{J}_n)^{-1}$ the matrix defined as in Lemma 2.4, i.e., $\gamma_T = 1$ if $\det \boldsymbol{\Phi}_T = 0$, and $\gamma_T = 0$ otherwise; Λ_{1T} is the smallest eigenvalue of $\boldsymbol{\Phi}_T$, and $c > 0$, and the elements of the matrix $\boldsymbol{\Phi}_T$ are defined by (2.27). As follows from the proof of Lemma 2.4, we have

$$P\{\tilde{\boldsymbol{\Phi}}_T\boldsymbol{\Phi}_T = \mathbf{J}_n\} \geq P\{\gamma_T = 0\} = P\{\det \boldsymbol{\Phi}_T \neq 0\} \geq 1 - \delta_0, \quad T > T_0',$$

where δ_0 is an arbitrary positive number. From the expression above and (2.39) we obtain

$$P\{\tilde{\boldsymbol{\Phi}}_T\boldsymbol{\Phi}_T = \mathbf{J}_n\} = P\Big\{\mathbf{U}_T = \tilde{\boldsymbol{\Phi}}_T\mathbf{Q}_T - \tilde{\boldsymbol{\Phi}}_T\mathbf{G}'_1(\boldsymbol{\alpha}_T)\mathbf{V}_{1T} - \tilde{\boldsymbol{\Phi}}_T\mathbf{G}'_2(\boldsymbol{\alpha}_T)\mathbf{V}_{2T}\Big\} \geq 1 - \delta_0,\ T > T_0'. \tag{2.47}$$

Set

$$\mathbf{B}_T = \boldsymbol{\Psi}_1(\boldsymbol{\alpha}_T)\tilde{\boldsymbol{\Phi}}_T\mathbf{G}'_1(\boldsymbol{\alpha}_T), \mathbf{q}_T = \boldsymbol{\Psi}_1(\boldsymbol{\alpha}_T)\tilde{\boldsymbol{\Phi}}_T\mathbf{Q}_T,$$
$$\mathbf{h}_T = \boldsymbol{\Psi}_1(\boldsymbol{\alpha}_T)\tilde{\boldsymbol{\Phi}}_T\mathbf{G}'_2(\boldsymbol{\alpha}_T)\mathbf{V}_{2T}. \tag{2.48}$$

Then $\mathbf{V'}_{1T}\mathbf{B}_T\mathbf{V}_{1T} = \mathbf{V'}_{1T}\mathbf{A}_T\mathbf{V}_{1T}$, where $\mathbf{A}_T$ is a symmetric matrix with elements $a_{ij}^T = \frac{1}{2}(b_{ij}^T + b_{ji}^T)$, $i, j = \overline{1,n}$ (where b_{ij}^T are the elements of $\mathbf{B}_T$). Thus, using (2.40), (2.44), (2.47), and (2.48), we obtain

$$P\{\tilde{\mathbf{\Phi}}_T\mathbf{\Phi}_T = \mathbf{J}_n\} = P\{\mathbf{V'}_{1T}\mathbf{A}_T\mathbf{V}_{1T} + \mathbf{V'}_{1T}(\mathbf{h}_T - \mathbf{q}_T) = 0\} \geq 1 - \delta_0, \quad T > T_0'. \tag{2.49}$$

For the matrix $\mathbf{A}_T$ we have

$$\mathbf{C'}_T\mathbf{A}_T\mathbf{C}_T = \mathbf{N}_T, \tag{2.50}$$

where $\mathbf{C}_T$ is orthogonal matrix, $\mathbf{N}_T = \text{diag}\,(\nu_{1T}, \cdots, \nu_{nT})$, ν_{iT} is the ith eigenvalue of $\mathbf{A}_T$. Applying Lemma 2.2 to $\mathbf{\Phi}_T$ and using (2.30), we obtain

$$p\lim_{T\to\infty} \tilde{\mathbf{\Phi}}_T = \mathbf{R}^{-1}(\alpha^0). \tag{2.51}$$

By consistency of $\boldsymbol{\alpha}_T$ and Assumption 2.2C, we get from (2.43) to (2.35)

$$p\lim_{T\to\infty} \mathbf{\Psi}_1(\boldsymbol{\alpha}_T) = p\lim_{T\to\infty} \mathbf{G}_1(\boldsymbol{\alpha}_T) = \mathbf{G}_1(\alpha^0). \tag{2.52}$$

Then from (2.48) to (2.51),

$$p\lim_{T\to\infty} \mathbf{B}_T = p\lim_{T\to\infty} \mathbf{A}_T = \mathbf{G}_1(\alpha^0)\mathbf{R}^{-1}(\alpha^0)\mathbf{G'}_1(\alpha^0) = \mathbf{A}.$$

Therefore, taking into account Assumptions 2.4 and 2.5, we derive that $\mathbf{A}$ is positive definite. According to Lemma 2.3, we see that

$$p\lim_{T\to\infty} \mathbf{N}_T = \mathbf{N}, \quad p\lim_{T\to\infty} \mathbf{C}_T = \mathbf{C}, \tag{2.53}$$

where $\mathbf{N} = \text{diag}\,(\nu_1, \ldots, \nu_n)$, ν_i is the ith eigenvalue of matrix $\mathbf{A}$, and $\mathbf{C}$ is the orthogonal matrix such that $\mathbf{C'AC} = \mathbf{N}$.

Set $\tilde{\nu}_{iT} = \nu_{iT}$ if $\nu_{iT} > 0$, and $\tilde{\nu}_{iT} = 1$ otherwise. Using the first relation in (2.53), we arrive at

$$p\lim_{T\to\infty} \tilde{\mathbf{N}}_T = \mathbf{N}, \tag{2.54}$$

where $\tilde{\mathbf{N}}_T = \text{diag}(\tilde{\nu}_{1T}, \ldots, \tilde{\nu}_{nT})$. By statement 3 of Lemma 2.3 for given number $\delta' > 0$ there exists $T_1' > 0$ such that

$$p\{\mathbf{N}_T = \tilde{\mathbf{N}}_T\} = P\{\nu_{iT} > 0, i = \overline{1,n}\} > 1 - \frac{\delta'}{3}, \quad T > T_1'.$$

Denote

$$\mathbf{Y}_T = \tilde{\mathbf{N}}_T^{1/2}\mathbf{C}_T^{-1}\mathbf{V}_{1T}. \tag{2.55}$$

Substituting $\mathbf{V}_{1T}$ from (2.55) in (2.49), we obtain, taking into account (2.50), the equality above and the inequality

$$P\{\det \mathbf{\Phi}_T = 0\} \leq \delta_0, \quad T > T_0',$$

that

$$P\{\mathbf{Y}'_T\mathbf{Y}_T + 2\mathbf{Y}'_T\mathbf{K}_T = 0\} \geq P\{\nu_{iT} > 0,\ i = \overline{1,n};\ \det \mathbf{\Phi}_T \neq 0\}$$
$$= P\{\nu_{iT} > 0,\ i = \overline{1,n}\} - P\{\nu_{iT} > 0,\ i = \overline{1,n};\ \det \mathbf{\Phi}_T = 0\}$$
$$\geq P\{\nu_{iT} > 0,\ i = \overline{1,n}\} - P\{\det \mathbf{\Phi}_T = 0\} \geq 1 - \frac{\delta}{3},$$

where $\delta = 3\delta_0 + \delta'$, i.e.

$$P\{Y_T'Y_T + 2Y_T'K_T = 0\} \geq 1 - \frac{\delta}{3}, \quad T > T_1. \tag{2.56}$$

Here

$$\mathbf{K}_T = \frac{1}{2}\mathbf{N}_T^{1/2}\mathbf{C}'_T(\mathbf{h}_T - \mathbf{q}_T). \tag{2.57}$$

As it was shown above,

$$\mathbf{Q}_T \overset{p}{\Rightarrow} \mathbf{Q}, \quad T \to \infty, \tag{2.58}$$

where $\mathbf{Q}_T$ is defined in (2.38), and Q is a normally distributed random variable.

From (2.48), according to (2.35), (2.45), (2.51), (2.52), and (2.58), we obtain that $\mathbf{q}_T \overset{p}{\Rightarrow} \mathbf{q}$, $T \to \infty$; and $p\lim_{T\to\infty} \mathbf{h}_T = \mathbf{O}_{m_1}$. It follows from (2.57), (2.53), (2.54) and the last two expressions, that the limit distribution of $\mathbf{K}_T$ coincides with the distribution of the random variable $\mathbf{K} = -\frac{1}{2}\mathbf{N}^{-1/2}\mathbf{C}'\mathbf{q}$. One can show that components of the m_1-dimensional random variable $\mathbf{K}$ are independent, centered, normally distributed with variance equal to 1. Therefore the limit distribution of $4||\mathbf{K}_T||^2$ is χ^2. Set $\mathbf{Y}_T = \tilde{\mathbf{Y}}_T - \mathbf{K}_T$. Then we obtain from (2.56)

$$P\{||\tilde{\mathbf{Y}}_T|| = ||\mathbf{K}_T||\} \geq 1 - \frac{\delta}{3}, \quad T > T_1. \tag{2.59}$$

The random variable K_T has a limit distribution, which implies that for a given $\delta > 0$ there exist $T_2 > 0$ and $\varepsilon_1 > 0$ such that

$$\frac{\delta}{6} > P\left\{||\mathbf{K}_T|| \geq \frac{\varepsilon_1}{2}\right\} \geq P\left\{||\tilde{\mathbf{Y}}_T|| \geq \frac{\varepsilon_1}{2},\ ||\mathbf{K}_T|| \geq \frac{\varepsilon_1}{2}\right\}$$
$$\geq P\left\{||\tilde{\mathbf{Y}}_T|| \geq \frac{\varepsilon_1}{2}\right\} - P\{||\tilde{\mathbf{Y}}_T|| \neq ||\mathbf{K}_T||\}, \quad T > T_2.$$

Then, taking into account (2.59), we obtain $P\{\|\tilde{\mathbf{Y}}_{\mathbf{T}}\| \geq (\varepsilon_1/2)\} < \delta/2$, $T > T_3 = \max(T_1, T_2)$. Since $||\mathbf{Y}_T|| \leq ||\tilde{\mathbf{Y}}_T|| + ||\mathbf{K}_T||$, for an arbitrary $\delta > 0$ there exists $\varepsilon_1 > 0$ such that

$$P\{||\mathbf{Y}_T|| < \varepsilon_1\} \geq P\left\{||\tilde{\mathbf{Y}}_T|| < \frac{\varepsilon_1}{2}\right\} - P\left\{||\mathbf{K}_T|| \geq \frac{\varepsilon_1}{2}\right\} > 1 - \frac{2\delta}{3}, \quad T > T_3. \tag{2.60}$$

We have from (2.55)

$$||\mathbf{V}_{1T}|| \leq ||\mathbf{C}_{1T}|| \cdot ||\mathbf{Y}_T||, \tag{2.61}$$

where $\mathbf{C}_{1T} = \mathbf{C}_T\mathbf{N}_T^{-1/2}$. It follows from (2.53) to (2.54) that $p\lim_{T\to\infty} \mathbf{C}_{1T} = \mathbf{C}_1 = \mathbf{C}\mathbf{N}^{-1/2}$. Hence, for a given $\delta > 0$ there exists $\varepsilon > 0$ such that

$$P\{\|\mathbf{C}_{1T}\|\,\varepsilon_1 < \varepsilon\} \geq 1 - \frac{\delta}{3}, \quad T > T_4. \tag{2.62}$$

Obviously, (2.60) and (2.62) are simultaneously satisfied for $T_0 = \max(T_3, T_4)$. Inserting in the inequality

$$P\{a \geq b\} \geq P\{a \geq \zeta\} - P\{b \geq \zeta\} \tag{2.63}$$

the values $a = ||\mathbf{C}_{1T}|| \cdot ||\mathbf{Y}_T||$, $b = ||\mathbf{C}_{1T}||\varepsilon_1$, and $\zeta = \varepsilon$, we obtain from (2.60) to (2.62)

$$\frac{2}{3}\delta > P\{\|\mathbf{C}_{1T}\| \cdot \|\mathbf{Y}_T\| \geq \|\mathbf{C}_{1T}\|\,\varepsilon_1\} \geq P\{\|\mathbf{C}_{1T}\| \cdot \|\mathbf{Y}_T\| \geq \varepsilon\} - \frac{\delta}{3}, \quad T > T_0.$$

Then (2.64) follows from the inequalities above and (2.61). The lemma is proved. □

Lemma 2.9. *Suppose that Assumptions 2.1–2.7 hold true. Then for given $\delta > 0$ there exist $\varepsilon > 0$ and $T_0 > 0$ such that*

$$P\{\|\mathbf{U}_T\| \geq \varepsilon\} < \delta, \quad T > T_0. \tag{2.64}$$

Proof. Set

$$\boldsymbol{\Sigma}_T = \tilde{\boldsymbol{\Phi}}_T\mathbf{Q}_T - \tilde{\boldsymbol{\Phi}}_T\mathbf{G}'_1(\boldsymbol{\alpha}_T)\mathbf{V}_{1T} - \tilde{\boldsymbol{\Phi}}_T\mathbf{G}'_2(\boldsymbol{\alpha}_T)\mathbf{V}_{2T}.$$

For some number $\varepsilon_3 > 0$ we have

$$P\{\|\mathbf{U}_T\| \geq \varepsilon_3,\, \|\mathbf{U}_T\| = \|\boldsymbol{\Sigma}_T\|\} \geq P\{\|\mathbf{U}_T\| \geq \varepsilon_3\} - P\{\|\mathbf{U}_T\| \neq \|\boldsymbol{\Sigma}_T\|\}.$$

Then, taking into account (2.47) it follows that

$$P\{\|\mathbf{U}_T\| \geq \varepsilon_3\} \leq P\{\|\mathbf{U}_T\| \geq \varepsilon_3, \|\mathbf{U}_T\| = \|\boldsymbol{\Sigma}_T\|\} + \delta_0 \leq P\{\|\boldsymbol{\Sigma}_T\| \geq \varepsilon_3\} + \delta_0, \quad T > T'_0.$$

From the inequalities above we obtain

$$P\{\|\mathbf{U}_T\| \geq \varepsilon_3\} \leq P\{\|\mathbf{\Sigma}_T\| \geq \varepsilon_3\} + \delta_0 \leq P\left\{||\tilde{\mathbf{\Phi}}_T\mathbf{G}'_1(\boldsymbol{\alpha}_T)|| \cdot ||\mathbf{V}_{1T}|| \geq \frac{\varepsilon_3}{3}\right\}$$
$$+P\left\{||\tilde{\mathbf{\Phi}}_T\mathbf{G}'_2(\boldsymbol{\alpha}_T)|| \cdot ||\mathbf{V}_{2T}|| \geq \frac{\varepsilon_3}{3}\right\} + P\left\{||\tilde{\mathbf{\Phi}}_T\mathbf{Q}_T|| \geq \frac{\varepsilon_3}{3}\right\} + \delta_0, \quad T > T'_0. \tag{2.65}$$

Let us estimate the terms in the right-hand side of (2.65). According to (2.51) and (2.52), for given $\delta > 0$ and $\varepsilon_1 > 0$ we have

$$P\{||\tilde{\mathbf{\Phi}}_T\mathbf{G}'_1(\boldsymbol{\alpha}_T)|| - ||\mathbf{R}^{-1}(\boldsymbol{\alpha}^0)\mathbf{G}'_1(\boldsymbol{\alpha}^0)|| < \varepsilon_1\} \geq P\{||\tilde{\mathbf{\Phi}}_T$$
$$\mathbf{G}'_1(\boldsymbol{\alpha}_T) - \mathbf{R}^{-1}(\boldsymbol{\alpha}^0)\mathbf{G}'_1(\boldsymbol{\alpha}^0)|| < \varepsilon_1\} \geq 1-\delta, \quad T > T_1.$$

From above, setting $\varepsilon_2 = \varepsilon||\mathbf{R}^{-1}(\boldsymbol{\alpha}^0)\mathbf{G}'_1(\boldsymbol{\alpha}^0)|| + \varepsilon\varepsilon_1$, where $\varepsilon > 0$ is arbitrary, we obtain

$$P\{\varepsilon||\tilde{\mathbf{\Phi}}_T\mathbf{G}'_1(\boldsymbol{\alpha}_T)|| < \varepsilon_2\} \geq 1 - \delta, \quad T > T_1. \tag{2.66}$$

Multiply both sides of the inequality in curly brackets in the expression (2.66) by $||\tilde{\mathbf{\Phi}}_T\mathbf{G}'_1(\boldsymbol{\alpha}_T)||$. Set in (2.63) $a = ||\tilde{\mathbf{\Phi}}_T\mathbf{G}'_1(\boldsymbol{\alpha}_T)|| \cdot ||\mathbf{V}_{1T}||$, $b = ||\tilde{\mathbf{\Phi}}_T\mathbf{G}'_1(\boldsymbol{\alpha}_T)||\varepsilon$, $\zeta = \varepsilon_2$. According to (2.46) and (2.63), we have

$$\delta > P\{||\tilde{\mathbf{\Phi}}_T\mathbf{G}'_1(\boldsymbol{\alpha}_T)|| \cdot ||\mathbf{V}_{1T}|| \geq \varepsilon_2\} - P\{||\tilde{\mathbf{\Phi}}_T\mathbf{G}'_1(\boldsymbol{\alpha}_T)||\varepsilon \geq \varepsilon_2\}, \quad T > T_2.$$

From the line above and (2.66) we obtain, setting $\delta = \delta_1/6$,

$$P\{||\tilde{\mathbf{\Phi}}_T\mathbf{G}'_1(\boldsymbol{\alpha}_T)|| \cdot \|\mathbf{V}_{1T}\| \geq \varepsilon_2\} < \frac{\delta_1}{3}, \quad T > \max(T_1, T_2). \tag{2.67}$$

According to (2.51), (2.35), and (2.45), $\tilde{\boldsymbol{\phi}}_T\mathbf{G}'_2(\boldsymbol{\alpha}_T)\mathbf{V}_{2T}$ converges in probability to 0. Thus, for given $\varepsilon_2 > 0$ and $\delta_1 > 0$ we can find $T_3 > 0$ such that

$$P\{||\tilde{\mathbf{\Phi}}_T\mathbf{G}'_2(\boldsymbol{\alpha}_T)\mathbf{V}_{2T}|| \geq \varepsilon_2\} < \frac{\delta_1}{3}, \quad T > T_3. \tag{2.68}$$

It follows from (2.58) and (2.51) that for given $\delta_1 > 0$ one can find $\varepsilon_4 > 0$ and $T_4 > 0$ for which

$$P\{||\tilde{\mathbf{\Phi}}_T\mathbf{Q}_T|| \geq \varepsilon_4\} < \frac{\delta_1}{3}, \quad T > T_4.$$

Varying ε_1 one can always achieve that $\varepsilon_2 = \varepsilon_4$; therefore, setting in the right-hand side of (2.65) $\varepsilon_3 = 3\varepsilon_4 = 3\varepsilon_2$, we obtain (2.64) from (2.65), (2.67), (2.68) and the last inequality setting $\varepsilon = \varepsilon_3$, $\delta = \delta_1 + \delta_0$, $T_0 = \max(T'_0, T_1, T_2, T_3, T_4)$.

Lemma is proved. □

2.2.3 Fundamental Results

In this subsection we determine the limit of the sequence of random variables U_T. To do this we consider the convex programming problem:

$$\begin{cases} \varphi_T(\mathbf{X}) = \frac{1}{2}\mathbf{X}'\Phi_T(\omega)\mathbf{X} - \mathbf{Q}'_T(\omega)\mathbf{X} \to \min, \\ \beta_i(\mathbf{X}) = \nabla \mathbf{g}'_i(\boldsymbol{\alpha}^0)\mathbf{X} + l_i(\mathbf{X}) \leq \mathbf{O}_{m_1}, \quad i \in I_1^0, \\ \beta_i(\mathbf{X}) = \sqrt{T} g_i(\boldsymbol{\alpha}^0) + \nabla g'_i(\boldsymbol{\alpha}^0)\mathbf{X} + l_i(\mathbf{X}) \leq \mathbf{O}_{m_2}, \quad i \in I_2^0, \end{cases} \tag{2.69}$$

where $X \in \Re^n$, $\mathbf{\Phi}_T(\omega) = \mathbf{\Phi}_T$, $\mathbf{Q}_T(\omega) = \mathbf{Q}_T$, $\omega \in \Omega$, Ω is a sample space.

The last two expressions in (2.69) can be rewritten as

$$\beta_i(\mathbf{X}) = \sqrt{T} g_i\left(\frac{\mathbf{X}}{\sqrt{T}} + \boldsymbol{\alpha}^0\right), \quad i \in I, \tag{2.70}$$

$$l_i(\mathbf{X}) = \frac{1}{2\sqrt{T}}\mathbf{X}' g_{i2}\left(\boldsymbol{\alpha}^0 + \theta_{T1}(\omega)\frac{\mathbf{X}}{\sqrt{T}}\right)\mathbf{X}, \quad i \in I, \tag{2.71}$$

where functions g_{i2} and $\theta_{T1} = \theta_{T1}(\omega)$ are defined in the same way as in (2.43).

By the solution to the problem (2.69) we understand the vector $\mathbf{U}_T^*(\omega)$, calculated for fixed ω. Since $\mathbf{\Phi}_T(\omega)$ is a symmetric matrix, the necessary conditions for the existence of an extremum in (2.69) can be formulated as follows:

$$\mathbf{\Phi}_T(\omega)\mathbf{U}_T^*(\omega) - \mathbf{Q}_T(\omega) + \sum_{i=1}^{m} \nabla\beta_i(\mathbf{U}_T^*(\omega))v_{iT}^*(\omega) = \mathbf{O}_n,$$

$$v_{iT}^*(\omega)\beta_i(\mathbf{U}_T^*(\omega)) = \mathbf{O}_n, \quad v_{iT}^*(\omega) \geq 0, \quad i \in I. \tag{2.72}$$

The necessary conditions for the existence of an extremum in (2.12) can be written, according to (2.37) and (2.39), in the following form:

$$\mathbf{\Phi}_T(\omega)\mathbf{U}_T^*(\omega) - \mathbf{Q}_T(\omega) + \sum_{i=1}^{m} \nabla g_i(\boldsymbol{\alpha}_T(\omega))v_{iT}(\omega) = \mathbf{O}_n,$$

$$v_{iT}(\omega)g_i(\boldsymbol{\alpha}_T(\omega)) = 0, \quad v_{iT}(\omega) \geq 0,\ i \in I. \tag{2.73}$$

Setting $\mathbf{X} = \mathbf{U}_T(\omega)$, we obtain according to (2.70)

$$\beta_i(\mathbf{U}_T(\omega)) = \sqrt{T} g_i(\boldsymbol{\alpha}_T(\omega)), \quad \nabla\beta_i(\mathbf{U}_T(\omega)) = \nabla g_i(\boldsymbol{\alpha}_T(\omega)),$$

which implies together with (2.72) and (2.73) that vectors $\mathbf{U}_T(\omega)$ and $\mathbf{V}_T(\omega)$ are the solutions to the system of equations (2.72), i.e. $\mathbf{U}_T(\omega)$ satisfies the necessary conditions of extremum in (2.69). This observation leads to Theorem 2.7.

Theorem 2.7. *Under Assumptions 2.1–2.7, the random variable* $\mathbf{U}_T(\omega) = \sqrt{T}(\boldsymbol{\alpha}_T(\omega) - \boldsymbol{\alpha}^0)$*, is the solution to (2.12), converging in distribution as* $T \to \infty$ *to the random variable* $\mathbf{U}(\omega)$*, which is the solution to the quadratic programming problem*

$$\varphi(X) = \frac{1}{2}\mathbf{X}'\mathbf{R}(\boldsymbol{\alpha}^0)\mathbf{X} - \mathbf{Q}'(\omega)\mathbf{X} \to \min, \quad \nabla g_i'(\boldsymbol{\alpha}^0)\mathbf{X} \le 0,\ i \in I_1^0. \tag{2.74}$$

Here $\mathbf{Q}(\omega)$ *is a normally distributed centered random variable with covariance matrix* $\sigma^2\mathbf{R}(\boldsymbol{\alpha}^0)$.

Proof. Consider the quadratic programming problem:

$$\tilde{\varphi}_T(\mathbf{X}) = \frac{1}{2}\mathbf{X}'\mathbf{R}(\boldsymbol{\alpha}^0)\mathbf{X} - \mathbf{Q}'_T(\omega)\mathbf{X} \to \min, \quad \nabla g_i'(\boldsymbol{\alpha}^0)\mathbf{X} \le 0,\ i \in I_1^0. \tag{2.75}$$

Denote its solution by $\tilde{\mathbf{U}}_T(\omega)$. According to Lemma 2.5, $\tilde{\mathbf{U}}_T(\omega)$ is a continuous function of $\mathbf{Q}_T : \tilde{\mathbf{U}}_T(\omega) = f(\mathbf{Q}_T(\omega))$. Then by (2.58) we have $f(\mathbf{Q}_T(\omega)) \overset{p}{\Rightarrow} f(\mathbf{Q}(\omega))$. On the other hand, according to (2.74), $\mathbf{U}(\omega) = f(\mathbf{Q}(\omega))$. Thus,

$$\tilde{\mathbf{U}}_T(\omega) \overset{p}{\Rightarrow} \mathbf{U}(\omega), \quad T \to \infty. \tag{2.76}$$

Let

$$\mathbf{O}_T = \{\mathbf{X} : \beta_i(\mathbf{X}) \le 0, i \in I\}, \quad \mathbf{O} = \{\mathbf{X} : \nabla g_i'(\boldsymbol{\alpha}^0)\mathbf{X} \le 0, i \in I_1^0\}.$$

By Assumption 2.6 and (2.70), $\mathbf{O}_T$ is a convex set. According to Assumption 2.6, $l_i(\mathbf{X}) \ge 0$, $\mathbf{X} \in \mathfrak{R}^n$. Therefore, (2.69) implies that $\nabla g_i'(\boldsymbol{\alpha}^0)\mathbf{U}_T(\omega) \le 0$, $i \in I_1^0$, since $\beta_i(\mathbf{U}_T(\omega)) \le 0, i \in I_1^0$ (see (2.70)). Thus, $\mathbf{U}_T(\omega) \in \mathbf{O}$.

Since $R(\boldsymbol{\alpha}^0)$ is positive definite (see Assumption 2.5), $\tilde{\varphi}_T(\mathbf{X})$ is a strongly convex function of $\mathbf{X}$. For such a function we have, since $\mathbf{U}_T(\omega) \in O$, the following relation with some constant $\mu > 0$ (Karmanov 1975):

$$||\mathbf{U}_T(\omega) - \tilde{\mathbf{U}}_T(\omega)||^2 \le \frac{2}{\mu}[\tilde{\varphi}_T(\mathbf{U}_T(\omega)) - \tilde{\varphi}_T(\tilde{\mathbf{U}}_T(\omega))].$$

To shorten the notation we omit below the argument ω.

For arbitrary $\varepsilon > 0$,

$$\begin{aligned} P\{||\mathbf{U}_T - \tilde{\mathbf{U}}_T||^2 < \varepsilon^2\} &\ge P\left\{\frac{2}{\mu}[\tilde{\varphi}_T(\mathbf{U}_T) - \tilde{\varphi}_T(\tilde{\mathbf{U}}_T)] < \varepsilon^2\right\} \\ &\ge P\left\{|\tilde{\varphi}_T(\mathbf{U}_T) - \varphi_T(\mathbf{U}_T)| < \frac{\varepsilon_1}{2}\right\} + P\left\{[\varphi_T(\mathbf{U}_T) - \tilde{\varphi}_T(\tilde{\mathbf{U}}_T)] < \frac{\varepsilon_1}{2}\right\} - 1, \end{aligned} \tag{2.77}$$

where $\varepsilon_1 = \mu\varepsilon^2/2$.

We estimate the probability in the right-hand side of (2.77). From (2.69), and (2.75) we have for arbitrary $\varepsilon_2 > 0$ and $\delta > 0$

$$P\{|\tilde{\varphi}_T(\mathbf{U}_T)-\varphi_T(\mathbf{U}_T)| < \varepsilon_2\} \geq 1-P\{\|\mathbf{U}_T\| \geq \sqrt{b}\}-P\left\{||\mathbf{R}(\boldsymbol{\alpha}^0)-\boldsymbol{\Phi}_T|| \geq \frac{2\varepsilon_2}{b}\right\} > 1-\delta, \quad T>T_1, \tag{2.78}$$

where b is a positive number (to derive (2.78) we used Lemma 2.9 and (2.30)).

Based on (2.30) and (2.76), we obtain in a similar manner

$$P\{|\tilde{\varphi}_T(\tilde{\mathbf{U}}_T) - \varphi_T(\tilde{\mathbf{U}}_T)| < \varepsilon_2\} > 1 - \delta, \quad T > T_2. \tag{2.79}$$

Let us estimate the second term in the right-hand side of (2.77). After some transformations we arrive at

$$P\left\{\varphi_T(\mathbf{U}_T) - \tilde{\varphi}_T(\tilde{\mathbf{U}}_T) < \frac{\varepsilon_1}{2}\right\} \geq P\{\varphi_T(\mathbf{U}_T) - \varphi_T(\tilde{\mathbf{U}}_T) \leq 0\} \tag{2.80}$$
$$+P\left\{|\varphi_T(\tilde{\mathbf{U}}_T) - \tilde{\varphi}_T(\tilde{\mathbf{U}}_T)| < \frac{\varepsilon_1}{2}\right\} - 1.$$

Consider the first term in the right-hand side of (2.81). If the matrix $\boldsymbol{\Phi}_T$ is positive definite, problem (2.69) has a unique solution, because $\mathbf{O}_T$ is a convex set and Assumption 2.7 holds true. Then, on the basis of the analysis of expressions (2.72) and (2.73) carried out above, we conclude that $\mathbf{U}_T$ satisfies the necessary conditions for the existence of the minimum in (2.69). Therefore, $\mathbf{U}_T^* = \mathbf{U}_T$ if $\Lambda_{iT} > 0$, $i = \overline{1,n}$, where $\boldsymbol{\Lambda}_{iT}$ is the ith eigenvalue of $\boldsymbol{\Phi}_T$. Thus,

$$P\{\varphi_T(\mathbf{U}_T) - \varphi_T(\tilde{\mathbf{U}}_T) \leq 0\} \geq P\{\tilde{\mathbf{U}}_T \in \mathbf{O}_T, \Lambda_{iT} > 0, i = \overline{1,n}\}$$
$$= P\{\beta_i(\tilde{\mathbf{U}}_T) \leq 0, i \in I; \Lambda_{iT} > 0, i = \overline{1,n}\}$$

implying that

$$P\{\varphi_T(\mathbf{U}_T) - \varphi_T(\tilde{\mathbf{U}}_T) \leq 0\} \geq P\{\Lambda_{iT} > 0, i = \overline{1,n}\} - m + \sum_{i=1}^{m} P\{\beta_i(\tilde{\mathbf{U}}_T) \leq 0\}. \tag{2.81}$$

From (2.71) to (2.76), continuity of the function g_{i2} and its boundedness in $\boldsymbol{\alpha} = \boldsymbol{\alpha}^0$, it follows that

$$p \lim_{T\to\infty} l_i(\tilde{\mathbf{U}}_T) = 0. \tag{2.82}$$

Using (2.76) and (2.82), as well as the inequalities $\nabla g_i'(\boldsymbol{\alpha}^0)\mathbf{U} \leq 0$, $i \in I_1^0$, and $g_i(\boldsymbol{\alpha}^0) < 0$ $i \in I_2^0$, it is easy to show that for arbitrary $\eta_i > 0$ there exists $T_{3i} > 0$ such that

$$P\{\beta_i(\tilde{\mathbf{U}}_T) \leq 0\} \geq 1 - \eta_i, \quad T > T_{3i}, \, i \in I. \tag{2.83}$$

Applying the statement 3 of Lemma 2.3 to $\boldsymbol{\Phi}_T$ and taking into account (2.30) and Assumption 2.5, we have for arbitrary $\eta > 0$ $P\{\Lambda_{iT} > 0, i = \overline{1,n}\} > 1 - (\eta/2)$, $T > T_5$.

From the last inequality, (2.83) and (2.81), setting $\eta = 2\sum_{i=1}^{m} \eta_i$, $T_4 = \max_{i \in I} T_{3i}$, we obtain

$$P\{\varphi_T(\mathbf{U}_T) - \varphi_T(\tilde{\mathbf{U}}_T) \leq 0\} > 1 - \eta, \quad T > T_6 = \max(T_4, T_5). \tag{2.84}$$

From (2.84) and (2.81) with $\varepsilon_2 = \varepsilon_1/2$, it follows from (2.79) that

$$P\{\varphi_T(\mathbf{U}_T) - \varphi_T(\tilde{\mathbf{U}}_T) \leq \varepsilon_2\} > 1 - \delta - \eta, \quad T > \max(T_2, T_6). \tag{2.85}$$

Inserting in the right-hand side of (2.77) with $\varepsilon_2 = \varepsilon_1/2$ the estimates (2.85) and (2.78), we obtain

$$p \lim_{T \to \infty} ||\mathbf{U}_T - \tilde{\mathbf{U}}_T|| = 0. \tag{2.86}$$

The statement of the theorem follows from (2.86) and (2.76). □

Corollary 2.1. *If $I_1^0 = \emptyset$ (i.e., all constraints for $\boldsymbol{\alpha} = \boldsymbol{\alpha}^0$ are inactive), then by (2.74) we have $\mathbf{U} = \mathbf{R}^{-1}(\boldsymbol{\alpha}^0)\mathbf{Q}$, i.e., the vector $\mathbf{U}_T = \sqrt{T}(\boldsymbol{\alpha}_T - \boldsymbol{\alpha}^0)$ is asymptotically normal. It is known that $\mathbf{U}_T$ has the same asymptotic distribution when there are no constraints ($I = \emptyset$). Obviously, both cases $I = \emptyset$ and $I_1^0 = \emptyset$ (for $I \neq \emptyset$) are asymptotically equivalent.*

Theorem 2.8. *If Assumptions 2.1–2.7 hold true, then the random variable $\mathbf{V}_{1T}$ with components $v_{iT} = \sqrt{T}\lambda_{iT}$, $i \in I_1^0$ where λ_{iT} are the Lagrange multipliers from problem (2.12), converge in distribution as $T \to \infty$ to the random variable $\mathbf{V}_1 = \mathbf{V}_1(\omega)$, where $\mathbf{V}_1 = \mathbf{V}_1(\omega)$ is a vector of Lagrange multipliers from problem (2.74).*

Proof. Consider the dual problems to (2.74) and (2.75), respectively:

$$\frac{1}{2}\mathbf{Y}'\mathbf{A}\mathbf{Y} - \mathbf{H}'\mathbf{Y} \to \min, \quad \mathbf{Y} \geq \mathbf{O}_n, \tag{2.87}$$

$$\frac{1}{2}\mathbf{Y}'\mathbf{A}\mathbf{Y} - \mathbf{H}'_T\mathbf{Y} \to \min, \quad \mathbf{Y} \geq \mathbf{O}_n, \tag{2.88}$$

where $\mathbf{A} = \mathbf{G}_1(\boldsymbol{\alpha}^0)\mathbf{R}^{-1}(\boldsymbol{\alpha}^0)\mathbf{G}'_1(\boldsymbol{\alpha}^0)$ is a positive definite matrix,

$$\mathbf{H} = \mathbf{H}(\omega) = \mathbf{G}_1(\boldsymbol{\alpha}^0)\mathbf{R}^{-1}(\boldsymbol{\alpha}^0)\mathbf{Q}(\omega),$$

$$\mathbf{H}_T = \mathbf{H}_T(\omega) = \mathbf{G}_1(\boldsymbol{\alpha}^0)\mathbf{R}^{-1}(\boldsymbol{\alpha}^0)\mathbf{Q}_T(\omega).$$

The solutions to (2.87) and (2.88) are $\mathbf{V}_1(\omega) = \bar{f}(\mathbf{H}(\omega))$ and $\tilde{\mathbf{V}}_{1T}(\omega) = \bar{f}(\mathbf{H}_T(\omega))$, respectively, where $\bar{f}$ is a continuous function (see Lemma 2.5). By (2.58) and continuity of $\bar{f}$ we have

$$\bar{f}(\mathbf{H}_T(\omega)) \stackrel{p}{\Rightarrow} \bar{f}(\mathbf{H}(\omega)), \quad T \to \infty,$$

i.e.,

$$\tilde{\mathbf{V}}_{1T}(\omega) \overset{p}{\Rightarrow} \mathbf{V}_1(\omega), \quad T \to \infty. \tag{2.89}$$

Consider one of the equations which give the necessary conditions for the existence of an extremum in (2.75):

$$\mathbf{R}(\boldsymbol{\alpha}^0)\tilde{\mathbf{U}}_T(\omega) - \mathbf{Q}_T(\omega) + G_1(\boldsymbol{\alpha}^0)\tilde{\mathbf{V}}_{1T}(\omega) = \mathbf{O}_n.$$

Subtracting this equation from (2.39) we obtain after some transformations

$$(\mathbf{G}'_1(\boldsymbol{\alpha}_T) - \mathbf{G}'_1(\boldsymbol{\alpha}^0))\mathbf{V}_{1T} + \mathbf{G}'_1(\boldsymbol{\alpha}^0)(\mathbf{V}_{1T} - \tilde{\mathbf{V}}_{1T}) + \boldsymbol{\Gamma}_T = \mathbf{O}_n, \tag{2.90}$$

where $\boldsymbol{\Gamma}_T = \boldsymbol{\Phi}_T\mathbf{U}_T - \mathbf{R}(\boldsymbol{\alpha}^0)\tilde{\mathbf{U}}_T + \mathbf{G}'_2(\boldsymbol{\alpha}_T)\mathbf{V}_{2T}$. As above, we omit the argument ω. Using (2.30), (2.35), (2.45), (2.86), and Theorem 2.7, one can show that

$$p \lim_{T\to\infty} \boldsymbol{\Gamma}_T = \mathbf{O}_n. \tag{2.91}$$

According to Assumption 2.4, the matrix $\mathbf{G}_1(\boldsymbol{\alpha}^0)$ is of full rank. Therefore after some rearrangements we obtain from (2.90)

$$||\mathbf{V}_{1T} - \tilde{\mathbf{V}}_{1T}|| \le ||\tilde{\mathbf{G}}(\boldsymbol{\alpha}^0)\mathbf{G}_1(\boldsymbol{\alpha}^0)\boldsymbol{\Gamma}_T|| + ||\tilde{\mathbf{G}}(\boldsymbol{\alpha}^0)\mathbf{G}_1(\boldsymbol{\alpha}^0)(\mathbf{G}'_1(\boldsymbol{\alpha}_T) - \mathbf{G}'_1(\boldsymbol{\alpha}^0))\mathbf{V}_{1T}||, \tag{2.92}$$

where $\tilde{\mathbf{G}}(\boldsymbol{\alpha}^0) = (\mathbf{G}_1(\boldsymbol{\alpha}^0)\mathbf{G}'_1(\boldsymbol{\alpha}^0))^{-1}$.

According to (2.91), for arbitrary $\varepsilon > 0$ and $\delta > 0$ we have

$$P\left\{||\tilde{\mathbf{G}}(\boldsymbol{\alpha}^0)\mathbf{G}_1(\boldsymbol{\alpha}^0)\boldsymbol{\Gamma}_T|| < \varepsilon\right\} \ge P\left\{\boldsymbol{\Gamma}_T < \frac{\varepsilon}{||\tilde{\mathbf{G}}(\boldsymbol{\alpha}^0)\mathbf{G}_1(\boldsymbol{\alpha}^0)||}\right\} \ge 1 - \delta, \quad T > T_1.$$

By (2.52) and Lemma 2.8 for the same ε and δ, one can find $b > 0$ and $T_2 > 0$ for which

$$P\left\{||\tilde{\mathbf{G}}(\boldsymbol{\alpha}^0)\mathbf{G}_1(\boldsymbol{\alpha}^0)(\mathbf{G}'_1(\boldsymbol{\alpha}_T) - \mathbf{G}'_1(\boldsymbol{\alpha}^0))\mathbf{V}_{1T}|| < \varepsilon\right\}$$

$$\ge 1 - P\left\{||\mathbf{G}'_1(\boldsymbol{\alpha}_T) - \mathbf{G}'_1(\boldsymbol{\alpha}^0)|| \ge \frac{\varepsilon_1}{b}\right\} - P\{||\mathbf{V}_{1T}|| \ge b\} \ge 1 - \delta, \quad T > T_2,$$

where $\varepsilon_1 = \varepsilon/||\tilde{\mathbf{G}}(\boldsymbol{\alpha}^0)\mathbf{G}_1(\boldsymbol{\alpha}^0)||$.

From the last two inequalities and (2.92) it follows that $||\mathbf{V}_{1T} - \tilde{\mathbf{V}}_{1T}||$ converges in probability to 0 as $T \to \infty$, which together with (2.89) proves the theorem. □

Using Theorems 2.7 and 2.8 one can determine in the case of large samples the accuracy of estimation of parameters and check statistical hypotheses. We show that the distribution of $\mathbf{V}_1$ is not concentrated at 0. Assume the opposite: $\mathbf{V}_1 = \mathbf{O}_{m_1}$ almost surely. According to the necessary conditions for the extremum in (2.74), we have $\mathbf{R}(\boldsymbol{\alpha}^0)\mathbf{U} - \mathbf{Q} + \mathbf{G}_1(\boldsymbol{\alpha}^0)\mathbf{V}_1 = \mathbf{O}_n$. Hence, if the assumption holds true, the vector $\mathbf{U}$ is normally distributed. In this case, by (2.44), Theorem 2.7 and the equality $p\lim_{T\to\infty} \boldsymbol{\Psi}_1(\boldsymbol{\alpha}_T) = \mathbf{G}_1(\boldsymbol{\alpha}^0)$ (see (2.52)) we obtain that the distribution of the vector $\mathbf{W}_{1T}$, defined in (2.44), converges as $T \to \infty$ to the normal distribution, which is not possible since $\mathbf{W}_{1T} \geq \mathbf{O}_{m_1}$. Thus, $P\{\mathbf{V}_1 \neq \mathbf{O}_{m_1}\} > 0$. Taking into account Lemma 2.7 the obtained result can be formulated as below.

Theorem 2.9. *If Assumptions 2.1–2.7 hold true, then* $P\{v_i > 0\} > 0,\ i \in I_1^0$; $P\{v_i = 0\} = 1,\ i \in I_2^0$, *where* v_i *is a limit of* v_{iT} *as* $T \to \infty$.

In mathematical programming the following result is known. If the so-called strict complementary slackness conditions hold, then the Lagrange multiplier, corresponding to an active constraint, is greater than 0, and the multiplier corresponding to an inactive constraint is equal to 0. The statement of the theorem can be regarded as an analogue of this property for Lagrange multipliers.

2.3 Asymptotic Properties of Nonlinear Regression Parameters Estimates by the Least Squares Method Under a Priory Inequality Constraints (Non-Convex Case)

2.3.1 Assumptions and Auxiliary Results

The results described above are based on the assumption of convexity of the admissible region **M**, given by restrictions (2.12). In order to make the restrictions on the region less strict, we change Assumption 2.2B by putting $\mathbf{M} = \Re^n$. This means that the first and second derivatives of the regression function are bounded on whole space of regression parameters. In what follows we use the assumption below.

Assumption 2.2′. A. Functions $f_t(\boldsymbol{\alpha}),\ t = \overline{1,T}$ and $g_i(\boldsymbol{\alpha}),\ i \in I$ are twice continuously differentiable on $\Re^n$.

B. For all $\boldsymbol{\alpha} \in \Re^n$ and all possible values of independent variables x_t there exist constants c_1 and c_2 such that

$$\left|\frac{\partial f_t(\boldsymbol{\alpha})}{\partial \alpha_i}\right| \leq c_1,\ \left|\frac{\partial^2 f_t(\boldsymbol{\alpha})}{\partial \alpha_i\,\partial \alpha_j}\right| \leq c_2,\quad t = \overline{1,T},\ i,j = \overline{1,n},$$

C. Functions $g_i(\boldsymbol{\alpha}),\ i \in I$ and their first and second derivatives are bounded in the neighborhood of $\boldsymbol{\alpha} = \boldsymbol{\alpha}^0$.

Furthermore, we replace Assumptions 2.6 and 2.7 by the assumption below.

Assumption 2.6′. Gradients $\nabla g_i\ (\boldsymbol{\alpha}_T),\ i \in I_a$ are linearly independent.

Thus, we consider Assumptions 2.1, 2.2′, 2.3–2.5, and 2.6′. These assumptions allow after some changes in the restrictions on $f_t(\boldsymbol{\alpha})$ and $g_i(\boldsymbol{\alpha})$, $i \in I$, to get rid of the convexity of constraints. Let us discuss how the new assumptions will influence the auxiliary results obtained before.

Lemmas 2.3–2.5 remain unchanged, since Assumptions 2.2–2.7 are not used in the proofs. Lemmas 2.6–2.9 rely on these assumptions. Note that Assumptions 2.2 and 2.2′ differ only in property B. One can check that the new Assumption 2.2′B includes Assumption 2.2B as a particular case. Assumptions 2.6 and 2.7 imply linear independence of gradients in left-hand side parts of active restrictions at $\boldsymbol{\alpha} = \boldsymbol{\alpha}_T$, which belongs to the admissible set. Recall that linear independence of gradients was used in the proofs of Lemmas 2.6–2.9. Assumption 2.6′ provides such a property for gradients for any $\boldsymbol{\alpha}_T \in \Re^n$. Therefore, Lemmas 2.6–2.9 hold true also for the case considered in this subsection.

2.3.2 *Fundamental Result*

We prove the theorem about the limit distribution of $\sqrt{T}(\boldsymbol{\alpha}_T - \boldsymbol{\alpha}^0)$ in the non-convex case. In the proof we use Lemmas 2.3–2.9, which also hold true in the non-convex case.

Theorem 2.10. *Under Assumptions 2.1, 2.2′, 2.3–2.5, and 2.6′, the random variable* $\mathbf{U}_T = \sqrt{T}(\boldsymbol{\alpha}_T - \boldsymbol{\alpha}^0)$, *where* $\boldsymbol{\alpha}_T$ *is the solution of (2.12), converges in distribution as* $T \to \infty$ *to a random variable* $\mathbf{U}$ *which is the solution to quadratic programming problem (2.74).*

Proof. In the proof we follow the same scheme as in the proof of Theorem 2.7.

Consider the quadratic programming problem

$$\begin{cases} \varphi_T(\mathbf{X}) = \frac{1}{2}\mathbf{X}'\boldsymbol{\Phi}_T(\omega)\mathbf{X} - \mathbf{Q}'_T(\omega)\mathbf{X} \to \min, \\ \beta_i(\mathbf{X}) = \nabla g_i'(\boldsymbol{\xi}_T(\omega))\mathbf{X} \le \mathbf{O}_{m_1}, \quad i \in I_1^0, \\ \beta_i(\mathbf{X}) = \sqrt{T} g_i(\boldsymbol{\alpha}^0) + \nabla g_i'(\boldsymbol{\xi}_T(\omega))\mathbf{X} \le \mathbf{O}_{m_2}, \quad i \in I_2^0, \end{cases} \tag{2.93}$$

where $\mathbf{X} \in \Re^n$; $\mathbf{Q}_T(\omega) = \mathbf{Q}_T$ and $\boldsymbol{\Phi}_T(\omega) = \boldsymbol{\Phi}_T$ have the same meaning as in (2.71), the function $\beta_i(\mathbf{X})$ is determined in (2.70), $\boldsymbol{\xi}_T = \boldsymbol{\xi}_T(\omega) = \boldsymbol{\alpha}^0 + \theta(\omega)(\boldsymbol{\alpha}_T(\omega) - \boldsymbol{\alpha}^0)$ and $\theta(\omega) \in [0, 1]$, $\omega \in \Omega$, Ω is the sample space. The solution $\mathbf{U}_T^* = \mathbf{U}_T^*(\omega)$ to problem (2.93) satisfies the minimum conditions (2.72), as well as the vector $\mathbf{U}_T = \mathbf{U}_T(\omega)$ (compare the expression (2.72) with necessary conditions for the minimum (2.73) in the problem (2.12)).

Consider the quadratic programming problem

$$\tilde{\varphi}_T(\mathbf{X}) = \frac{1}{2}\mathbf{X}'\mathbf{R}(\boldsymbol{\alpha}^0)\mathbf{X} - \mathbf{Q}'_T(\omega)\mathbf{X} \to \min,$$

$$\nabla g_i'(\boldsymbol{\xi}_T(\omega))\mathbf{X} \le 0, \quad i \in I_1^0. \tag{2.94}$$

Denote its solution by $\tilde{\mathbf{U}}_T = \tilde{\mathbf{U}}_T(\omega)$. According to Lemma 2.5, $\tilde{\mathbf{U}}_T$ is the continuous function of $\mathbf{Q}_T$ and $\nabla g_i(\boldsymbol{\xi}_T)$, $i \in I_1^0$:

$$\tilde{\mathbf{U}}_T = q(\mathbf{Q}_T; \nabla g_i(\boldsymbol{\xi}_T), i \in I_1^0).$$

It follows by the definition of $\boldsymbol{\xi}_T$ that $p\lim_{T\to\infty} \boldsymbol{\xi}_T = \boldsymbol{\alpha}^0$, which implies, by continuity of $\nabla g_i(\boldsymbol{\alpha})$, $i \in I$ in $\Re^n$ (see Assumption 2.2′), that

$$p \lim_{T\to\infty} \nabla g_i(\boldsymbol{\xi}_T) = \nabla g_i(\boldsymbol{\alpha}^0), \quad i \in I_1^0.$$

Then by the continuity of the function $q(\cdot)$ we have

$$q(\mathbf{Q}_T; \nabla g_i(\boldsymbol{\xi}_T), i \in I_1^0) \overset{p}{\Rightarrow} q(\mathbf{Q}; \nabla g_i(\boldsymbol{\alpha}^0), i \in I_1^0), \quad T \to \infty. \tag{2.95}$$

Here we used that from the convergence in probability to a constant we get its convergence in distribution.

The solution to problem (2.74) can be written in the form

$$\mathbf{U} = q(\mathbf{Q}; \nabla g_i(\boldsymbol{\alpha}^0),\ i \in I_1^0).$$

Thus, from this expression and (2.95) we get (2.76).

Define

$$\mathbf{O}_T = \{\mathbf{X} : \beta_i(\mathbf{X}) \leq 0, i \in I\}, \quad \mathbf{O} = \{\mathbf{X} : \nabla g_i'(\boldsymbol{\xi}_T)\mathbf{X} \leq 0, i \in I_1^0\}.$$

In terms of the notation (2.70) and the constraints in (2.93), we have $\nabla g_i'(\boldsymbol{\xi}_T)\mathbf{U}_T = \beta_i(\mathbf{U}_T) = \sqrt{T} g_i(\boldsymbol{\alpha}_T) \leq 0$, $i \in I_1^0$, which implies $\mathbf{U}_T \in \mathbf{O}$. Then all the considerations used for obtaining (2.77)–(2.81) will remain true (see the proof of Theorem 2.7).

Let us estimate (2.81). Consider the first term in the right-hand side of this expression. If the matrix $\boldsymbol{\Phi}_T$ is positive definite, the square programming problem (2.93) has a unique solution. Therefore, comparing (2.72) and (2.73), we derive that $\mathbf{U}_T^* = \mathbf{U}_T$, where $\mathbf{U}_T^*$ is the solution to (2.93). We obtain

$$\begin{aligned} P\{\varphi_T(\mathbf{U}_T) - \varphi_T(\tilde{\mathbf{U}}_T) \leq 0\} &\geq P\{\tilde{\mathbf{U}}_T \in \mathbf{O}_T, \boldsymbol{\Lambda}_{iT} > 0, i = \overline{1,n}\} \\ &= P\{\beta_i(\tilde{\mathbf{U}}_T) \leq 0, i \in I_2^0; \boldsymbol{\Lambda}_{iT} > 0, i = \overline{1,n}\} \\ &\geq P\{\boldsymbol{\Lambda}_{iT} > 0, i = \overline{1,n}\} - m_2 + \sum_{i \in I_2^0} P\{\beta_i(\tilde{\mathbf{U}}_T) \leq 0\}, \end{aligned} \tag{2.96}$$

where m_2 is the number of elements in $\mathbf{I}_2^0$, and $\boldsymbol{\Lambda}_{iT}$ is the ith eigenvalue of the matrix $\boldsymbol{\Phi}_T$.

We find the limit of $P\{\beta_i(\tilde{\mathbf{U}}_T) \leq 0\}$ as $T \to \infty$ based on (2.76), inequalities $g_i(\boldsymbol{\alpha}^0) < 0$, $i \in I_2^0$ and the limit $p\lim_{T\to\infty} \boldsymbol{\xi}_T = \boldsymbol{\alpha}^0$.

For $\varepsilon > 0$ we have

$$
\begin{aligned}
P\{\beta_i(\tilde{\mathbf{U}}_T) \leq 0\} &= P\left\{-\nabla g_i'(\boldsymbol{\xi}_T)\tilde{\mathbf{U}}_T \geq \sqrt{T} g_i(\boldsymbol{\alpha}^0)\right\} \\
&= P\left\{\nabla g_i'(\boldsymbol{\alpha}^0)\tilde{\mathbf{U}} - \nabla g_i'(\boldsymbol{\xi}_T)\tilde{\mathbf{U}}_T \geq \sqrt{T} g_i(\boldsymbol{\alpha}^0) + \nabla g_i'(\boldsymbol{\alpha}^0)\tilde{\mathbf{U}}\right\} \\
&\geq P\left\{\nabla g_i'(\boldsymbol{\alpha}^0)\tilde{\mathbf{U}} - \nabla g_i'(\boldsymbol{\xi}_T)\tilde{\mathbf{U}}_T \geq \sqrt{T} g_i(\boldsymbol{\alpha}^0) + |\nabla g_i'(\boldsymbol{\alpha}^0)\tilde{\mathbf{U}}|\right\} \\
&\geq P\left\{\nabla g_i'(\boldsymbol{\alpha}^0)\tilde{\mathbf{U}} - \nabla g_i'(\boldsymbol{\xi}_T)\tilde{\mathbf{U}}_T \geq -\varepsilon, \sqrt{T} g_i(\boldsymbol{\alpha}^0) + |\nabla g_i'(\boldsymbol{\alpha}^0)\tilde{\mathbf{U}}| \geq -\varepsilon\right\} \\
&\geq P\{\nabla g_i'(\boldsymbol{\alpha}^0)\tilde{\mathbf{U}} - \nabla g_i'(\boldsymbol{\xi}_T)\tilde{\mathbf{U}}_T \geq -\varepsilon\} - 1 + P\left\{\sqrt{T} g_i(\boldsymbol{\alpha}^0) + |\nabla g_i'(\boldsymbol{\alpha}^0)\tilde{\mathbf{U}}| < -\varepsilon\right\} \\
&\geq P\{||\nabla g_i'(\boldsymbol{\alpha}^0)\tilde{\mathbf{U}} - \nabla g_i'(\boldsymbol{\xi}_T)\tilde{\mathbf{U}}_T|| \leq -\varepsilon\} - 1 + P\left\{\sqrt{T} g_i(\boldsymbol{\alpha}^0) + |\nabla g_i'(\boldsymbol{\alpha}^0)\tilde{\mathbf{U}}| < -\varepsilon\right\} \\
&\geq 1 - \eta_{1i} - \eta_{2i},\ T > T_{3i},\ i \in I_2^0, \qquad (2.97)
\end{aligned}
$$

where $\eta_{ki} > 0$ are arbitrary. For such values and $\varepsilon > 0$ there exists T_{3i} for which (2.97) holds true. Then, taking $\eta_i = \eta_{1i} + \eta_{2i}$, we obtain

$$
P\{\beta_i(\tilde{\mathbf{U}}_T) \leq 0\} \geq 1 - \eta_i, \quad T > T_{3i},\ i \in I_2^0. \qquad (2.98)
$$

From the proof of Theorem 2.7 we derive for any $\eta > 0$

$$
P\{\boldsymbol{\Lambda}_{iT} > 0, i = \overline{1, n}\} > 1 - \frac{\eta}{2}, \quad T > T_5.
$$

Further, from the expression above and (2.96), (2.98), we get

$$
P\{\varphi_T(\mathbf{U}_T) - \varphi_T(\tilde{\mathbf{U}}_T) \leq 0\} > 1 - \eta, \quad T > T_6, \qquad (2.99)
$$

where

$$
\eta = 2\sum_{i \in I_2^0} \eta_i, \quad T_6 = \max(T_4, T_5), \quad T_4 = \max_{i \in I_2^0} T_{3i}.
$$

From above, (2.81) and (2.79) hold true provided that assumptions of Theorem 2.10 are satisfied. Then from (2.81), (2.79) and inequality (2.99) we derive (2.85).

By the same arguments as in the proof of Theorem 2.7 and the fact that in our settings inequalities (2.77), (2.78), (2.85) and the limit (2.76) hold true, we get the statement of the theorem. □

2.4 Limit Distribution of the Estimate of Regression Parameters Which Are Subject to Equality Constraints

The described methodology of derivation the limit distribution of a multidimensional parameter regression estimate can be modified for the case of equality constraints

$$g(\boldsymbol{\alpha}) = \mathbf{O}_m, \quad g(\boldsymbol{\alpha}) \in \Re^m. \tag{2.100}$$

Namely, one can consider the problem

$$S_T(\boldsymbol{\alpha}) \to \min, \quad g_i(\boldsymbol{\alpha}) = 0, \, i \in I = \{1, 2, \ldots, m\}, \tag{2.101}$$

where $S_T(\boldsymbol{\alpha})$ is determined in (2.13).

For this case we need Assumptions 2.1, 2.2′, and 2.3. Assumptions 2.4 (Sect. 2.2) and 2.6′ (Sect. 2.3) need to be changed in order to take into account the absence of constraints in (2.100).

Assumption 2.4′ Gradients $\nabla g_i(\boldsymbol{\alpha}^0)$, $i \in I = \{1, \ldots, m\}$ are linearly independent in the neighborhood $\boldsymbol{\alpha} = \boldsymbol{\alpha}^0$.

Assumption 2.6″ Gradients $\nabla g_i(\boldsymbol{\alpha}_T)$, $i \in I$ are linearly independent.

Taking into account these assumptions, consider the lemmas from Sect. 2.2. Obviously, Lemmas 2.3–2.5 will remain true. Using Assumptions 2.1, 2.2′, 2.3, 2.4′, 2.5 and 2.6″, we can proceed to the proof of Lemma 2.6.

Proof of Lemma 2.6. The necessary conditions for the existence of the extremum in problems (2.101) and (2.13) are given by expressions (2.24). According to Sects. 2.2 and 2.3, for Assumptions 2.1, 2.2′, 2.3, 2.4′, 2.5′, and 2.6″, expressions (2.26)–(2.32) hold true. The assertion $p\lim_{T\to\infty} \lambda_{iT}=0$, $i\in I$ follows from relations (2.24) and (2.32), since the gradients $\nabla g_i(\boldsymbol{\alpha}_T)$, $i=\overline{1,m}$ are linearly independent.
The proof of Lemmas 2.8 and 2.9 can be simplified since for the considered case $m_1 = m$, implying

$$\mathbf{V}_{1T} = \mathbf{V}, \quad \mathbf{V}_{2T} = \mathbf{O}_{m_2}, \quad \mathbf{G}_1(\boldsymbol{\alpha}_T) = \mathbf{G}(\boldsymbol{\alpha}_T), \quad \mathbf{G}_2(\boldsymbol{\alpha}_T) = \mathbf{O}_{m_2}, \quad \mathbf{W}_{1T} = \mathbf{O}_{m_1}.$$

□

Lemma 2.9 allows to prove a theorem on the limit distribution of estimates of regression parameters.

Theorem 2.11. *If Assumptions 2.1, 2.2′, 2.3, 2.4′, 2.5′, and 2.6″ are satisfied, then the random variable* $\mathbf{U}_T = \sqrt{T}(\boldsymbol{\alpha}_T - \boldsymbol{\alpha}^0)$, *where* $\boldsymbol{\alpha}_T$ *is a solution to (2.101), converges in distribution to a random variable* $\mathbf{U}$, *which is normally distributed with the expectation* $E\{\mathbf{U}\} = \mathbf{O}_n$ *and the covariance matrix*

$$\mathbf{K} = \mathbf{R}^{-1}(\boldsymbol{\alpha}^0)\sigma^2[\mathbf{J}_n - \mathbf{G}'(\boldsymbol{\alpha}^0)(\mathbf{G}(\boldsymbol{\alpha}^0)\mathbf{R}^{-1}(\boldsymbol{\alpha}^0)\mathbf{G}'(\boldsymbol{\alpha}^0))^{-1}\mathbf{G}(\boldsymbol{\alpha}^0)\mathbf{R}^{-1}(\boldsymbol{\alpha}^0)].$$

The proof is analogous to the proof of Theorem 2.10.

Consider the optimization problem

$$\varphi_T(\mathbf{X}) = \frac{1}{2}\mathbf{X}'\mathbf{\Phi}_T\mathbf{X} - \mathbf{Q}'_T\mathbf{X} \to \min, \quad \beta_i(\mathbf{X}) = \nabla g_i'(\boldsymbol{\xi}_T)\mathbf{X} = 0, \quad i \in I, \quad (2.102)$$

where $\mathbf{X} \in \mathfrak{R}^n$, $\mathbf{Q}_T$ is determined by (2.38), $\beta_i(\mathbf{X})$ is determined by (2.70), and $\boldsymbol{\xi}_T = \boldsymbol{\alpha}^0 + \theta(\boldsymbol{\alpha}_T - \boldsymbol{\alpha}^0)$, $\theta \in [0, 1]$. To define the constraints in (2.102) we used the series expansion

$$\beta_i(\mathbf{X}) = \sqrt{T} g_i\left(\frac{\mathbf{X}}{\sqrt{T}} + \boldsymbol{\alpha}^0\right) = \sqrt{T} g(\boldsymbol{\alpha}^0) + \nabla g_i'(\boldsymbol{\xi}_T)\mathbf{X}, \quad i \in I \quad (2.103)$$

and the fact that $g_i(\boldsymbol{\alpha}^0) = 0, i \in I$.

Each of the constraints $g_i(\boldsymbol{\alpha}) = 0$ can be represented as two constraints-inequalities:

$$g_i(\boldsymbol{\alpha}) \leq 0, \quad -g_i(\boldsymbol{\alpha}) \leq 0, \quad i \in I. \quad (2.104)$$

Then it is easy to see that the expression (2.102) coincides for $I_2^0 = \emptyset$ with (2.93) if $I = I_1^0$. Then, taking into account inequalities (2.104), the first system of inequalities in (2.93) becomes

$$\nabla g_i'(\boldsymbol{\xi}_T)\mathbf{X} \leq 0, \quad -\nabla g_i'(\boldsymbol{\xi}_T)\mathbf{X} \leq 0, \quad i \in I = I_1^0 \quad (2.105)$$

and

$$\mathbf{U} = [\mathbf{J}_n - \mathbf{R}^{-1}(\boldsymbol{\alpha}^0)\mathbf{G}'(\boldsymbol{\alpha}^0)(\mathbf{G}(\boldsymbol{\alpha}^0)\mathbf{R}^{-1}(\boldsymbol{\alpha}^0)\mathbf{G}'(\boldsymbol{\alpha}^0))^{-1}]\mathbf{R}^{-1}(\boldsymbol{\alpha}^0)\mathbf{Q} \quad (2.106)$$

is the solution to the problem

$$\varphi(\mathbf{X}) = \frac{1}{2}\mathbf{X}'\mathbf{R}(\boldsymbol{\alpha}^0)\mathbf{X} - \mathbf{Q}'\mathbf{X} \to \min, \quad \nabla g_i'(\boldsymbol{\alpha}^0)\mathbf{X} = 0, \quad i \in I, \quad (2.107)$$

where $\mathbf{Q} \sim N(\mathbf{O}_n, \sigma^2\mathbf{R}(\boldsymbol{\alpha}^0))$. The solution $\mathbf{U}_T^*$ to (2.102) satisfies the necessary conditions for the existence of extremum in this problem, i.e.,

$$\mathbf{\Phi}_T\mathbf{U}_T^* - \mathbf{Q}_T + \sum_{i \in I} \nabla\beta_i(\mathbf{U}_T^*)v_{iT}^* = \mathbf{O}_n, \quad \beta_i(\mathbf{U}_T^*) = 0, \quad i \in I, \quad (2.108)$$

where $v_{iT}^* = \sqrt{T}\lambda_{iT}^*$, λ_{iT}^* is the Lagrange coefficient for problem (2.102).

The necessary conditions for the existence of the extremum in (2.101) and (2.13) are of the form:

$$\mathbf{\Phi}_T\mathbf{U}_T + \sum_{i \in I} \nabla g_i(\boldsymbol{\alpha}_T)v_{iT} - \mathbf{Q}_T = \mathbf{O}_n, \quad g_i(\boldsymbol{\alpha}_T) = \mathbf{O}_n, \quad i \in I, \quad (2.109)$$

where $v_{iT} = \sqrt{T}\lambda_{iT}$.

According to (2.103) we have:

$$\beta_i(\mathbf{U}_T) = \sqrt{T} g_i(\boldsymbol{\alpha}_T),$$

$$\nabla\beta_i(\mathbf{X}) = \nabla g_i\left(\frac{\mathbf{X}}{\sqrt{T}} + \boldsymbol{\alpha}^0\right), \quad i \in I. \quad (2.110)$$

Hence

$$\nabla\beta_i(\mathbf{U}_T) = \nabla g_i(\boldsymbol{\alpha}_T), \quad i \in I. \tag{2.111}$$

Comparing equalities (2.108) and (2.109), and taking into account expressions (2.110) and (2.111), we see that $\mathbf{U}_T$ and $\mathbf{V}_T = [v_{1T} \dots v_{mT}]'$ satisfies (2.108). By (2.105), the problem (2.94) from Sect.2.3 can be reformulated as

$$\tilde{\varphi}_T(\mathbf{X}) = \frac{1}{2}\mathbf{X}'\mathbf{R}(\boldsymbol{\alpha}^0)\mathbf{X} - \mathbf{Q}'_T\mathbf{X} \to \min, \quad \nabla g_i'(\xi_T)\mathbf{X} = 0, \quad i \in I \tag{2.112}$$

and its solution is

$$\tilde{\mathbf{U}}_T = [\mathbf{J}_n - \mathbf{R}^{-1}(\boldsymbol{\alpha}^0)\mathbf{G}'(\xi_T)(\mathbf{G}(\xi_T)\mathbf{R}^{-1}(\boldsymbol{\alpha}^0)\mathbf{G}'(\xi_T))^{-1}]\mathbf{R}^{-1}(\boldsymbol{\alpha}^0)\mathbf{Q}_T. \tag{2.113}$$

We obtain the expression (2.76) from (2.106) and (2.113), according to (2.58) and Assumption 2.2′A. From expressions (2.102) and (2.103) we get $\beta_i(\mathbf{U}_T)$=0, $i \in \mathbf{O}$, where $\mathbf{O} = \{\mathbf{X} : \nabla g_i(\boldsymbol{\xi}_T)\mathbf{X} = 0, i \in I\}$; hence, $\mathbf{U}_T \in \mathbf{O}$.
Further, repeating all arguments used in the proof of Theorem 2.7, we obtain (2.79) and (2.81).

Let us estimate the first term in the right-hand side of (2.81). If $\boldsymbol{\Phi}_T$ is the positive definite matrix and T is large enough so that $\boldsymbol{\xi}_T$ belongs to the neighborhood of $\boldsymbol{\alpha}^0$ where the Assumption 2.4′ holds true, then the solution to (2.102) is unique. If $\boldsymbol{\Phi}_T$ is positive definite and Assumption 2.6″ holds true, the solution to (2.101), (2.13) is also unique. Consequently, (2.109) determines the unique vector $\mathbf{U}_T$, which is also the solution to (2.108). Therefore we have $\mathbf{U}_T^* = \mathbf{U}_T$, under the conditions that $\boldsymbol{\Lambda}_{iT} > 0$, $i = \overline{1,n}$, $||\xi_T - \boldsymbol{\alpha}^0|| < \varepsilon_3$, where $\boldsymbol{\Lambda}_{iT}$ is the ith eigenvalue of $\boldsymbol{\Phi}_T$, and $\varepsilon_3 > 0$ determines the neighborhood $\boldsymbol{\alpha}^0$, in which Assumptions 2.4′ holds true. Thus

$$\begin{aligned}
&P\{\varphi_T(\mathbf{U}_T) - \varphi_T(\tilde{\mathbf{U}}_T) \le 0\} \\
&\quad \ge P\{\Lambda_{iT} > 0, i = \overline{1,n}; ||\xi_T - \boldsymbol{\alpha}^0|| < \varepsilon_3\} \\
&\quad = P\{\Lambda_{iT} > 0, i = \overline{1,n}\} - P\{\Lambda_{iT} > 0, i = \overline{1,n}; ||\xi_T - \boldsymbol{\alpha}^0|| \ge \varepsilon_3\} \\
&\quad \ge P\{\Lambda_{iT} > 0, i = \overline{1,n}\} - P\{||\xi_T - \boldsymbol{\alpha}^0|| \ge \varepsilon_3\} \\
&\quad = P\{\Lambda_{iT} > 0, i = \overline{1,n}\} - 1 + P\{||\xi_T - \boldsymbol{\alpha}^0|| < \varepsilon_3\}.
\end{aligned}$$

Since $\boldsymbol{\xi}_T$ converges in probability to $\boldsymbol{\alpha}^0$, we have for arbitrary $\eta > 0$

$$P\{||\xi_T - \boldsymbol{\alpha}^0|| < \varepsilon_3\} > 1 - \frac{\eta}{2}, \quad T > T_4.$$

According to the proof of Theorem 2.7, we have for arbitrary $\eta > 0$

$$P\{\boldsymbol{\Lambda}_{iT} > 0, i = \overline{1,n}\} > 1 - \frac{\eta}{2}, \quad T > T_5.$$

Thus (2.84) follows from the last three inequalities. Repeating the arguments from the proof of Theorem 2.10, we obtain from (2.84), (2.79), and (2.81) with $\varepsilon_2 = \varepsilon_1/2$, the expressions (2.85) and (2.86) (see Sect. 2.2). Thus the statement of the theorem follows from (2.86) and (2.76).

Theorem 2.11, which is a special case of the results obtained into Sect. 2.2, is given in Lutkepohl (1983). However, the proof given in this subsection is in our opinion more complete.

2.5 Asymptotic Properties of the Least Squares Estimates of Parameters of a Linear Regression with Non-Stationary Variables Under Convex Restrictions on Parameters

2.5.1 Settings

Unlike the previous subsections we consider here a case when independent variables can have a trend, and the regression is linear. Such a situation is quite common in many econometric problems.

We assume that the admissible domain of the multidimensional parameter is a convex set. In this situation, the asymptotical distribution for such a parameter cannot be obtained as a special case of the results from Sects. 2.2 and 2.3 due to the presence of the trend in variables. We find the limit distribution based on proofs given in these subsections.

Consider the estimation problem

$$S_T(\alpha) = \frac{1}{2}\sum_{t=1}^{T}(y_t - \mathbf{x}'_t\alpha)^2 \to \min, \quad g_i(\alpha) \leq 0,\ i \in I, \tag{2.114}$$

where the values x_t, y_t, $t = \overline{1,T}$ and functions $g_i(\alpha)$, $i \in I$, are known.

One can see that problem (2.114) is a particular case of problems (2.12) and (2.13). However, this problem significantly differs from (2.12) to (2.13) because the restrictions on the regression function $f_t(\alpha) = \mathbf{x}'_t\alpha$ are less strong, i.e. it might tend to $\pm\infty$ when t increases.

We solve the problem based on rather general assumptions about the regressors and the noise.

Assumption 2.8. Assume that the random variables ε_t are centered, independent, do not depend on $\mathbf{x}_t$, $t = 1, 2, \ldots$, and have the same distribution $\boldsymbol{\Phi}_t(u)$, $t = 1, 2, \ldots$, such that $\sup_{t=1,2,\ldots} \int_{|u|>0} u^2 d\boldsymbol{\Phi}_t(u) \to 0$ as $c \to \infty$. We denote by σ^2 the variance of ε_t.

Assumption 2.1 from Sect. 2.2 is a particular case of Assumption 2.8.

Denote $\mathbf{P}_T = \sum_{t=1}^{T} \mathbf{x}_t \mathbf{x}'_t$ and $\mathbf{E}_T = \mathrm{diag}(\sqrt{\rho_{11}^T}, \sqrt{\rho_{22}^T}, ..., \sqrt{\rho_{nn}^T})$, where ρ_{ii}^T are the elements on the main diagonal of $\mathbf{P}_T$.

Assumption 2.9. For all T, the matrix $\mathbf{P}_T$ is non-degenerate. The matrix $\mathbf{R}_T = \mathbf{E}_T^{-1}\mathbf{P}_T\mathbf{E}_T^{-1} \to \mathbf{R}$ as $T \to \infty$, where $\mathbf{R}$ is some positive definite matrix.

Assumption 2.10. Assume that constraints $g_i(\boldsymbol{\alpha})$ are twice continuously differentiable convex functions. Further, assume that the matrix $\mathbf{G}$ whose rows are $\nabla g_i'(\boldsymbol{\alpha}^0)$, $i \in I$, is of full rank.

Assumption 2.11. Assume that $\rho_{ii}^T \to \infty$, $x_{T+1,i}^2/\rho_{ii}^T \to 0$, $i = \overline{1,n}$, as $T \to \infty$.

Assumption 2.12. For some $(m \times m)$ diagonal matrix $\bar{\mathbf{E}}_T$ with positive elements on the main diagonal $\bar{e}_{Ti}$, $i = \overline{1,m}$, there exists the limit $\lim_{T\to\infty} \tilde{\mathbf{G}}_T = \tilde{\mathbf{G}}$ of the matrix $\tilde{\mathbf{G}}_T = \bar{\mathbf{E}}_T\mathbf{G}\mathbf{E}_T^{-1}$, where $\mathbf{G}$ is a matrix composed from the rows $\nabla g_i'(\boldsymbol{\alpha}^0)$, $i = \overline{1,n}$. At the same time, (1) $\tilde{\mathbf{G}}_1 = \lim_{T\to\infty} \tilde{\mathbf{G}}_{T1}$, where $\tilde{\mathbf{G}}_{T1} = \bar{\mathbf{E}}_{T1}\mathbf{G}_1\mathbf{E}_T^{-1}$, $\bar{\mathbf{E}}_{T1} = \mathrm{diag}(\bar{e}_{Ti})$, $i \in I_1^0$, and $\mathbf{G}_1$ is a matrix composed of $\nabla g_i'(\boldsymbol{\alpha}^0)$, $i \in I_1^0$, is of full rank, (2) there exists $\lim_{T\to\infty} \bar{e}_{Ti}e_{Tj}^{-1} < \infty$, $i \in I$, $j = \overline{1,n}$, where $e_{Tj} = \sqrt{\rho_{jj}^T}$.

Assumption 2.12 holds true, in particular, when the regressors are bounded, see previous subsections. Then $\bar{\mathbf{E}}_T = \sqrt{T}\mathbf{J}_m$, $E_T = \sqrt{T}\mathbf{J}_n$.

Here and everywhere in this subsection we assume that the regression parameter satisfies (2.14).

2.5.2 *Consistency of Estimator*

We start with the proof of the consistency of the solution to problem (2.114).

Theorem 2.12. *If Assumptions 2.8–2.12 are satisfied, then the solution* $\boldsymbol{\alpha}_T$ *to problem (2.114) is a consistent estimate of* $\boldsymbol{\alpha}^0$.

Proof. After some transformation problem (2.114) can be written as

$$\frac{1}{2}\boldsymbol{\alpha}'\mathbf{P}_T\boldsymbol{\alpha} - \boldsymbol{\alpha}'\mathbf{X}'_T\mathbf{Y}_T \to \min, \quad g_i(\boldsymbol{\alpha}) \le 0,\ i \in I, \tag{2.115}$$

where $\mathbf{X}_T$ is the $T \times n$ matrix.

According to Assumption 2.9, the matrix $\mathbf{P}_T$ is positive defined. Therefore one can always find a non-degenerate matrix $\mathbf{H}_T$ such that $\mathbf{P}_T = \mathbf{H}'_T\mathbf{H}_T$. We assume that $\boldsymbol{\beta} = \mathbf{H}_T\boldsymbol{\alpha}$. Taking into account these transformations, the quadratic programming problem (2.115) can be written as

$$\frac{1}{2}\boldsymbol{\beta}'\boldsymbol{\beta} - \boldsymbol{\beta}'(\mathbf{H}_T^{-1})'\mathbf{X}'_T\mathbf{Y}_T \to \min, \quad h_i(\boldsymbol{\beta}) \le 0,\ i \in I, \tag{2.116}$$

where $h_i(\boldsymbol{\beta}) = g_i(\mathbf{H}_T^{-1}\boldsymbol{\beta})$ is convex.

Put $\mathbf{K}_\beta = \{\boldsymbol{\beta} : h_i(\boldsymbol{\beta}) \leq 0, \boldsymbol{\beta} \in \Re^n\}$. The set $\mathbf{K}_\beta$ is convex by convexity of functions $h_i(\boldsymbol{\beta})$, $i \in I$. Then we transform the problem (2.116) to

$$||\boldsymbol{\beta} - \boldsymbol{\beta}_T^*||^2 \to \min, \quad \boldsymbol{\beta} \in \mathbf{K}_\beta, \tag{2.117}$$

where $\boldsymbol{\beta}_T^* = (\mathbf{H}_T^{-1})'\mathbf{X}'_T\mathbf{Y}_T$ is the solution to (2.116) without taking into account the constraint $\boldsymbol{\beta} \in \mathbf{K}_\beta$.

It follows from (2.117) that its solution $\boldsymbol{\beta}_T$ is the projection of $\boldsymbol{\beta}_T^*$ onto $\mathbf{K}_\beta$. This projection is unique by convexity of $\mathbf{K}_\beta$. It is known that the distance from an arbitrary point a that does not belong to some convex set A, to the projection of a onto this set, does not exceed the distance from a to an arbitrary point of A. Therefore, we have

$$||\boldsymbol{\beta}_T - \boldsymbol{\beta}^0||^2 \leq ||\boldsymbol{\beta}_T^* - \boldsymbol{\beta}^0||^2, \tag{2.118}$$

where $\boldsymbol{\beta}^0 = \mathbf{H}_T\boldsymbol{\alpha}^0 \in \mathbf{K}_\beta$ since by condition (2.14) we obtain

$$h_i(\boldsymbol{\beta}^0) = h_i(\mathbf{H}_T\boldsymbol{\alpha}^0) = g_i(\boldsymbol{\alpha}^0) \leq 0, \quad i \in I.$$

From inequality (2.118) we get

$$||\mathbf{H}_T(\boldsymbol{\alpha}_T - \boldsymbol{\alpha}^0)||^2 \leq ||\mathbf{H}_T(\boldsymbol{\alpha}_T^* - \boldsymbol{\alpha}^0)||^2$$

which implies

$$(\boldsymbol{\alpha}_T - \boldsymbol{\alpha}^0)'\mathbf{E}_T\mathbf{R}_T\mathbf{E}_T(\boldsymbol{\alpha}_T - \boldsymbol{\alpha}^0) \leq (\boldsymbol{\alpha}_T^* - \boldsymbol{\alpha}^0)'\mathbf{E}_T\mathbf{R}_T\mathbf{E}_T(\boldsymbol{\alpha}_T^* - \boldsymbol{\alpha}^0).$$

Then we obtain

$$\mu_{\min}(\mathbf{R}_T)||\mathbf{E}_T(\boldsymbol{\alpha}_T - \mathbf{R}^0)||^2 \leq \mu_{\max}(\mathbf{R}_T)||\mathbf{U}_T^*||^2, \tag{2.119}$$

where $\mu_{\max}(\mathbf{R}_T)$ and $\mu_{\min}(\mathbf{R}_T)$ are, respectively, the maximal and the minimal eigenvalues of $\mathbf{R}_T$, and $\mathbf{U}_T^* = \mathbf{E}_T(\boldsymbol{\alpha}_T^* - \boldsymbol{\alpha}^0)$.

Denote $e_{T,\min}^2 = \min_{i=\overline{1,n}} e_{Ti}^2$, where $e_{Ti}^2 = \rho_{ii}^T$. Then the inequality $||\mathbf{E}_T(\boldsymbol{\alpha}_T - \boldsymbol{\alpha}^0)||^2 \geq e_{T,\min}^2||\boldsymbol{\alpha}_T - \boldsymbol{\alpha}^0||^2$ holds true. Taking into account this inequality, we obtain from (2.119)

$$||\boldsymbol{\alpha}_T - \boldsymbol{\alpha}^0||^2 \leq \left(\frac{\mu_{\max}(\mathbf{R}_T)}{e_{T,\min}^2\mu_{\min}(\mathbf{R}_T)}\right)||\mathbf{U}_T^*||^2. \tag{2.120}$$

According to Anderson (1971), under Assumptions 2.8, 2.9 and 2.11 of Theorem 2.6.1, the distribution of $\mathbf{U}_T^*$ converges as $T \to \infty$ to the distribution of the variable $\mathbf{U}^* \sim N(\mathbf{O}_n, \sigma^2\mathbf{R}^{-1})$. By Rao (1965, Theorem XII, Chap. 2), we have $||\mathbf{U}_T^*||^2 \overset{p}{\Rightarrow} ||\mathbf{U}^*||^2$.

According to Assumption 2.9, the eigenvalues of the matrix $\mathbf{R}_T$ converge as $T \to \infty$ to non-zero values. Moreover, according to Assumption 2.11, $e_{T,\min}^2 \to \infty$

as $T \to \infty$. Therefore, the first term in the right-hand side of inequality (2.120) converges in probability to zero. Then the right-hand side of (2.120) also converges in probability to zero, which implies the statement of the theorem. □

Corollary 2.2. *If Assumptions 2.8–2.12 hold true, then the solution to (2.114) with linear constraints* $g_i(\boldsymbol{\alpha}) = \mathbf{g}'_i\boldsymbol{\alpha} - b_i \leq 0,\ i \in I$, *converges in probability to* $\boldsymbol{\alpha}^0$.

Corollary 2.3. *If Assumptions 2.8–2.12 are satisfied and, at the same time, we have* $\bar{\mathbf{E}}_T = \sqrt{T}\mathbf{J}_m$ *and* $\mathbf{E}_T = \sqrt{T}\mathbf{J}_n$, *then the solution* $\boldsymbol{\alpha}_T$ *to problem (2.114) with linear constraints converges to* $\boldsymbol{\alpha}^0$ *in the mean-square sense.*

Proof. Under the conditions of Corollary 2.3, $\boldsymbol{\alpha}^*_T$ converges in the mean-square sense to $\boldsymbol{\alpha}^0$ (Demidenko 1981, Sect. 1.7), $\mu_{\max}(\mathbf{R}_T) \to \mu_{\max}(\mathbf{R})$, and $\mu_{min}(\mathbf{R}_T) \to \mu_{\min}(\mathbf{R}) > 0$ as $T \to \infty$. This directly implies the statement of corollary. □

2.5.3 Limit Distribution of the Parameter Estimate

Consider two auxiliary results.

Lemma 2.10. *Assume that the conditions of Theorem 2.12 are satisfied. Then, for given* $\delta > 0$, *there exist* $\varepsilon > 0$ *and* $T_0 > 0$, *such that*

$$P\{||\mathbf{U}_T|| \geq \varepsilon\} < \delta, \quad T > T_0,\ \mathbf{U}_T = \mathbf{E}_T(\boldsymbol{\alpha}_T - \boldsymbol{\alpha}^0). \tag{2.121}$$

Proof. Transforming inequality (2.119), we obtain

$$||\mathbf{U}_T||^2 \leq \mu_T ||\mathbf{U}^*_T||^2, \tag{2.122}$$

where $\mu_T = \mu_{\max}(\mathbf{R}_T)/\mu_{\min}(\mathbf{R}_T)$. According to Assumption 2.9, we have $\mu_T \to \mu \neq 0$ as $T \to \infty$, and, as in the proof of Theorem 2.12, $||\mathbf{U}^*_T||^2 \Rightarrow^p ||\mathbf{U}^*||^2$. Therefore, taking into account inequality (2.122), for given $\delta > 0$, there exist some $\varepsilon > 0$ and $T_0 > 0$ for which

$$\delta > P\{\mu_T ||\mathbf{U}^*_T||^2 \geq \varepsilon\} > P\{||\mathbf{U}_T||^2 ge\varepsilon\}, \quad T > T_0.$$

Lemma is proved. □

Lemma 2.11. *Suppose that Assumptions 2.9–2.12 are satisfied. Then the solution to the minimization problem with respect to* $\mathbf{X} = \mathbf{E}_T(\boldsymbol{\alpha} - \boldsymbol{\alpha}^0) \in \Re^n$

$$\begin{cases} \varphi_T(\mathbf{X}) = \frac{1}{2}\mathbf{X}'\mathbf{R}_T\mathbf{X} - \mathbf{Q}'_T\mathbf{X} \to \min, \\ \beta_i(\mathbf{X}) = (\bar{e}_{Ti}\nabla g'_i(\boldsymbol{\alpha}^0)\mathbf{E}_T^{-1})\mathbf{X} + l_i(\mathbf{X}) \leq \mathbf{O}_{m_1}, \quad i \in I_1^0, \\ \beta_i(\mathbf{X}) = \bar{e}_{Ti}g_i(\boldsymbol{\alpha}^0) + (\bar{e}_{Ti}\nabla g'_i(\boldsymbol{\alpha}^0)\mathbf{E}_T^{-1})\mathbf{X} + l_i(\mathbf{X}) \leq \mathbf{O}_{m_2}, \quad i \in I_2^0 \end{cases} \tag{2.123}$$

is $\mathbf{U}_T = \mathbf{E}_T(\boldsymbol{\alpha}_T - \boldsymbol{\alpha}^0)$.

In (2.123),

$$\mathbf{Q}_T = \mathbf{E}_T^{-1} \sum_{t=1}^{T} \varepsilon_t \mathbf{x}_t,$$

$$\beta_i(\mathbf{X}) = \bar{e}_{Ti} g_i(\mathbf{E}_T^{-1}\mathbf{X} + \boldsymbol{\alpha}^0), \quad i \in I, \tag{2.124}$$

$$l_i(\mathbf{X}) = \frac{1}{2}\mathbf{X}'(\bar{e}_{Ti}\mathbf{E}_T^{-1} g_{i2}(\boldsymbol{\alpha}^0 + \theta_{T1}\mathbf{E}_T^{-1}\mathbf{X})\mathbf{E}_T^{-1})\mathbf{X}, \quad i \in I, \tag{2.125}$$

where functions g_{i2} and θ_{T1} are defined in (2.43).

Proof. The cost function in the problem (2.123) is strictly convex (according to Assumption 2.9) with convex domain, since the related constraints are convex functions. Hence, the solution to (2.123) (which we denote by $\mathbf{U}_T^*$) is unique. It satisfies the following necessary and sufficient conditions for the existence of the minimum:

$$\mathbf{R}_T\mathbf{U}_T^* - \mathbf{Q}_T + \sum_{i=1}^{m} \nabla\beta_i(\mathbf{U}_T^*) v_{Ti}^* = \mathbf{O}_n,$$

$$v_{Ti}^*\beta_i(\mathbf{U}_T^*) = 0, \quad v_{Ti}^* \geq 0,\ i \in I, \tag{2.126}$$

where v_{Ti}^* is the Lagrange multiplier corresponding to the ith constraint.

Problem (2.114) is also a convex programming problem, such that $\boldsymbol{\alpha}_T$ satisfies the necessary and sufficient conditions for the existence of its minimum:

$$\mathbf{P}_T\boldsymbol{\alpha}_T - \boldsymbol{\alpha}_T'\mathbf{X}'_T\mathbf{Y}_T + \sum_{i=1}^{m} \nabla g_i(\boldsymbol{\alpha}_T)\lambda_{Ti} = \mathbf{O}_n,$$

$$\lambda_{Ti} g_i(\boldsymbol{\alpha}_T) = 0, \quad \lambda_{Ti} \geq 0,\ i \in I, \tag{2.127}$$

where λ_{Ti}, $i \in I$ are Lagrange multipliers for problem (2.114).

After some transformations of the first equality in (2.127), multiplying it by $\mathbf{E}_T^{-1}$ from the left, and putting $v_{Ti} = \bar{e}_{Ti}^{-1}\lambda_{Ti}$, $i \in I$, we obtain

$$\mathbf{R}_T\mathbf{U}_T - \mathbf{Q}_T + \sum_{i=1}^{m} \bar{e}_{Ti}\mathbf{E}_T^{-1}\nabla g_i(\boldsymbol{\alpha}_T) v_{Ti} = \mathbf{O}_n,$$

$$v_{Ti}\bar{e}_{Ti} g_i(\boldsymbol{\alpha}_T) = 0, \quad v_{Ti} \geq 0,\ i \in I. \tag{2.128}$$

Comparing the expressions (2.126) and (2.128), we obtain that condition (2.126) holds true when $\mathbf{U}_T^* = \mathbf{U}_T$ and $v_{Ti}^* = v_{Ti} \geq 0$, $i \in I$, because according to (2.124) we have $\beta_i(\mathbf{U}_T) = \bar{e}_{Ti} g_i(\boldsymbol{\alpha}_T)$ and $\nabla\beta_i(\mathbf{U}_T) = \bar{e}_{Ti}\mathbf{E}_T^{-1}\nabla g_i(\boldsymbol{\alpha}_T)$, $i \in I$. Since the conditions (2.126) determine a unique solution to (2.123), the statement of the lemma holds true.

□

Let us find the limit of $\mathbf{U}_T$.

Theorem 2.13. *If Assumptions 2.8–2.12 hold true, then the random variable* $\mathbf{U}_T = \mathbf{E}_T(\boldsymbol{\alpha}_T - \boldsymbol{\alpha}^0)$ *converges in distribution as* $T \to \infty$ *to a random variable* $\mathbf{U}$*, which is the solution to the problem*

$$\varphi_T(\mathbf{X}) = \frac{1}{2}\mathbf{X}'\mathbf{R}\mathbf{X} - \mathbf{Q}'\mathbf{X} \to \min, \quad \tilde{\mathbf{G}}_1\mathbf{X} \leq \mathbf{O}_{m_1}, \tag{2.129}$$

where the $(m_1 \times n)$ *matrix* $\tilde{\mathbf{G}}_1$ *consists of the rows of the matrix* $\tilde{\mathbf{G}}$ *with indices* $i \in I_1^0$.

Proof. Consider the quadratic programming problem

$$\varphi_T^0(\mathbf{X}) = \frac{1}{2}\mathbf{X}'\mathbf{R}\mathbf{X} - \mathbf{Q}'_T\mathbf{X} \to \min, \quad \tilde{\mathbf{G}}_{T1}\mathbf{X} \leq \mathbf{O}_{m_1}. \tag{2.130}$$

Denote its solution by $\mathbf{U}_T^0$. By Theorem 2.6.1 (Anderson 1971), for the vector $\mathbf{Q}_T$ we have

$$\mathbf{Q}_T \stackrel{p}{\Rightarrow} \mathbf{Q} \sim N(\mathbf{O}_n, \sigma^2\mathbf{R}), \quad T \to \infty. \tag{2.131}$$

According to Lemma 2.5, $\mathbf{U}_T^0$ is continuous in $\mathbf{Q}_T$ and $\tilde{\mathbf{G}}_{T1}$: $\mathbf{U}_T^0 = f(\mathbf{Q}_T, \tilde{\mathbf{G}}_{T1})$. From here, (2.131), and the fact that by Assumption 2.12 we have $\lim_{T\to\infty} \tilde{\mathbf{G}}_{T1} = \tilde{\mathbf{G}}_1$, we obtained $f(\mathbf{Q}_T, \tilde{\mathbf{G}}_{T1}) \stackrel{p}{\Rightarrow} f(\mathbf{Q}, \tilde{\mathbf{G}}_1)$. According to (2.129), we have $\mathbf{U} = f(\mathbf{Q}, \tilde{\mathbf{G}}_1)$. Thus,

$$\mathbf{U}_T^0 \stackrel{p}{\Rightarrow} \mathbf{U} \quad \text{as } T \to \infty. \tag{2.132}$$

Denote by $\mathbf{O}_T = \{\mathbf{X} : \beta_i(\mathbf{X}) \leq 0, i \in I\}$ and $\mathbf{O} = \{\mathbf{X} : (\bar{e}_{Ti}\nabla g_i'(\boldsymbol{\alpha}^0)\mathbf{E}_T^{-1})\mathbf{X} \leq 0, i \in I_1^0\}$ admissible domains in problems (2.123) and (2.130), respectively.

According to Assumption 2.10 functions $g_i(\boldsymbol{\alpha})$, $i \in I$ are convex. By the convexity of the constraints we have $l_i(\mathbf{X}) \geq 0$, $\mathbf{X} \in \Re^n$. Therefore, since $\beta_i(\mathbf{U}_T) \leq 0$, $i \in I_1^0$, it follows from the first constraint in (2.123) that $(\bar{e}_{Ti}\nabla g_i'(\boldsymbol{\alpha}^0)\mathbf{E}_T^{-1})\mathbf{U}_T \leq 0$, $i \in I_1^0$. This implies $\mathbf{U}_T \in \mathbf{O}$.

According to Assumption 2.9, $\varphi_T^0(\mathbf{X})$ is a strongly convex function of $\mathbf{X}$. Then see Karmanov (1975),

$$||\mathbf{U}_T - \mathbf{U}_T^0||^2 \leq \frac{2}{\mu}[\varphi_T^0(\mathbf{U}_T) - \varphi_T^0(\mathbf{U}_T^0)], \quad \mu > 0 \tag{2.133}$$

because $\mathbf{U}_T \in \mathbf{O}$, i.e., $\mathbf{U}_T$ satisfies the constraints of the problem (2.130).

According to (2.133), we have for arbitrary $\varepsilon > 0$

$$\begin{aligned} P\{||\mathbf{U}_T - \mathbf{U}_T^0||^2 < \varepsilon^2\} &\geq P\left\{\frac{2}{\mu}[\varphi_T^0(\mathbf{U}_T) - \varphi_T^0(\mathbf{U}_T^0)] < \varepsilon^2\right\} \\ &\geq P\left\{|\varphi_T^0(\mathbf{U}_T) - \varphi_T(\mathbf{U}_T)| < \frac{\varepsilon_1}{2}\right\} + P\left\{\varphi_T(\mathbf{U}_T) - \varphi_T^0(\mathbf{U}_T^0) < \frac{\varepsilon_1}{2}\right\} - 1, \end{aligned} \tag{2.134}$$

where $\varepsilon_1 = \varepsilon^2\mu/2$.

To transform the right-hand side of (2.134) we need to derive some relations connecting the cost functions in problems (2.123) and (2.130). For arbitrary $\varepsilon_2 > 0$, we have

$$P\{|\varphi_T^0(\mathbf{X}) - \varphi_T(\mathbf{X})| < \varepsilon_2\} = P\left\{\left|\frac{1}{2}\mathbf{X}'(\mathbf{R} - \mathbf{R}_T)\mathbf{X}\right| < \varepsilon_2\right\}. \tag{2.135}$$

Substituting $\mathbf{X} = \mathbf{U}_T$ in (2.135) we obtain

$$P\{|\varphi_T^0(\mathbf{U}_T) - \varphi_T(\mathbf{U}_T)| < \varepsilon_2\} \geq 1 - P\left\{|\mathbf{U}_T| \geq \sqrt{a}\right\} - P\left\{|\mathbf{R} - \mathbf{R}_T| \geq \frac{2\varepsilon_2}{a}\right\} \tag{2.136}$$

where a is some positive number. According to Assumption 2.9, for sufficiently large T the third term in the right-hand side of (2.136) is equal to 0. Then, applying Lemma 2.10, we obtain from (2.136)

$$P\{|\varphi_T^0(\mathbf{U}_T) - \varphi_T(\mathbf{U}_T)| < \varepsilon_2\} \geq 1 - \delta, \quad T > T_1. \tag{2.137}$$

Put $\mathbf{X} = \mathbf{U}_T^0$ in (2.135). After performing some necessary transformations and taking into account that $\mathbf{U}_T^0$ has a limit distribution (cf. (2.132)), we obtain for arbitrary $\varepsilon_2 > 0$ and $\delta > 0$

$$P\{|\varphi_T^0(\mathbf{U}_T^0) - \varphi_T(\mathbf{U}_T^0)| < \varepsilon_2\} = D_{T2} \geq 1 - \delta, \quad T > T_2. \tag{2.138}$$

Consider the second term in the right-hand side of (2.134). We have

$$\begin{aligned} P\left\{\varphi_T(\mathbf{U}_T) - \varphi_T^0(\mathbf{U}_T^0) < \frac{\varepsilon_1}{2}\right\} &\geq P\{\varphi_T(\mathbf{U}_T) - \varphi_T(\mathbf{U}_T^0) \leq 0\} \\ +P\left\{\varphi_T(\mathbf{U}_T^0) - \varphi_T^0(\mathbf{U}_T^0) < \frac{\varepsilon_1}{2}\right\} &- 1 = D_{T1} + D_{T2} - 1, \end{aligned} \tag{2.139}$$

where D_{T2} is the probability defined in inequality (2.138) at $\varepsilon_2 = \varepsilon_1/2$.

To determine the probability D_{T1} in (2.139), we write the constraints of problem (2.123) for $\mathbf{X} = \mathbf{U}_T^0$ as follows:

$$\begin{aligned} &\tilde{\mathbf{G}}_{T1}\mathbf{U}_T^0 + \mathbf{L}_1(\mathbf{U}_T^0) \leq \mathbf{O}_{m_1}, \\ &\tilde{\mathbf{G}}_{T2}\mathbf{U}_T^0 + \bar{\mathbf{E}}_{T2}g^{(2)}(\boldsymbol{\alpha}^0) + \mathbf{L}_2(\mathbf{U}_T^0) \leq \mathbf{O}_{m_2}. \end{aligned} \tag{2.140}$$

Here, $\mathbf{L}_k(\mathbf{X})$ is vector of dimension m_k, $k = \overline{1,2}$, whose ith component is a function $l_i(\mathbf{X})$, $i \in I_k^0$, specified by the expression (2.125); $\tilde{\mathbf{G}}_{Tk} = \bar{\mathbf{E}}_{Tk}\mathbf{G}_k\mathbf{E}_T^{-1}$, $k = 1, 2$. In this case, we have by Assumption 2.12

$$\lim_{T\to\infty} \tilde{\mathbf{G}}_{Tk} = \tilde{\mathbf{G}}_k, \tag{2.141}$$

where $\bar{\mathbf{E}}_{Tk} = \operatorname{diag}(\bar{e}_{Ti})$.

According to (2.125), (2.132) and Assumption 2.12, we have

$$p \lim_{T\to\infty} l_i(\mathbf{U}_T^0) = \frac{1}{2}(\mathbf{U}_T^0)'(\bar{e}_{Ti}\mathbf{E}_T^{-1} g_{i2}(\boldsymbol{\alpha}^0 + \theta_{T1}\mathbf{E}_T^{-1}\mathbf{U}_T^0)\mathbf{E}_T^{-1})\mathbf{U}_T^0 = 0, \quad i \in I$$

which implies $p \lim_{T\to\infty} \mathbf{L}_2(\mathbf{U}_T^0) = \mathbf{O}_{m_2}$.

By (2.139) we obtain $D_{T1} \geq P\{\mathbf{U}_T^0 \in O_T\}$. Therefore, we have the sequence of inequalities:

$$\begin{aligned}
D_{T1} &= P\{\varphi_T(\mathbf{U}_T) - \varphi_T(\mathbf{U}_T^0) \leq 0\} \\
&\geq P\{\tilde{\mathbf{G}}_{T1}\mathbf{U}_T^0 + \mathbf{L}_1(\mathbf{U}_T^0) \leq \mathbf{O}_{m_1}, \tilde{\mathbf{G}}_{T2}\mathbf{U}_T^0 + \mathbf{L}_2(\mathbf{U}_T^0) + \bar{\mathbf{E}}_{T2}\mathbf{g}^{(2)}(\boldsymbol{\alpha}^0) \leq \mathbf{O}_{m_2}\} \\
&\geq P\{\tilde{\mathbf{G}}_{T1}\mathbf{U}_T^0 + \mathbf{L}_1(\mathbf{U}_T^0) \leq \mathbf{O}_{m_1}\} + P\{\tilde{\mathbf{G}}_{T2}\mathbf{U}_T^0 + \mathbf{L}_2(\mathbf{U}_T^0) + \bar{\mathbf{E}}_{T2}\mathbf{g}^{(2)}(\boldsymbol{\alpha}^0) \leq \mathbf{O}_{m_2}\} - 1 \\
&\geq \sum_{i\in I_1^0} P\left\{\psi_{Ti}^{(1)} + l_i(\mathbf{U}_T^0) \leq 0\right\} + \sum_{i\in I_2^0} P\left\{\psi_{Ti}^{(2)} + l_i(\mathbf{U}_T^0) + \bar{e}_{Ti} g_i(\boldsymbol{\alpha}^0) \leq 0\right\} + 1 - m_1 - m_2,
\end{aligned}$$

where $\psi_{Ti}^{(k)}$ is the ith component of the vector $\tilde{\mathbf{G}}_{Tk}\mathbf{U}_T^0$, $k = \overline{1,2}$.

According to (2.129), we have $\tilde{\mathbf{G}}_1\mathbf{U} \leq \mathbf{O}_{m_1}$. Then (2.132) and (2.141) imply that $\tilde{\mathbf{G}}_{T1}\mathbf{U}_T^0 \overset{p}{\Rightarrow} \tilde{\mathbf{G}}_1\mathbf{U} \leq \mathbf{O}_{m_1}$. Since $p \lim_{T\to\infty} l_i(\mathbf{U}_T^0) = 0$, $i \in I$, one can see that for an arbitrary $\delta_1 > 0$ there exists $T_3 > 0$ for which

$$P\left\{\psi_{Ti}^{(1)} + l_i(\mathbf{U}_T^0) \leq 0\right\} > 1 - \delta_1, \quad i \in I_1^0,\ T > T_3.$$

Since $l_i(\mathbf{U}_T^0)$, $i \in I$, converges to zero in probability, we have by (2.131), (2.140) and Assumption 2.12, that for an arbitrary $\delta_2 > 0$

$$P\{\bar{e}_{Ti}^{-1}(|\psi_{Ti}^{(2)}| + l_i(\mathbf{U}_T^0)) \leq -g_i(\boldsymbol{\alpha}^0)\} > 1 - \delta_2, \quad i \in I_2^0,\ T > T_4.$$

Substituting this and the previous inequalities in the lower estimate obtained for D_{1T}, we get

$$D_{T1} \geq 1 - m_1\delta_1 - m_2\delta_2, \quad T > \max(T_3, T_4). \tag{2.142}$$

Put $\varepsilon_2 = \varepsilon_1/2$ in the expression (2.138). Then, taking into account (2.142), we obtain from (2.139)

$$P\left\{\varphi_T(\mathbf{U}_T) - \varphi_T^0(\mathbf{U}_T^0) < \frac{\varepsilon_1}{2}\right\} > 1 - \delta - m_1\delta_1 - m_2\delta_2, \quad T > \max(T_2, T_4). \tag{2.143}$$

Consider problem (2.134). Substituting inequality (2.137) with $\varepsilon_2 = \varepsilon_1/2$ and inequality (2.143) into (2.134), we get

$$P\{||U_T - U_T^0||^2 < \varepsilon^2\} > 1 - 2\delta - m_1\delta_1 - m_2\delta_2, \quad T > \max_{1\leq i\leq 4} T_i.$$

Thus, we obtain $p \lim_{T\to\infty} ||\mathbf{U}_T - \mathbf{U}_T^0||^2 = 0$. The statement of the theorem follows from this expression and (2.132). □

Chapter 3
Method of Empirical Means in Nonlinear Regression and Stochastic Optimization Models

In stochastic optimization and identification problems (Ermoliev and Wets 1988; Pflug 1996), it is not always possible to find the explicit extremum for the expectation of some random function. One of the methods for solving this problem is the method of empirical means, which consists in approximation of the existing cost function by its empiric estimate, for which one can solve the corresponding optimization problem. In addition, it is obvious that many problems in mathematical statistics (for example, estimation of unknown parameters by the least squares, the least modules, the maximum likelihood methods, etc.) can be formulated as special stochastic programming problems with specific constraints for unknown parameters which stresses the close relation between stochastic programming and estimation theory methods. In such problems the distributions of random variables or processes are often unknown, but their realizations are known. Therefore, one of the approaches for solving such problems consists in replacing the unknown distributions with empiric distributions, and replacing the corresponding mathematical expectations with their empiric means. The difficulty is in finding conditions under which the approximating problem converges in some probabilistic sense to the initial one. We discussed this briefly in Sect. 2.1. Convergence conditions are of course essentially dependent on the cost function, the probabilistic properties of random observations, metric properties of the space, in which the convergence is investigated, a priori constraints on unknown parameters, etc. In the notation used in statistical decision theory the problems above are closely related with the asymptotic properties of unknown parameters estimates, i.e. their consistency, asymptotic distribution, rate of convergence, etc.

It should be noted that there are a lot of publications devoted to the method of empirical means. Among them we can name the works of E. Yubi, J. Dupacjva, Yu. Ermoliev, V. Kankova, A. King, P.S. Knopov, V.I. Norkin, G. Salinetti, F. Shapiro, R.J. Wets and others. Some of these publications are given in the references. There are many approaches for solving this problem. We pay principal attention to the approach based on the general theory of convergence of the extremum points of a random function to the limit point. Here we will rely on The-

P.S. Knopov and A.S. Korkhin, *Regression Analysis Under A Priori Parameter Restrictions*, Springer Optimization and Its Applications 54,
DOI 10.1007/978-1-4614-0574-0_3,

orem 2.1. We shall briefly discuss one of the well-known approaches, based on the notions of epi-distance and epi-convergence (Dupacova and Wets 1986; King 1986, 1988; Salinetti and Wets 1986; Shapiro 1989, 1991; Shapiro et al. 2009; Wets 1979, 1983). These approaches are widely used in modern asymptotic estimation theory and proved to be useful for solving the problems below. It should also be noted, that in classical estimation theory one usually considers estimation problems without constraints on unknown parameters, or only with equality-constraints. Inequality type constraints make the estimation problem much more complicated, especially when we need to find the asymptotic distribution of obtained estimates. We discuss some results concerning above-mentioned problems herein.

We begin with the simplest but important stochastic programming problem and demonstrate the basic ideas of the proposed approach. Then we consider more sophisticated models which are investigated by similar methods.

In this chapter, we consider random functions which have continuous and discrete parameters. Therefore, for convenience we use the notation s, t, u, etc. for continuous arguments, and i, j, k, etc., for discrete. As a rule, by $||\cdot||$ we denote the norm in $\Re^l$ or in some functional space, if it will not lead to misunderstanding.

3.1 Consistency of Estimates Obtained by the Method of Empirical Means with Independent Or Weakly Dependent Observations

Let $\{\xi_i, i \in N\}$ be independent identically distributed observations of a random variable defined on a probability space $(\mathbf{\Omega}, \Im, P)$ with values in some metric space $(Y, L(Y))$, ξ be a random value with the same distribution and taking values in the same metric space with norm $||\cdot||$, and let $L(Y)$ be the minimal σ-algebra on Y. We assume that I is a closed subset in $\Re^l, l \geq 1$, possibly $I = \Re^l$. We assume that $f : I \times Y \to \Re$ is a nonnegative function satisfying the following conditions:

1. $f(\mathbf{u}, z), \mathbf{u} \in I$, is continuous for all $z \in Y$
2. For any $\mathbf{u} \in I$, the mapping $f(\mathbf{u}, z), z \in Y$, is $L(Y)$-measurable

The problem consists in finding the minimum point of the function

$$F(\mathbf{u}) = E(f(\mathbf{u}, \xi)), \mathbf{u} \in I \tag{3.1}$$

and its minimal value.

This problem is approximated by the following one: find the minimum points of the function

$$F_n(\mathbf{u}) = \frac{1}{n} \sum_{i=1}^{n} f(\mathbf{u}, \xi_i) \tag{3.2}$$

and its minimal value.

Theorem 3.1 (Knopov and Kasitskaya 1995, 2002). *Let the following conditions be satisfied:*

1. *For any $c > 0$, $E(\max_{||\mathbf{u}|| \le c} f(\mathbf{u}, \xi)) < \infty$, where $|| \cdot ||$ is a norm in $\Re^l$;*
2. *If $P\{\xi \in Y'\} = 1$, then for all $z \in Y'$ we have $f(\mathbf{u}, z) \to \infty$ as $||z|| \to \infty$;*
3. *There is a unique point $\mathbf{u}_0$, at which the function $F(\mathbf{u})$ attains its minimum.*

Then, for any n and $\omega \in \mathbf{\Omega}'$, when $P(\mathbf{\Omega}')=1$, there is at least one vector $\mathbf{u}_n = \mathbf{u}_n(\omega) \in I$ for which the minimum value of $F_n(\mathbf{u})$ is attained and, for any $n \ge 1$, the vector $\mathbf{u}_n$ can be chosen to be G'_n -measurable, where $G'_n = G_n \cap \mathbf{\Omega}'$ and $G_n = \sigma\{\xi_i, i = \overline{1,n}\}$. In this case, with probability 1, $\mathbf{u}_n \to \mathbf{u}_0$ and $F(\mathbf{u}_n) \to F(\mathbf{u}_0)$.

The proof of this statement is given in Knopov and Kasitskaya (1995, 2002). The proof is based on Theorem 2.1 and Remark 2.2, where the universal approach for proving consistency and strong consistency of the estimates of unknown parameters for the case of random variables with values in some metric space, is presented.

Let us briefly discuss some key steps.

1. Taking into account condition 2 of Theorem 2.1, one can prove that there exists $n_0 \ge 1$ such that with probability 1 all $\mathbf{u}_n, n \ge n_0$, belong to the compact set $K = \{\mathbf{u} \in I : ||\mathbf{u}|| \le c\}$.
2. By ergodic theorem, it is easy to see that $P\{\lim_{n\to\infty} F_n(\mathbf{u}) = F(\mathbf{u})\} = 1$.
3. Choosing $c(\gamma) = 2E\{\sup_{\{\mathbf{u},\mathbf{u}':||\mathbf{u}-\mathbf{u}'||<\gamma\}} ||f(\mathbf{u}, s) - f(\mathbf{u}', s)\}$, one can prove that $P\{\lim_{n\to\infty} \sup_{||\mathbf{u}-\mathbf{u}'||<\gamma} ||F_n(\mathbf{u}) - F_n(\mathbf{u}') < c(\gamma)\} = 1$.

Therefore Condition 4 of Theorem 2.1 is satisfied too. Using Theorem 2.1, we immediately obtain the statement of Theorem 3.1.

Remark 3.1. Using the ergodic theorem, it is possible to prove that Theorem 3.1 also holds true in the case when the random sequence $\{\xi_i, i \in N\}$ is stationary in the narrow sense. In other words, the problem

$$F(\mathbf{u}) = E(f(\mathbf{u}, \xi)) \to \min_{\mathbf{u}\in I}$$

is approximated by the problem.

$$F_n(\mathbf{u}) = \frac{1}{n} \sum_{i=1}^{n} f(\mathbf{u}, \xi) \to \min_{\mathbf{u}\in I}.$$

When time is continuous, a statement similar to Theorem 3.1 is also true.

Theorem 3.2 (Knopov and Kasitskaya 2002). *Let $\{\xi(t), t \in \Re\}$ be a random process stationary in the narrow sense and defined on the probability space $(\mathbf{\Omega}, \Im, P)$ with values in $\Re$.*

Let the following conditions be satisfied:

1. *For any $c > 0$, $E\{\max_{||\mathbf{u}|| \le c} f(\mathbf{u}, \xi(0))\} < \infty$*
2. *If I is an unbounded set for any $z \in Y'$, and $P\{\xi(t) \in Y' \forall t \ge 0\} = 1$, then $f(\mathbf{u}, z) \to \infty$ as $||\mathbf{u}|| \to \infty$*
3. *There is a unique element $\mathbf{u}_0 \in I$ for which the minimal value of the function $F(\mathbf{u}) = Ef(\mathbf{u}, \xi(0))$ is attained*

Then, for all $T > 0$ and $\omega \in \mathbf{\Omega}'$, $P(\mathbf{\Omega}') = 1$, there is at least one vector $\mathbf{u}\ (T) \in I$ for which the minimal value of the function

$$F_T(\mathbf{u}) = \frac{1}{T}\int_0^T f(\mathbf{u}, \xi(t))\mathrm{d}t$$

is attained and, for each $T > 0$, the variable $\mathbf{u}(T)$ is G'_T-measurable, where $G'_T = G_T \cap \mathbf{\Omega}'$ and $G_T = \sigma\{\xi(t), 0 \leq t \leq T\}$.

Let $\mathbf{u}_0 = \arg\min F(\mathbf{u})$, where $F(\mathbf{u}) = Ef(\mathbf{u}, \xi(0))$. Then we have

$$P\left\{\lim_{T\to\infty} \mathbf{u}(T) = \mathbf{u}\right\} = 1, \quad P\left\{\lim_{T\to\infty} F_T(\mathbf{u}_T) = F(\mathbf{u}_0)\right\} = 1.$$

Taking into account Remark 3.1, the proof follows by the same arguments as in the discrete time case.

Further generalization of the results on the strong consistency or convergence with probability 1 of an approximating stochastic programming problem to the original one consists in considering instead of $f(\mathbf{u}, \xi)$ the functions of more general form. We take a function of the form

$$F_n(\mathbf{u}) = \frac{1}{n}\sum_{i=1}^{n} f(i, \mathbf{u}, \xi_i), \tag{3.3}$$

where $\xi_i \in \mathfrak{R}$ are, generally speaking, dependent observations with some restrictions on the character of dependence. More precisely, the following result takes place.

Theorem 3.3 (Knopov and Kasitskaya 2002). *Let the stochastic function $f(i, \mathbf{u}, \xi_i)$ satisfy the conditions:*

1. *For any $\mathbf{u} \in I$, there exists a function F(x) such that $F(\mathbf{u}) = \lim_{||\mathbf{u}||\to\infty} F_n(\mathbf{u})$, and a point $\mathbf{u}_0 \in I$ such that $F(\mathbf{u}_0) < F(\mathbf{u})$ if $\mathbf{u} \neq \mathbf{u}_0$*
2. *The function $f(i, \mathbf{u}, z)$ is continuous with respect to the second argument uniformly in i and z*
3. *If the set I is unbounded, then for fixed i and z, $f(i, \mathbf{u}, z) \to \infty$ as $||\mathbf{u}|| \to \infty$*
4. *There exists a function $c(\gamma) > 0, c(\gamma) \to 0$ as $\gamma \to 0$, such that for any $\delta > 0$, there exists some γ_0 such that for any element $\mathbf{u} \in I$ and $0 < \gamma < \gamma_0$, the following relation holds true:*

$$\overline{\lim_{n\to\infty}} \frac{1}{n}\sum_{i=1}^{n} E \sup_{\substack{||\mathbf{u}-\mathbf{u}'|| < \gamma \\ \|\mathbf{u}-\mathbf{u}_0\| > \delta}} |f(i, \mathbf{u}, \xi_i) - f(i, \mathbf{u}', \xi_i)| < c(\gamma).$$

5. *The function* $f(i, \mathbf{u}, \xi_i)$ *satisfies the strong intermixing condition*

$$\alpha(j) = \sup_i \sup_{\substack{A \in \sigma_{-\infty}^{i} \\ B \in \sigma_{i+j}^{\infty}}} |P(AB) - P(A)P(B)| \le \frac{c}{1 + j^{1+\varepsilon}}, \ \varepsilon > 0,$$

$$\sigma_n^m = \sigma\{f(i, \mathbf{u}, \xi_i), n \le i \le m, \mathbf{u} \in I\},$$

6. $E(f(i, \mathbf{u}, \xi_i))^{2+\delta} < \infty, \varepsilon\delta > 2.$

Let $\mathbf{u}_n = \arg\min_{\mathbf{u} \in I} F_n(\mathbf{u})$. Then

$$P\left\{\lim_{n\to\infty} ||\mathbf{u}_n - \mathbf{u}_0|| = 0\right\} = 1, \quad P\left\{\lim_{n\to\infty} F_n(\mathbf{u}_n) = F(\mathbf{u}_0)\right\} = 1.$$

Similar statement is also true for the case when a continuous stochastic function $f(t, \mathbf{u}, \xi(t))$ is observed on the interval $[0, T]$, i.e., if we take

$$F_T(\mathbf{u}) = \frac{1}{T} \int_0^T f(t, \mathbf{u}, \xi(t)) \,\mathrm{d}t,$$

where $\xi(t)$ is a random process, stationary in the narrow sense. In this case, the minimization problem $\min_{\mathbf{u} \in I} F_T(\mathbf{u})$ is considered, as well as the asymptotic behavior of $\mathbf{u}_T = \arg\min_{\mathbf{u} \in I} F_T(\mathbf{u})$ and $F_T(\mathbf{u}_T)$ as $T \to \infty$.

Consider one more example. Let $\{\boldsymbol{\xi}_t, t \ge 0\}$ be a random field with continuous time defined on the probability space $(\boldsymbol{\Omega}, \Im, P)$, $\boldsymbol{\xi}_t \in \Re^m, m \ge 1$; J is a closed subset in $\Re^p$, $\{f(t, \mathbf{x}, \mathbf{y}) : \Re^+ \otimes J \otimes \Re^p \to \Re^+\}, \Re^+ = [0, \infty)$ is continuous in all parameters function. Let us minimize the functional

$$F_T(\mathbf{x}) = \frac{1}{T} \int_0^T f(t, \mathbf{x}, \boldsymbol{\xi}_t)\mathrm{d}t, \ \mathbf{x} \in J.$$

Theorem 3.4 (Knopov and Kasitskaya 2002). *Let the random function* $f(t, \mathbf{x}, \boldsymbol{\xi}_t)$ *satisfy the conditions:*

1. *For any* $\mathbf{x} \in J$ *there exists a function* $F(\mathbf{x})$ *such that*

$$F(\mathbf{x}) = \lim_{T\to\infty} EF_T(\mathbf{x})$$

and a point $\mathbf{x}^* \in J$ *such that* $F(\mathbf{x}^*) < F(\mathbf{x})$ *if* $\mathbf{x} \ne \mathbf{x}^*$
2. *If* J *is an unbounded set then* $f(t, \mathbf{x}, \mathbf{y}) \to \infty$ *as* $||\mathbf{x}|| \to \infty$ *for any fixed* t *and* $\mathbf{y}$
3. *There exists a function* $c(\gamma) > 0, c(\gamma) \to 0$ *as* $\gamma \to \infty$, *such that for any* $\delta > 0$ *there exists* $\gamma_0 > 0$ *such that for any element* $\mathbf{x}' \in J$, $0 < \gamma < \gamma_0$:

$$\overline{\lim_{T\to\infty}} \frac{1}{T} \int_0^T E \sup_{\substack{||\mathbf{x}-\mathbf{x}'|| < \gamma \\ ||\mathbf{x}-\mathbf{x}^*|| > \delta}} |f(t, \mathbf{x}, \boldsymbol{\xi}_t) - f(t, \mathbf{x}', \boldsymbol{\xi}_t)|\mathrm{d}t < c(\gamma).$$

4. *Strongly mixing condition for the random process* $\boldsymbol{\xi}_t$ *is fulfilled:*

$$\sup_{t\geq 0} \sup_{\substack{A \in \sigma_{-\infty}^{t} \\ B \in \sigma_{t+\tau}^{\infty}}} |P(AB) - P(A)P(B)| \leq \frac{C}{1+\tau^{1+\varepsilon}},$$

where $\tau > 0, \varepsilon > 0, \sigma_a^b$ *is the* σ *-algebra generated by* $\{\boldsymbol{\xi}_t, a \leq t \leq b\}$

5. $\sup_{t\geq 0} Ef(t, \mathbf{x}, \boldsymbol{\xi}_t)^{2+\delta} < \infty$, *where* $\varepsilon\delta > 2, ||\mathbf{x}||_p < \infty$.

Then $P\{\lim_{t\to\infty} ||\mathbf{x}_T - \mathbf{x}^*|| = 0\} = 1$, where $\mathbf{x}_T = [\arg\min_{\mathbf{x}\in J} F_T(\mathbf{x})]$.

Consider the case when the unknown parameter u is an element of some compact set K from some function space. Let us formulate the problem and the obtained results. First we consider the case when $K \subset C_{[0,1]}$ is the space of continuous on $[0, 1]$ functions with the uniform metric $||\cdot||$.

Let ξ be a random variable defined on the probability space $(\boldsymbol{\Omega}, \Im, P)$, and let $f(t, y, z) : [0, 1] \times \Re \times \Re \to \Re^+$ be a function continuous on $[0, 1] \times \Re$ for fixed z and measurable in z for fixed t and y. The problem consists in finding

$$\min_{u\in K} E\left\{\int_0^1 f(t, u(t), \xi)\mathrm{d}t\right\} = \min_{u\in K} F(u).$$

This problem is approximated with the following one: Find

$$\min_{u\in K} \frac{1}{n+1}\sum_{i=1}^{n} f\left(\frac{i}{n}, u\left(\frac{i}{n}\right), \xi_i^n\right) = \min_{u\in K} F_n(u),$$

where $\{\xi_i^n, 0 \leq i \leq n, n \geq 1\}$ is a sequence of series of independent observations of a random variable ξ.

Let $u_n = \arg\min_{u\in K} F_n(u)$, $u_0 = \arg\min_{u\in K} F(u)$. We find the conditions under which $u_n \to u_0$, $F_n(u_n) \to F(u_0)$ as $n \to \infty$ in some probabilistic sense.

Theorem 3.5 (Knopov and Kasitskaya 2002). *Let the following conditions hold true:*

1. $E\{\max_{u\in K} f^4(t, u(t), \xi)\} \leq L, t \in [0, 1]$ *with some constant* $L > 0$
2. *There exists a unique point* $u_0 \in K$ *such that* $F(u) > F(u_0), u \neq u_0$

Then, for any $n\geq 1$, one can chose a function u_n such that for any $t \in [0, 1] u_n(t, \omega)$, $\omega \in \boldsymbol{\Omega}$ is $\Im_n$-measurable, where $\Im_n$ is the σ-algebra generated by random variables $\{\xi_i^n, 0 \leq i \leq n\}$. Moreover,

$$P\left\{\lim_{n\to\infty} ||u_n - u_0|| = 0\right\} = 1,$$

$$P\left\{\lim_{n\to\infty} F_n(u_n) = F(u_0)\right\} = 1.$$

As above, the proof consists in checking the conditions of Theorem 2.1.

Remark 3.2. The choice of $C_{[0,1]}$ as the function space on which the compact set K is given is undoubtedly not unique. For instance, the cases when K belongs to some Hilbert space with corresponding metric, to space of continuous functions with multidimensional argument, etc., seem to be equally interesting.

Let us consider the observation model 4 from Sect. 2.1 with criteria

$$Q_n(\alpha) = \frac{1}{n}\sum_{i=1}^{n}\left[y_i - \alpha\left(\frac{i}{n}\right)\right]^2 \tag{3.4}$$

or

$$\tilde{Q}_n(\alpha) = \frac{1}{n}\sum_{i=1}^{n}\left|y_i - \alpha\left(\frac{i}{n}\right)\right| \tag{3.5}$$

and let α_n and $\breve{\alpha}_n$ be, respectively, least squares and least modules estimates. Then by Theorem 3.4 α_n and $\breve{\alpha}_n$ are strongly consistent estimates of unknown function α^0 from model 4, Sect. 2.1. To show this, put $f(x, y, z) = |\alpha^0(x) - y + z|^2$ for criterion (3.4) and $f(x, y, z) = |\alpha^0(x) - y + z|$ for criterion (3.5). In case of criterion (3.5) the conditions of Lemma 2.3 must be fulfilled.

It should be noted, that the properties of estimates for criteria of type (3.2) for independent observations are investigated in detail in frames of the theory of robust estimates (according to the terminology of Hampel et al. 1986; Huber 1981) and there is no need to review these results in detail. We only note that in these monographs and in works of other authors (Van de Geer 1995; Ivanov 1984a, b, 1997; Liese and Vajda 1994; Yubi 1977) consistency conditions, rates of convergence, asymptotic distributions and other important properties of estimates are investigated.

We would like to discuss briefly one more class of regression models, the study of which was initiated in Hampel et al. (1986), Huber (1981), Vapnik (1982, 1996) and some others. We quote one of these results which concern the properties of estimates in the case when there are some a priori constraints on the range of unknown parameters. We will follow the terminology used in original monographs.

In what follows, the parameter set Θ is a locally compact space with a countable base, $(\mathfrak{X}, \mathfrak{A}, P)$ is a probability space, and $\rho(x, \theta)$ is some real-valued function on $\mathfrak{X} \times \mathbf{\Theta}$.

Assume that $x_1, x_2, ...$ are independent random variables with values in $\mathfrak{X}$ having the same probability distribution P. Let $T_n(x_1, ..., x_n)$ be any sequence of functions $T_n : \mathfrak{X}^n \to \mathbf{\Theta}$ such that

$$\frac{1}{n}\sum_{i=1}^{n}\rho(x_i, T_n) - \inf_{\theta}\frac{1}{n}\sum_{i=1}^{n}\rho(x_i, \theta) \to 0 \tag{3.6}$$

almost surely (or in probability). We want to give the sufficient conditions ensuring that every such sequence T_n converges almost surely (or in probability) to some constant θ_0.

Convergence of T_n will be proved under the following set of assumptions.

(A-1). For each fixed $\theta \in \boldsymbol{\Theta}$, $\rho(x,\theta)$ is $\mathfrak{A}$-measurable, and $\rho(x,\theta)$ is separable in the sense of Doob: there is a P-null set N and a countable subset $\theta' \subset \theta$ such that for every open set $U \subset \theta$ and every closed interval A, the sets

$$\{x|\rho(x,\theta) \in A, \forall\theta \in U\}, \quad \{x|\rho(x,\theta) \in A, \forall\theta \in U \cap \theta'\}$$

differ by at most a subset of N.

This assumption ensures measurability of the infinia and limits occurring below. For fixed P, the function ρ can always be replaced by a separable version.

(A-2). The function ρ is a.s. lower semicontinuous in θ, that is,

$$\inf_{\theta' \in U} \rho(x,\theta') \to \rho(x,\theta) \quad \text{a.s.}$$

as the neighborhood U of θ shrinks to $\{\theta\}$.

(A-3). There is a measurable function $a(x)$ such that

$$E\{\rho(x,\theta) - a(x)\}^- < \infty \text{ for all } \theta \in \boldsymbol{\Theta},$$

$$E\{\rho(x,\theta) - a(x)\}^+ < \infty \text{ for some } \theta \in \boldsymbol{\Theta}.$$

Thus, $\gamma(\theta) = F = E\{\rho(x,\theta) - a(x)\}$ is well-defined.

(A-4). There is $\theta_0 \in \boldsymbol{\Theta}$ such that $\gamma(\theta) > \gamma(\boldsymbol{\Theta}_0)$ for all $\theta \neq \theta_0$.

If θ is not compact, we make one-point compactification by adding the infinity point ∞.

(A-5). There is a continuous function $b(\theta) > 0$ such that

1. $\inf\limits_{\theta \in \boldsymbol{\Theta}} \frac{\rho(x,\theta) - a(x)}{b(\theta)} \geq h(x)$

   for some integrable h
2. $\liminf\limits_{\theta \to \infty} b(\theta) > \gamma(\theta_0)$,
3. $E\left\{\liminf\limits_{\theta \to \infty} \frac{\rho(x,\theta) - a(x)}{b(\theta)}\right\} \geq 1.$

If θ is compact, then (2) and (3) are redundant.

Example 3.1. Let $\boldsymbol{\Theta} = \mathfrak{X}$ be the real axis, and let P be any probability distribution, having a unique median θ_0. Then (A-1)–(A-5) are satisfied for $\rho(x,\theta) = |x - \theta|$, $a(x) = |x|, b(\theta) = |\theta| + 1$. (This implies that the sample median is a consistent estimate of the median.)

Taken together, (A-2), (A-3) and (A-5) (1) imply by monotone convergence the following strengthened version of (A-2):

(A-2″) As the neighborhood U of θ shrinks to $\{\theta\}$,

$$E\left\{\inf_{\theta' \in U} \rho(x,\theta') - a(x)\right\} \to E\{\rho(x,\theta) - a(x)\}.$$

Note that the set $\{\theta \in \mathbf{\Theta} | E(|\rho(x,\theta) - a(x)|) < \infty\}$ is independent of the particular choice of $a(x)$; if there is $a(x)$ satisfying (A-3), then one can choose $a(x) = \rho(x, \theta_0)$.

Lemma 3.1 (Huber 1981). *If (A-1), (A-3) and (A-5) hold, then there is a compact set $C \subset \mathbf{\Theta}$ such that every sequence T_n satisfying (3.6) a.s. stays in C forever.*

Theorem 3.6 (Huber 1981). *If (A-1), (A-2″), (A-3), and (A-4) hold, then every sequence T_n satisfying (3.6), converges (by Lemma 3.1) to θ_0 almost surely. Analogous statement is true for the convergence in probability.*

3.2 Regression Models for Long Memory Systems

Before we considered the cases when the observations are dependent, and the corresponding random sequences or processes satisfy strong mixing condition. This imposes rather strong restrictions on the rate with which the dependence between observations decreases while the distance between them increases. During recent years a lot of investigations were devoted to making the conditions of weak dependence less restrictive. Further, the notion of strong dependence was introduced. The system satisfying the condition of strong dependence were named as "systems with long memory". Let us quote one of the typical results for such systems on consistency of the unknown parameter estimates.

Let

$$S_T(\boldsymbol{\beta}) = \frac{1}{T}\int_0^T f(t,\boldsymbol{\beta},\eta(t))\,\mathrm{d}t,$$

where $f(t,\boldsymbol{\beta},\eta(t)) : [0,\infty) \times J \times \Re^1 \to [0,\infty)$, $s \geq 1$, is a known function, J is a closed subset in $\Re^p$, $p \geq 1$, $||\cdot||$ is a norm in $\Re^p$, $\eta(t), t \in \Re$, is a random noise.

Let the following conditions be fulfilled:

1. $f(t,\boldsymbol{\beta},y)$ is a function, continuous on the whole set of variables
2. $\eta(t), t \in \Re$ is a stationary second-order process with null expectation, represented in form $\eta(y) = G(\xi(t))$, where $\xi(t), t \in \Re$, is real-valued, measurable, mean square continuous Gaussian random process with null expectation, variance 1 and correlation function $B(t) = \mathrm{cov}(\xi(0),\xi(t)) = L(t)|t|^{-\alpha}, 0 < \alpha < 1$, where $L(t) = L'(|t|)$, $t > 0$, is a non-negative function slowly varying at infinity, that is $\forall s > 0 \lim_{t\to\infty} L(ts)/L(t) = 1$, and bounded on every finite interval. Moreover, $\lim_{t\to\infty}\sup_{t\in[0,T)} L(ts)/L(t) < \infty, G(u), u \in \Re$, is real-valued, measurable, non-random function with $EG^2(\xi(0)) < \infty$.

Suppose that there exists $t_0 > 0$ such that for $t > t_0$ the function $B(t), t \in \Re$, is decreasing.

Theorem 3.7 (Moldavskaya 2002). *Let conditions 1 and 2 be satisfied. Moreover, assume that*

1. *For any $\boldsymbol{\beta} \in J$ there exist a function $S(\boldsymbol{\beta})$ such that $S(\boldsymbol{\beta}) = \lim_{T\to\infty} ES_T(\boldsymbol{\beta})$, and a point $\boldsymbol{\beta}^* \in J$, in which this function attains its unique minimum*
2. *If the set J is unbounded, then $f(t,\boldsymbol{\beta},y) \to \infty$ as $||\boldsymbol{\beta}|| \to \infty$, for fixed t, y*
3. *There exists a function $c(\gamma) > 0$ such that $c(\gamma) \to 0$ as $\gamma \to 0$, and for any $\varepsilon > 0$ there exists such $\gamma_0 > 0$ that for $0 < \gamma < \gamma_0$ and any $\boldsymbol{\beta}' \in J$*

$$\overline{\lim_{T\to\infty}} \frac{1}{T} \int_0^T E \sup_{\substack{||\boldsymbol{\beta}-\boldsymbol{\beta}'|| < \gamma \\ ||\boldsymbol{\beta}-\boldsymbol{\beta}^*|| > \varepsilon}} |f(t,\boldsymbol{\beta},\eta(t)) - f(t,\boldsymbol{\beta}',\eta(t))| \mathrm{d}t < c(\gamma),$$

4. *$Ef(t,\boldsymbol{\beta},G(\xi(0))) < \infty$*
5. *$\sup_{\substack{t\geq 0 \\ \boldsymbol{\beta}\in J}} \sum_{i\geq 1} C_i^2(t,\boldsymbol{\beta})/i! < \infty$, where $C_i^j(t,\boldsymbol{\beta})$ are the coefficients in the expansion of the function f into the series of Chebyshev-Hermite polynomials*
6. *$\sup_{T\geq 0} \int_0^1 \int_0^1 \frac{L(T|t-s|)}{L(T)|t-s|^\alpha} \mathrm{d}t \; \mathrm{d}s < \infty$.*

Then the estimate of the parameter $\boldsymbol{\beta}^*$, determined as the minimum point of the functional $S_T(\boldsymbol{\beta})$, is strongly consistent.

We note that in works of A.V. Ivanov and N.N. Leonenko series of fundamental results concerning asymptotic behavior of nonlinear regression estimates for stochastic systems with strongly dependent observations were obtained. We quote one of these results concerning strong consistency of the least squares estimates, obtained in Ivanov and Leonenko (2001, 2002).

Let $(\boldsymbol{\Omega}, \Im, P)$ be a complete probability space, and $\xi(t) = \xi(\omega,t) : \boldsymbol{\Omega} \times \Re \to \Re$ a stochastic process satisfying the following condition:

A. $\xi(t), t \in \Re$, is a real measurable mean-square continuous stationary Gaussian process with $E\xi(t) = 0$ and the covariance function

$$B(t) = \mathrm{cov}(\xi(0),\xi(t)) = \frac{1}{(1+t^2)^{\alpha/2}}, \quad 0 < \alpha < 1, t \in \Re.$$

Under condition A, the process $\xi(t), t \in \Re$, admits the representation

$$\xi(t) = \int_{R^1} \mathrm{e}^{i\lambda t} \sqrt{f_\alpha(\lambda)} W(\mathrm{d}\lambda), \tag{3.7}$$

where $W(\cdot)$ is the complex Gaussian white noise on the measurable space $(\Re, \mathfrak{B}(\Re))$. Spectral density $f_\alpha(\lambda), \lambda \in \Re$ of the process $\xi(t)$ is of the form (see, for example Donoghue 1969),

$$f_\alpha(\lambda) = \frac{2^{(1-\alpha)/2}}{\Gamma(\alpha/2)\pi^{1/2}} K_{(\alpha-1)/2}(|\lambda|)|\lambda|^{(\alpha-1)/2}, \quad \lambda \in \Re,$$

where

$$K_\nu(z) = \frac{1}{2}\int_0^\infty s^{\nu-1}\exp\left\{-\frac{1}{2}z\left(s+\frac{1}{s}\right)\right\}\mathrm{d}s, \quad z > 0,$$

is the modified Bessel function of the third kind of order ν (Watson 1944).

B. A nonlinear Borel function $G : \Re \to \Re$ is such that

$$\int_{-\infty}^{\infty} G^2(u)\varphi(u)\,\mathrm{d}u < \infty,$$

where

$$\varphi(u) = \frac{1}{\sqrt{2\pi}}\mathrm{e}^{-u^2/2}, \quad u \in \Re.$$

Under condition B, the function $G(u), u \in \Re$, can be expanded into the series

$$G(u) = \sum_{k=0}^{\infty}\frac{C_k}{k!}H_k(u), \quad C_k = \int_{-\infty}^{\infty} G(u)H_k(u)\varphi(u)\mathrm{d}u, \quad k = 0, 1, ... \tag{3.8}$$

of orthogonal Chebyshev-Hermite polynomials

$$H_k(u) = (-1)^k e^{u^2/2}\frac{d^k e^{-u^2/k}}{\mathrm{d}u^k}, \qquad k = 0, 1, ...$$

in Hilbert space $L_2(\Re, \varphi(u)\,\mathrm{d}u)$.

Additionally we assume that the function G satisfies the condition

B′. There exists an integer $m \geq 1$ such that $C_1 = \cdots = C_{m-1} = 0$ and $C_m \neq 0$.

The integer $m \geq 1$ is called the Hermite rank of G (Taqqu 1979; Dobrushin and Mayor 1972).

Let $\mathbf{\Theta} \subset \Re$ be a bounded open interval. Under conditions A and B we consider the regression model

$$y(t) = y_\theta(t) = g(t, \theta) + G(\xi(t)), \quad t \in \Re,$$

where $g(t, \theta) : \Re \times \mathbf{\Theta}^c \to \Re$ is a measurable function depending on unknown parameter $\theta \in \mathbf{\Theta}$, and the random noise $\varsigma(t) = g(\xi(t)), t \in \Re$, is such that $E\varsigma(0) = 0$ (or $C_0 = 0$) and $E\varsigma^2(0) = 1$. We want to estimate the single parameter θ from observations of the stochastic process $y_\theta(t), t \in [0, T]$, as $T \to \infty$.

Any random variable $\hat{\theta}_T \in \mathbf{\Theta}^c$ satisfying the property

$$Q_T(\hat{\theta}_T) = \inf_{\tau \in \mathbf{\Theta}^c} Q_T(\tau), \quad Q_T(\tau) = \int_0^T [y_\theta(t) - g(t, \tau)]^2 \mathrm{d}t$$

is called the least squares estimate of unknown parameter $\theta \in \boldsymbol{\Theta}$ obtained from the observations $y_\theta(t), t \in [0, T]$, where $\boldsymbol{\Theta}^c$ is the closure of $\boldsymbol{\Theta}$. Set

$$b_T(\tau_1, \tau_2) = \int_0^T (g(t, \tau_1) - g(t, \tau_2))^2 \mathrm{d}t,$$

$$\eta_T(\tau) = \int_0^T G(\xi(t))(g(t, \tau) - g(t, \theta))\, \mathrm{d}t,$$

$$\gamma(T) = \int_0^T G^2(\xi(t))\mathrm{d}t.$$

By definition of least squares estimate, we have almost surely

$$Q_T(\theta) = \gamma(T) \geq Q_T(\hat{\theta}_T) = \gamma(T) - 2\eta_T(\hat{\theta}_T) + b_T(\hat{\theta}_T, \theta)$$

or

$$b_T(\hat{\theta}_T, \theta) \leq 2\eta_T(\hat{\theta}_T).$$

If $g(t, \theta)$ is differentiable with respect to θ, then we write

$$\frac{\partial^r g(t, \theta)}{\partial \theta^r} = g_r(t, \theta), \quad d_{rT}^2(\tau) = \int_0^T g_r^2(t, \theta)\mathrm{d}t, \quad r = 1, 2, 3.$$

Suppose that the function $g(t, \tau)$ can be extended to some interval $\boldsymbol{\Theta}^* \supset \boldsymbol{\Theta}$ in such a way that $g(t, \cdot) \in C^3(\boldsymbol{\Theta}^*)$, the functions $g_r(t, \tau), r = 1, 2$, are bounded on $[0, T] \times \boldsymbol{\Theta}^c$ for any $T > 0$, and $g_3(t, \tau)$ is locally square integrable in t.

Note that the function

$$A_\theta^2(T) = \frac{1}{T} d_{1T}^2(\theta) = \int_0^1 g_1^2(Ts, \theta)\mathrm{d}s \tag{3.9}$$

is used in limit theorems below as a part of normalizing factors.

The symbols $k_i, i = 0, 1, 2, \ldots$ denote positive constants whose particular values are not essential for the purposes of this work.

The following condition holds for a wide class of regression functions.

D. Suppose that

$$\sup_{0 \leq s \leq 1} \frac{|g_1(Ts, \tau)|}{A_\theta(T)} \leq k_1 < \infty$$

uniformly in $T > 0$ and $\tau \in \boldsymbol{\Theta}^c$.

The following condition is called the contact condition or condition for distinguishability of parameters.

E. For any $\rho > 0$, there exists a number $r = r(\rho) > 0$ such that

$$\inf_{|u| > \rho (T^{-1} d_{1T}^2(\theta))^{-1/4}} \frac{b_T(u + \theta, \theta)}{T[T^{-1} d_{1T}^2(\theta)]^{1/2}} > r.$$

For the least squares method we have

Theorem 3.8 (Ivanov and Leonenko 2001). *Let conditions A, B, B′, D, and E hold for $\alpha \in (0, 1/m)$, where m is the Hermitian rank of the function G. Then*

$$T^{-1/4} d_{1T}^{1/2}(\theta)(\hat{\theta}_T - \theta) \xrightarrow{P} 0$$

as $T \to \infty$.

For the asymptotic expansion of these estimates we refer to Ivanov and Leonenko (2001, 2002). Such results will be considered in Sect. 3.4.3.

3.3 Statistical Methods in Stochastic Optimization and Estimation Problems

We have noted already that nonlinear regression parameter estimation problems can be formulated as special stochastic programming problems based on the method of empirical means. Meanwhile, the objective function may not be continuous, and often it turns out that it is enough to demand that it is upper semicontinuous. We shall first briefly consider one of the most well-known and advanced approaches for investigation of the asymptotic properties of estimates developed in works of R. Wets, H. Attouch, J. Dupacova, A. King, G. Salinetti, A. Shapiro and others. This approach is based on the notion of epi-convergence, since it is equivalent to the set-convergence of the epigraphs. For the precise definition of epi-convergence and its basic properties we refer to Attouch (1984). First we introduce some conditions which are needed later on.

Let $(\mathbf{\Omega}, \Im, P)$ be a probability space where $\mathbf{\Omega}-$ the support of $P-$ is a closed subset of a Polish space X, and $\Im$ is the Borel sigma-field related to $\mathbf{\Omega}$; we may regard $\mathbf{\Omega}$ as the set of possible values of the random element ξ defined on the probability space $(\mathbf{\Omega}', \Im', \tilde{P}')$. If P is known, our problem can be formulated as follows:

$$\text{find } \mathbf{x}^* \in \Re^n \text{that minimizes } Ef(\mathbf{x}), \tag{3.10}$$

where $Ef(\mathbf{x}) := \int_{\mathbf{\Omega}} f(\mathbf{x}, \xi) P(\mathrm{d}\xi) = E\{f(\mathbf{x}, \xi)\}$ and $f : \Re^n \times \mathbf{\Omega} \to \Re \cup \{\infty\} = (-\infty, \infty)$ is a lower semicontinuous function; we set $Ef(\mathbf{x}) = \infty$ whenever $\xi \mapsto f(\mathbf{x}, \xi)$ is not bounded above by a summable (extended real-valued) function. We refer to $\mathrm{dom} Ef := \{\mathbf{x} | Ef(\mathbf{x}) < \infty\}$ as to the effective domain of Ef. Points that do not belong to dom Ef cannot minimize Ef and thus are effectively excluded from the optimization problem (3.10). Hence, the constraints on **x** are already included in the model. Note that by definition of the integral we always have:

$$\mathrm{dom}\, Ef \subset \{\mathbf{x} | f(\mathbf{x}, \xi) < \infty \text{ a.s.}\}.$$

An extended real-valued function $h : \Re^n \times \mathbf{\Omega} \to \bar{\Re} = [-\infty, \infty]$ is said to be *proper* if $h > -\infty$ and is not identically $+\infty$; it is called *lower semicontinuous* (l.sc.) at $\mathbf{x}$ if for any sequence $\{\mathbf{x}^k\}_{k=1}^{\infty}$ converging to $\mathbf{x}$ we have $\liminf_{k\to\infty} h(\mathbf{x}^k) \geq h(\mathbf{x})$, including the values ∞ or $-\infty$. The extend real-valued function f defined on $\Re^n$ is called a *random lower semicontinuous function* if

1. For all $\xi \in \mathbf{\Omega}$ $f(\cdot, \xi)$ is lower semicontinuous
2. f is $B^n \times \Im$-measurable

where B^n is a Borel sigma-field on $\Re^n$.

Assumption 3.1. "Continuity" of f. The function $f : \Re^n \times Y \to (-\infty, \infty]$ with closed and non-empty $\text{dom} f := \{(\mathbf{x}, \xi) | f(\mathbf{x}, \xi) < \infty\} = S \times Y, S \subset \Re^n$, is such that for all $\mathbf{x} \in S \xi \mapsto f(\mathbf{x}, \xi)$ is continuous on Y, and for all $\xi \in Y$ $\mathbf{x} \mapsto f(\mathbf{x}, \xi)$ is lower semicontinuous on $\Re^n$, and locally lower Lipschitz on S in the following sense: for any $\mathbf{x}$ in S, there exists a neighborhood V of $\mathbf{x}$ and a bounded continuous function $\beta : Y \to R$ such that for all $\mathbf{x}' \in V \cap S$ and $\xi \in Y$

$$f(\mathbf{x}, \xi) - f(\mathbf{x}', \xi) \leq \beta(\xi) \cdot ||\mathbf{x} - \mathbf{x}'||.$$

Assumption 3.2. Convergence in distribution. Given a sample space (Z, F, μ) and an increasing sequence of sigma-fields $(F^{\nu})_{\nu=1}^{\infty}$ contained in F, let $P^{\nu} : \mathbf{\Omega} \times Z \to [0, 1], \nu = 1, ...$ be such that for all $\zeta \in Z$ $P^{\nu}(\cdot, \zeta$ is a probability measure on $(\mathbf{\Omega}, \Im)$, and for all $\mathrm{A} \in \Im$ $\zeta \mapsto P^{\nu}(\mathrm{A}, \zeta)$ is F^{ν}-measurable. For μ-almost all ζ in Z, the sequence $\{P^{\nu}(\cdot, \zeta)\}_{\nu=0}^{\infty}$ is $f(\mathbf{x}, \cdot)$-tight (asymptotically negligible), i.e. for each $\mathbf{x} \in S$ and $\varepsilon > 0$ there exists a compact set $K_{\varepsilon} \subset Y$ such that for $\nu = 0, 1, ...$

$$\int_{\mathbf{\Omega} \backslash K_{\varepsilon}} |f(\mathbf{x}, \xi)| P^{\nu}(\mathrm{d}\xi, \zeta) < \varepsilon$$

and

$$\int_{\mathbf{\Omega}} \inf_{x \in \Re^n} f(\mathbf{x}, \xi) P^{\nu}(d\xi, \zeta) > -\infty.$$

Theorem 3.9 (Dupacova and Wets 1986). *Under Assumptions 3.1 and 3.2 we have μ-almost surely*

$$\limsup_{\nu \to \infty} (\inf E^{\nu} f) \leq \inf Ef.$$

Moreover, there exists $Z_0 \in F$ with $\mu(Z \backslash Z_0) = 0$ such that

1. For all $\zeta \in Z_0$, any cluster point $\hat{\mathbf{x}}$ of any sequence $\{\mathbf{x}^{\nu}, \nu = 1, ...\}$ with $\mathbf{x}^{\nu} \in \arg\min E^{\nu} f^{\nu}(\cdot, \zeta)$ belongs to $\arg\min Ef$ (i.e. is an optimal estimate)
2. For $\nu = 1, \ldots, \zeta \mapsto \arg\min E^{\nu} f(\cdot, \zeta) : Z_0 \Rightarrow \Re^n$ is a closed-valued F^{ν}-measurable multifunction

In particular, if there is a compact set $D \subset \Re^n$ such that for $\nu = 1, ...(\arg\min E^{\nu} f) \cap D$ is non-empty μ-a.s. and $\{\mathbf{x}^*\} = \arg\min Ef \cap D$, then there exists a sequence

$\{\mathbf{x}^\nu : Z_0 \to \Re^n\}_{\nu=1}^\infty$ of F^ν-measurable selections of $\{\arg\min E^\nu f\}_{\nu=1}^\infty$ such that $\mathbf{x}^* = \lim_{\nu\to\infty} \mathbf{x}^\nu(\zeta)$ for μ-almost all ζ, and also $\inf Ef = \lim_{\nu\to\infty} (\inf E^\nu f)$ μ-a. s.

In stochastic programming problems there often arise situations where it is not necessary to prove the strong consistency of parameter estimates on which some functional minimum is reached. It is enough that the sequence of functionals in minimum points converges in one or another probability sense. In this case, we say that the functional convergence takes place. Below we present some interesting results in this direction obtained in Norkin (1989, 1992).

Consider the problem

$$F(x) = \int_{\mathbf{\Omega}} f(x, \xi) P(\mathrm{d}\xi) \to \min_{x\in K}, \tag{3.11}$$

where $x \in K \subset X$, K is a compact set in some topological space X, $\xi \in \mathbf{\Omega}$, $(\mathbf{\Omega}, \Im, P)$ is the probability space, the function $f : X \times \mathbf{\Omega} \to \Re$ is integrable for each fixed x, $F(x)$ is lower semicontinuos.

The approach for solving (3.11), when it is replaced by the sequence of problems of the type:

$$F_n(x) = \frac{1}{n}\sum_{i=1}^{n} f(x, \xi_i) \to \min_{x\in K}, \tag{3.12}$$

where ξ_i are independent identically distributed random variables whose distributions coincides with the distribution generated by the measure P, is considered in Yubi (1977), Kankova (1979), Vapnik (1979), Dupacova (1979), King (1988), King and Wets (1988), and Wets (1979). Solving the problem (3.12) depends on the observation set $(\xi_1, \xi_2, ..., \xi_n)$. Let us introduce the probability space $(\mathbf{\Omega}, \Im_\omega, P_\omega)$ which is the countable product of initial spaces, whose elementary outcomes are the sequences of observations $\omega = (\xi_1, \xi_2, ...)$. Note that the solutions to (3.12) depend on ω. There naturally arises the question about the convergence of optimal values $F_n^*(\omega) = \min_{x\in K} F_n(x)$ to $F^*(\omega) = \min_{x\in K} F(x)$ and of the solutions $X_n^*(\omega) = \arg\min_{x\in K} F_n(x)$ to $X^* = \arg\min_{x\in K} F(x)$ in one or another probability sense.

Write $F_n(x)$ as

$$F_n(x) = F(x) + \frac{1}{n}\sum_{i=1}^{n} (f(x, \xi_i) - F(x)) = F(x) + y_n(x, \omega).$$

Consider the functional

$$\Phi(y) = \min_{x\in K} \varphi(x, y(x)) = \min_{x\in K}(F(x) + y(x)),$$

where $F(x)$ is lower semicontinuous on the compact K function, $y(x) \in C(K; \Re)$, $C(K; \Re)$ is the Banach space of continuous on K functions with the norm $||y|| = \max_{x\in K} |y(x)|$.

The following statements hold true (Norkin 1989, 1992).

Theorem 3.10. *Functional $\Phi(y)$ is of Lipschitz type.*

Theorem 3.11. *If the function $F(x)$ is continuous on the metric compact K, then the multiple-valued mapping*

$$y(\cdot) \to X^*(y(\cdot)) = \arg\min_{x \in K}(F(x) + y(x))$$

is upper semicontinuous (and closed), and the deviation functional

$$y(\cdot) \to \Delta(X^*(y), X^*(\bar{y})) = \sup_{x \in X^*(y)} \inf_{x' \in X^*(\bar{y})} \rho(x, x')$$

is upper semicontinuous in $y(\cdot)$, and is continuous at the point $y(\cdot) \in C(X; \Re)$.

Thus, there exists a sequence of the Lipschitz functional $\Phi(y_n(\cdot, \omega))$, where $y_n(x, \omega) = (1/n)\sum_{i=1}^{n} f(x, \xi_i) - F(x)$. If the sequence $y_n(\cdot, \omega)$ of elements from $C(K; \Re)$ converges to $0 \in C(K; \Re)$ in one or another probability sense (almost surely, in probability, in distribution, in norm), then, since $\Phi(y)$ is a Lipschitz functional, the convergence $\Phi(y_n) \to \Phi(0) = \min_{x \in K} F(x)$, and $\Delta(X^*(y_n), X^*) \to 0$ takes place in the same sense. In other words, the convergence of the statistical decision method of stochastic programming problems takes place.

Thus, the investigation of convergence of the statistical decision method of the stochastic programming problems reduces to the study of convergence of random variables $y_n(\cdot, \omega)$ in $C(K; \Re)$ to $0 \in C(K; \Re)$ in one or another probabilistic sense. For every fixed $x\, y_n(x, \omega) \to 0$ P_ω-a.s. due to strong law of large numbers. Now we are looking for conditions under which the pointwise convergence in one or another probabilistic sense implies the uniform convergence in the same sense in $C(K; \Re)$.

Theorem 3.12. *Let $(\mathbf{\Omega}, \Im, P)$ be a probability space, D be a relatively open convex set in $\Re^p$. Suppose that the function $f : D \times \mathbf{\Omega} \to \Re$ is convex in $\mathbf{x}$ in D for almost all ξ, and integrable in ξ for all $\mathbf{x} \in D$. Then for any compact $K \subset D$*

$$||y_n(\cdot, \omega)|| = \max_{\mathbf{x} \in K} |y_n(\mathbf{x}, \omega)| \to 0 \quad P_\omega\text{-a.s.}$$

Theorem 3.13. *Let $(\mathbf{\Omega}, \Im, P)$ be the probability space, K be a compact set in the complete separable metric space. Suppose that the function $f : K \times \mathbf{\Omega} \to \Re$ is integrable in $\xi \in \mathbf{\Omega}$ for all $x \in K$, and of Lipschitz type in $x \in K$ uniformly in ξ:*

$$|f(x, \xi) - f(y, \xi)| \le L||y - x||, \quad x, y \in K.$$

Then P_ω-almost surely

$$||y_n(\cdot, \omega)|| = \max_{x \in K} \left| \frac{1}{n} \sum_{i=1}^{n} f(x, \xi_i) - F(x) \right| \to 0, \quad n \to \infty.$$

3.4 Empirical Mean Estimates Asymptotic Distribution

In this section we study limit distributions of unknown parameters estimates. In the first sub-subsection models with independent or weakly dependent observations are considered. In the second sub-subsection we consider long memory models.

3.4.1 *Asymptotic Distribution of Empirical Estimates for Models with Independent and Weakly Dependent Observations*

Assume that $\{\xi(i), i = 1, 2, ...\}$ is a sequence of independent identically distributed random variables defined on the probability space $(\boldsymbol{\Omega}, \Im, P)$, a random variable ξ has the same distribution as $\{\xi(i)\}$, the function $f(\mathbf{x}, \xi), \mathbf{x} \in R^p$, is continuous in the first argument and measurable in the second. Define the set of functionals

$$F_n(\mathbf{x}) = \frac{1}{n} \sum_{i=1}^{n} f(\mathbf{x}, \xi(i)), \quad \mathbf{x} \in J, \tag{3.13}$$

where

$$J = \{\mathbf{x} : g(\mathbf{x}) = (g_1(\mathbf{x}), ..., g_m(\mathbf{x})) \leq 0\}. \tag{3.14}$$

Here "$\leq$" is applied to each component of the vector.

We assume that the following conditions are satisfied.

A1. The function $f(\mathbf{x}, y)$ is twice continuously differentiable in the first argument, and $E \max_{||\mathbf{x}||_p \leq c} f(\mathbf{x}, \xi) < \infty, c > 0$. There exist $c > 0, \gamma_0 > 0$ such that for $||\mathbf{x} - \mathbf{x}^*|| \leq \gamma, \gamma < \gamma_0$, the inequality

$$E \left| \frac{\partial^2 f(\mathbf{x}, y)}{\partial x_i \partial x_k} \right| \leq c$$

holds true.

A2. The functions $g_i(\mathbf{x}), \mathbf{x} \in J$, are twice continuously differentiable, and for $||\mathbf{x} - \mathbf{x}^*|| \leq \gamma, \gamma < \gamma_0$,

$$\left| \frac{\partial g_i(\mathbf{x})}{\partial x_k} \right| \leq c, \qquad \left| \frac{\partial^2 g_i(\mathbf{x})}{\partial x_i \partial x_k} \right| \leq c,$$

where c is independent of x_i, x_k.

A3. The point $\mathbf{x}^*$ is the unique point for which $Ef(\mathbf{x}, \xi) > E(\mathbf{x}^*, \xi), \mathbf{x} \neq \mathbf{x}^*, \mathbf{x} \in J$, $\mathbf{x}^* \in J$, and

$$P\left\{ \lim_{n\to\infty} ||\mathbf{x}_n - \mathbf{x}^*|| = 0 \right\} = 1, \quad \mathbf{x}_n \in \arg\min_{\mathbf{x}\in J} F_n(\mathbf{x}).$$

A4. Let N_1 (respectively, N_2) be the set of arguments of indices i, for which $g_i(\mathbf{x}^*) = 0$ (respectively, $g_i(\mathbf{x}^*) < 0$). We assume that $\nabla g_j(\mathbf{x}^*)$ are linearly independent for $i \in N_1$.
A5. Functions $g_i(\mathbf{x})$ are convex.
A6. There exists a point $\mathbf{x}_*$ such that $g(\mathbf{x}_*) < 0$. Conditions A5 and A6 imply (see Korkhin 1985) that $\nabla g_i(\mathbf{x}_n)$ are linearly independent for those i for which $g_i(\mathbf{x}_n) = 0$.

Let us now investigate the asymptotic distribution of the variables $\mathbf{x}_n$. We introduce some relations following from the necessary conditions for the existence of the extremum. By definition,

$$\nabla F_n(\mathbf{x}_n) + \sum_{i=1}^{m} \lambda_i \nabla g_i(\mathbf{x}_n) = 0,$$

$$\lambda_{in} g_i(\mathbf{x}_n) = 0, \quad \lambda_{in} \geq 0,$$

∇g is the gradient of the function $g(\mathbf{x})$, $\lambda_{in} \geq 0$ are Lagrange multipliers. Denote $\nabla F_n(\mathbf{x}) = \tilde{F}_n(\mathbf{x})$. Applying the mean value theorem, we obtain

$$\tilde{F}_n(\mathbf{x}_n) = \tilde{F}_n(\mathbf{x}^*) + \tilde{\mathbf{\Phi}}_n(\varsigma_{1n}(\mathbf{x}_n - \mathbf{x}^*)), \tag{3.15}$$

where $\tilde{\mathbf{\Phi}}_n$ is a $p \times p$ matrix whose elements $\mathbf{\Phi}_n^{kl}$ are of the form:

$$\mathbf{\Phi}_n^{kl}(\varsigma_{1n}) = \frac{1}{n} \sum_{l=1}^{n} \left. \frac{\partial^2 f(\mathbf{x}, \xi(i))}{\partial x_k \partial x_l} \right|_{\mathbf{x}=\varsigma_{1n}},$$

$$\varsigma_{1n} = \mathbf{x}^* + \mathbf{\Theta}_1(\mathbf{x}_n - \mathbf{x}^*), \quad 0 \leq \mathbf{\Theta} \leq 1.$$

Thus we have

$$\tilde{F}_n(\mathbf{x}^*) + \tilde{\mathbf{\Phi}}_n(\varsigma_{1n})(\mathbf{x}_n - \mathbf{x}^*) + \sum_{i=1}^{n} \lambda_{in} \nabla g_i(\mathbf{x}_n) = 0.$$

We prove some auxiliary lemmas.

Lemma 3.2. *Let $\mathbf{\Phi}_n(s, \omega)$ be a sequence of functions satisfying Conditions 1 and 2 of Theorem 2.1, and assume that there exists a deterministic $\mathbf{\Phi}(s)$ such that for any $s \in K$*

$$P\left\{\lim_{n\to\infty} \mathbf{\Phi}_n(s, \omega) = \mathbf{\Phi}(s)\right\} = 1.$$

1. *Assume that for some sequence $s_n \in K$ there exists an element $s_0 \in K$ such that $P\{\lim_{n\to\infty} |s_n - s_0|| = 0\} = 1$.*

2. *There exists* $\gamma_0 > 0$ *such that for all* $0 < \gamma < \gamma_0$

$$P\left\{\overline{\lim_{n\to\infty}} \sup_{||s-s_0||<\gamma} |\boldsymbol{\Phi}_n(s) - \boldsymbol{\Phi}_n(s_0)| < c(\gamma)\right\} = 1,$$

where $c(\gamma) > 0$ *and* $c(\gamma) \to 0$ *as* $\gamma \to 0$.

Then $P\{\lim_{n\to\infty} \boldsymbol{\Phi}_n(s_n) = \boldsymbol{\Phi}(s_0)\} = 1$.

The proof is similar to the proof given in Dorogovtsev (1982). Since $P\{\overline{\lim}_{n\to\infty} \boldsymbol{\Phi}_n(s_n) = \boldsymbol{\Phi}(s_0)\} = 1$, it suffices to show that $P\{\overline{\lim}_{n\to\infty} |\boldsymbol{\Phi}_n(s_n) - \boldsymbol{\Phi}(s_0)| > \varepsilon\} = 0$ for any $\varepsilon > 0$. For any $\delta > 0$ we have

$$\begin{aligned} P\left\{\overline{\lim_{n\to\infty}} |\boldsymbol{\Phi}_n(s_n) - \boldsymbol{\Phi}(s_0)| \geq \varepsilon\right\} &= P\{\omega : ||s_n - s_0|| < \gamma, n \geq N(\omega)\}, \\ \overline{\lim_{n\to\infty}} |\boldsymbol{\Phi}_n(s_n) - \boldsymbol{\Phi}_n(s_0)| > \varepsilon, &\quad \overline{\lim_{n\to\infty}} \sup_{||s-s_0||<\gamma} |\boldsymbol{\Phi}_n(s_n) - \boldsymbol{\Phi}_n(s_0)| < c(\gamma) \\ = P\left\{\overline{\lim_{n\to\infty}} |\boldsymbol{\Phi}_n(s_n) - \boldsymbol{\Phi}_n(s_0)| < c(\gamma),\right. &\left. \overline{\lim_{n\to\infty}} |\boldsymbol{\Phi}_n(s_n) - \boldsymbol{\Phi}_n(s_0)| \geq \varepsilon\right\}. \end{aligned} \quad (3.16)$$

Taking $\gamma > 0$ such that $c(\gamma) < \varepsilon$, we obtain that the right-hand side of (3.16) vanishes.

Lemma 3.3. *Assume that condition A1 holds. Then for each element of the matrix of second derivatives we have*

$$P\left\{\lim_{n\to\infty} \boldsymbol{\Phi}_n^{kl}(\mathbf{x}_n) = \boldsymbol{\Phi}^{kl}(\mathbf{x}^*)\right\} = 1, \quad \boldsymbol{\Phi}_n^{kl}(\mathbf{x}) = \frac{\partial^2 F_n(\mathbf{x})}{\partial x_k \partial x_l},$$

$$\boldsymbol{\Phi}^{kl}(\mathbf{x}) = E\frac{\partial^2 f(\mathbf{x}, \xi)}{\partial x_k \partial x_l},$$

where $||\mathbf{x}_n - \mathbf{x}^*|| \to 0$ *as* $n \to \infty$ *with probability 1.*

Proof. Consider

$$\begin{aligned} &\sup_{\{\mathbf{x}:||\mathbf{x}-\mathbf{x}^*||<\gamma\}} |\boldsymbol{\Phi}_n^{kl}(\mathbf{x}) - \boldsymbol{\Phi}_n^{kl}(\mathbf{x}^*)| \\ &\leq \sup_{\{\mathbf{x}:||\mathbf{x}-\mathbf{x}^*||<\gamma\}} \frac{1}{n}\sum_{i=1}^{n} \sup_{\{\mathbf{x}:||\mathbf{x}-\mathbf{x}^*||<\gamma\}} \left|\frac{\partial^2 f(\mathbf{x}, \xi(i))}{\partial x_k \partial x_l} - \frac{\partial^2 f(\mathbf{x}^*, \xi(i))}{\partial x_k \partial x_l}\right| = \varsigma_n^{kl}(\gamma). \end{aligned}$$

Let

$$c_{kl}(\gamma) = 2E\left\{\sup_{\{\mathbf{x}:||\mathbf{x}-\mathbf{x}^*||<\gamma\}} \left|\boldsymbol{\Phi}_n^{kl}(\mathbf{x}) = \frac{\partial^2 f(\mathbf{x}, \xi)}{\partial x_k \partial x_l} - \boldsymbol{\Phi}_n^{kl}(\mathbf{x}) = \frac{\partial^2 f(\mathbf{x}^*, \xi)}{\partial x_k \partial x_l}\right|\right\}.$$

By the strong flow of large numbers, we have

$$P\left\{\lim_{n\to\infty}\varsigma_n^{kl}(\gamma)=\frac{1}{2}c_{kl}(\gamma)\right\}=1.$$

Since the function $\partial^2 f(\mathbf{x},\xi)/\partial x_k\partial x_l$ is continuous in $\mathbf{x}$, we pass to the limit under the expectation and obtain that $c(\gamma)\to 0$ as $\gamma\to 0$. Thus conditions of Lemma 3.2 are satisfied, implying $P\{\lim_{n\to\infty}\mathbf{\Phi}_n^{kl}(\mathbf{x}_n)=\mathbf{\Phi}^{kl}(\mathbf{x}^*)\}=1$. □

Now we find the asymptotic distribution of the optimal points. Multiplying the left-hand side of (3.15) by $\sqrt{n}$ and applying the mean value theorem to the function $g_i(\mathbf{x}_n)$, we get

$$\sqrt{n}\tilde{\mathbf{F}}_n(\mathbf{x}^*)+\tilde{\mathbf{\Phi}}_n\sqrt{n}(\mathbf{x}_n-\mathbf{x}^*)+\sum_{i=1}^{n}\sqrt{n}\lambda_{in}\nabla g_i(\mathbf{x}_n)=0,$$

$$\lambda_{in}g_i(\mathbf{x}^*)+\lambda_{in}\nabla g_i(\mathbf{x}^*+\mathbf{\Theta}_2\Delta\mathbf{x}_n)(\mathbf{x}_n-\mathbf{x}^*)=0.$$

Thus we arrive at the quadratic programming problem:

$$\frac{1}{2}\mathbf{x}^{\mathrm{T}}\tilde{\mathbf{\Phi}}_n\mathbf{x}+\tilde{\mathbf{F}}_n^{\mathrm{T}}(\mathbf{x}^*)\mathbf{x}\to\min, \tag{3.17}$$

$$b_i(\mathbf{x})=\sqrt{n g_i}(\mathbf{x}^*)+\nabla g_i^{\mathrm{T}}(\mathbf{x}^*)\mathbf{x}+l_i(\mathbf{x})\le 0,\quad \mathbf{x}\in\Re^p,$$

$$l_i(\mathbf{x})=\frac{1}{2\sqrt{n}}\mathbf{x}^{\mathrm{T}}g_{i2}\left(\mathbf{x}^*+\mathbf{\Theta}_1\frac{\mathbf{x}}{\sqrt{n}}\right)\mathbf{x},\quad \mathrm{T}i\in J, \tag{3.18}$$

$g_{i2}(\mathbf{x})$ is the $p\times p$ Hessian with elements $\partial^2 g_i(\mathbf{x})/\partial x_k\partial x_i$.

By the central limit theorem, $\tilde{\mathbf{F}}_n(\mathbf{x}^*)$ converges weakly to the normal variable $N(\mathbf{a},\mathbf{B})$, where vector $\mathbf{a}$ and matrix $\mathbf{B}$ are given by $\mathbf{a}=E\nabla f(\mathbf{x}^*,\xi_0)$, $\mathbf{B}=E[\nabla f(\mathbf{x}^*,\xi_0)-\mathbf{a}][\nabla f(\mathbf{x}^*,\xi_0)-\mathbf{a}]^{\mathrm{T}}$ (index "T" denotes transposing). By Lemma 3.3, $\tilde{\mathbf{\Phi}}_n\to\mathbf{\Phi}(\mathbf{x}^*)$ with probability 1 as $n\to\infty$. Besides (3.17) and (3.18), consider the following quadratic programming problem:

$$\tfrac{1}{2}\mathbf{x}^T\mathbf{\Phi}(\mathbf{x}^*)\mathbf{x}+\varsigma x\to\min, \tag{3.19}$$

$$\nabla g^{\mathrm{T}}(\mathbf{x}^*)\mathbf{x}\le 0,\quad i\in N_1, \tag{3.20}$$

where ς is $N(\mathbf{a},\mathbf{B})$.

Theorem 3.14. *Let conditions A1–A6 be satisfied. Then the random vector* $\boldsymbol{\eta}_n=\sqrt{n}(\mathbf{x}_n-\mathbf{x}^*)$ *which is the solution to problems (3.17) and (3.18), converges in distribution to the random vector* $\boldsymbol{\eta}$ *which is the solution to (3.19) and (3.20).*

The proof of Theorem 3.14 is analogous to the proof of Theorem 2.1, and is given in Knopov and Kasitskaya (2002).

Corollary 3.1. *If* $N_1 = \varnothing$, *then the vector* $\boldsymbol{\eta}$ *is normally distributed with parameters* $(\mathbf{0}, \boldsymbol{\Phi}^{-1}(\mathbf{x}^*)^{\mathrm{T}}\mathbf{B})$.

Now consider the case when $\{\boldsymbol{\xi}(i), i = 1, 2\}$ are stationary dependent random vectors.

Definition 3.1. We say that stationary vectors satisfy the strong mixing condition if

$$\alpha(n) = \sup_{\substack{A \in \mathbf{A}_{-\infty}^m \\ B \in \mathbf{A}_{n+m}^m}} |P(AB) - P(A)P(B)| \to 0 \tag{3.21}$$

as $n \to \infty$, where $\mathrm{A}_n^m = \sigma\{\boldsymbol{\xi}(i), n \le i \le m\}$ is the σ-algebra generated by the sequence $\{\boldsymbol{\xi}(i), n \le i \le m\}$.

For the theorem below we refer to Koroljuk et al. (1985).

Theorem 3.15. *Let* $\boldsymbol{\xi}(i) = \{\xi^1(i), \ldots, \xi^p(i)\}$, $E\boldsymbol{\xi}(i) = 0$, *be stationary in the narrow sense vectors satisfying conditions:*

1. *Strong mixing condition (3.21) with* $\alpha(n) \le c/n^{1+\varepsilon}, \varepsilon > 0$
2. $E||\boldsymbol{\xi}(i)||^{2+\delta} < \infty$, *where* $\varepsilon\delta > 2$

Then there exists a bounded spectral density $\mathbf{h}(\lambda) = \{h_{il}(\lambda)\}_{i,l=1}^p$, *continuous in 0. If, moreover,* $\mathbf{h}(\lambda)$ *is a non-degenerate matrix, then the vector* $(1/\sqrt{n})\sum_{i=1}^n \boldsymbol{\xi}(i)$ *is asymptotically normal as* $n \to \infty$ *with* $N(\mathbf{0}, 2\pi\mathbf{h}(0))$.

Using Theorem 3.15, we can prove similarly to Theorem 3.14 the following statement (Knopov and Kasitskaya 2002).

Theorem 3.16. *Assume that* $\nabla f(\mathbf{x}^*, \boldsymbol{\xi}(i))$ *satisfies the conditions:*

1. *Strong mixing condition with* $\alpha(n) \le c/n^{1+\varepsilon}, \varepsilon > 0$
2. $E||\nabla f(\mathbf{x}^*, \boldsymbol{\xi}(i))||^{2+\delta} < \infty$, *where* $\varepsilon\delta > 2$
3. *The spectral density* $\mathbf{h}(\lambda)$ *of the vector* $\nabla f(\mathbf{x}^*, \boldsymbol{\xi}(i))$ *is a non-degenerate in* $\lambda = 0$ *matrix*
4. *Conditions A1–A6 hold true*

Then the sequence of vectors $\boldsymbol{\eta}_n = \sqrt{n}(\mathbf{x}_n - \mathbf{x}^*)$ which are solutions to (3.17) and (3.18), converges weakly to the random vector $\boldsymbol{\eta}$, which is the solution to (3.19) and (3.20), where $\boldsymbol{\varsigma}$ is normally distributed $N(\mathbf{0}, 2\pi\mathbf{h}(0))$.

For dependent vectors which are difference martingales we have one more version of the central limit theorem.

Theorem 3.17. *Let* $\{\mathbf{u}_n, n \in N\}$ *be the stationary in the narrow sense metrically transitive process, satisfying* $E(\mathbf{u}_n/\Im_{n-1}) = 0$, *where* $\Im_n$ *is the* σ*-algebra generated by the random vectors* $\mathbf{u}_1, \ldots, \mathbf{u}_n$. *Then the distribution of the variable* $(1/\sqrt{n})\sum_{k=1}^n \mathbf{u}_k$ *converges weakly as* $n \to \infty$ *to the normal distribution* $N(\mathbf{0}, \mathbf{R})$, *where* $\mathbf{R} = E(\mathbf{u}_n\mathbf{u}_n^{\mathrm{T}})$.

Using Theorem 3.17, we obtain the following statement.

Theorem 3.18. *Assume that* $\nabla f(\mathbf{x}^*, \boldsymbol{\xi}(i))$ *in (3.13) satisfies the conditions:*

1. $E\{[\nabla f(\mathbf{x}^*, \boldsymbol{\xi}(i)) - \mathbf{a}]/G_{i-1}\} = 0$, *where* G_i *is the* σ*-algebra generated by the vectors* $f(\mathbf{x}^*, \boldsymbol{\xi}(1)), \ldots, f(\mathbf{x}^*, \boldsymbol{\xi}(i))$
2. *Conditions A1–A6 hold true*

Then the vector $\boldsymbol{\eta}_n = \sqrt{n}(\mathbf{x}_n - \mathbf{x}^*)$, which is the solution to the problems (3.17) and (3.18) converges weakly to the random vector $\boldsymbol{\eta}$, which is the solution to (3.13) and (3.14), where $\boldsymbol{\varsigma}$ is normally distributed random vector $N(\mathbf{a}, \mathbf{B})$.

Now we consider the case when the function f depends also on the time variable i. In this case we take $F_n(\mathbf{x}) = (1/n)\sum_{i=1}^{n} f(i, \mathbf{x}, \xi(i))$.

For simplicity, assume first that $\xi(i)$ is independent identically distributed random variables with $E\xi(i) = 0, E\xi^2(i) = \sigma^2$. As before, we assume that conditions A1–A6 hold true, together with Condition A7:

A7. The following limits exist:

$$\mathbf{a}(\mathbf{x}) = \lim_{n\to\infty} \sum_{i=1}^{n} E\nabla f(i, \mathbf{x}, \xi(i)),$$

$$\tilde{\mathbf{B}}(\mathbf{x}) = \lim_{n\to\infty} \frac{1}{n} \sum_{i=1}^{n} E[\nabla f(i, \mathbf{x}, \xi(i)) - \mathbf{a}(\mathbf{x})][\nabla f(i, \mathbf{x}, \xi(i)) - \mathbf{a}(\mathbf{x})]^{\mathrm{T}},$$

$$\boldsymbol{\Phi}_{ki}^{\gamma}(\mathbf{x}^*) = \overline{\lim_{n\to\infty}} \sum_{i=1}^{n} \sup_{\{\mathbf{x}:\|\mathbf{x}-\mathbf{x}^*\|<\gamma\}} E\left|\frac{\partial^2 \mathbf{f}(i, \mathbf{x}, \boldsymbol{\xi}(i))}{\partial x_k \partial x_i}\right|.$$

Under the Conditions A1–A7 one can prove Theorem 3.19 in the same way as Theorem 3.18.

Theorem 3.19. *Under Conditions A1–A7, the solution* $\boldsymbol{\eta}_n = \sqrt{n}(\mathbf{x}_n - \mathbf{x}^*)$ *to the problems (3.17) and (3.18) converges in distribution to the random vector* $\boldsymbol{\eta}$, *which is the solution to (3.19) and (3.20), where* $\boldsymbol{\varsigma}$ *is normally distributed* $N(\mathbf{a}(\mathbf{x}^*), \tilde{\mathbf{B}}(\mathbf{x}^*))$.

If $\xi(i)$ are not independent observations, but satisfy strong intermixing conditions, the following theorem about the asymptotic distribution of estimates takes place.

Theorem 3.20. *Assume that Conditions A1–A7 and the conditions below are satisfied.*

A8. The vector $\boldsymbol{\varsigma}_i = \nabla f(i, \mathbf{x}^*, \boldsymbol{\xi}_i)$ satisfies uniform strong mixing condition

$$\alpha(j) = \sup_{i} \sup_{\substack{A \in \sigma_{-\infty}^{i} \\ B \in \sigma_{i+j}^{\infty}}} |P(AB) - P(A)P(B)| \le \frac{C}{1 + j^{1+\varepsilon}},$$

where $\sigma_n^m = \sigma\{\boldsymbol{\varsigma}_i, n \le i \le m\}$.

A9. $E||\nabla f(i, \mathbf{x}^*, \boldsymbol{\xi}_i)||^{2+\delta} < \infty$ for $\varepsilon\delta > 2$.
Then Theorem 3.19 holds true.

Now we consider some specific regression models and find the asymptotic distribution of the unknown parameter estimates.

Model (2.4)

$$y_j = f(\boldsymbol{\alpha}, \mathbf{x}(j)) + \varepsilon(j),$$
$$\boldsymbol{\alpha}^0 \in J = \{\boldsymbol{\alpha} : g(\boldsymbol{\alpha}) \leq 0\},$$

where $\mathbf{x}(j)$ and $\varepsilon(j)$ are determined in model (2.4), $\mathbf{x}(j)$ are independent random vectors, $\varepsilon(j)$ are independent random variables, independent on $\mathbf{x}(j)$ for $j \geq 1$.

Let us examine criterion (2.7). Let

$$Q_n(\boldsymbol{\alpha}) = \frac{1}{n}\sum_{j=1}^{n}[y_j - f(\boldsymbol{\alpha}, \mathbf{x}(j))]^2.$$

We assume that the function $f(\boldsymbol{\alpha}, \mathbf{x})$ satisfies the following conditions:

1. $f(\boldsymbol{\alpha}, \mathbf{x})$ is twice continuously differentiable with respect to $\boldsymbol{\alpha}$, and for $\boldsymbol{\alpha} \in J$

$$\left|\frac{\partial f(\boldsymbol{\alpha}, \mathbf{x})}{\partial \alpha_i}\right| \leq c_1, \qquad \left|\frac{\partial^2 f(\boldsymbol{\alpha}, \mathbf{x})}{\partial x_k \partial x_i}\right| \leq c_2.$$

2. There exist

$$r_{kl} = E\frac{\partial f(\boldsymbol{\alpha}^0, \mathbf{x}(j))}{\partial \alpha_k} \cdot \frac{\partial f(\boldsymbol{\alpha}^0, \mathbf{x}(j))}{\partial \alpha_l},$$

where the matrix $\mathbf{R}$ with the elements r_{kl} is positive definite.

Theorem 3.21. *Assume that the function $f(\boldsymbol{\alpha}, \mathbf{x})$ satisfies Conditions 1 and 2, functions $g_i(\boldsymbol{\alpha})$ satisfy Conditions A2, A4–A6, $\boldsymbol{\alpha}_n = \arg\min_{\boldsymbol{\alpha}\in J} Q_n(\boldsymbol{\alpha})$ is strongly consistent estimate of $\boldsymbol{\alpha}^0$, and the spectral density $\mathbf{h}(\lambda)$ of the stationary sequence $\varepsilon(j)$ exists and is continuous and bounded at the point $\lambda = 0$.*

Let $\varepsilon(j)$ be independent identically distributed random variables with $E\varepsilon(j) = 0$, $E\varepsilon^2(j) = 1$. Then the vector $\sqrt{n}(\boldsymbol{\alpha}_n - \boldsymbol{\alpha}^0)$ is the solution to problems (3.17) and (3.18), where

$$\tilde{F}_n^i(\boldsymbol{\alpha}^0) = \frac{1}{n}\sum_{j=1}^{n}\frac{\partial f(\boldsymbol{\alpha}^0, \mathbf{x}(j))}{\partial \alpha_i}\varepsilon(j).$$

Here $\tilde{F}_n^i(\boldsymbol{\alpha}^0)$ is the element of $\tilde{\mathbf{F}}_n(\boldsymbol{\alpha}^0)$, $\boldsymbol{\Phi}_n(\boldsymbol{\alpha})$ is the $p \times p$ matrix with elements

$$\boldsymbol{\Phi}_n^{kl}(\boldsymbol{\alpha}) = \frac{1}{n}\sum_{j=1}^{n}\left\{\frac{\partial f(\boldsymbol{\alpha},\mathbf{x}(j))}{\partial\alpha_k}\cdot\frac{\partial f(\boldsymbol{\alpha},\mathbf{x}(j))}{\partial\alpha_l} + \frac{\partial^2 f(\boldsymbol{\alpha},\mathbf{x}(j))}{\partial\alpha_k\,\partial\alpha_l}[f(\boldsymbol{\alpha}^0,\mathbf{x}(j)) - f(\boldsymbol{\alpha},\mathbf{x}(j))]\right\}.$$

The vector $\sqrt{n}(\boldsymbol{\alpha}_n - \boldsymbol{\alpha}^0)$ converges weakly to the random vector $\boldsymbol{\eta}$, which is the solution to (3.19) and (3.20), where $\boldsymbol{\varsigma}$ is normal $N(\mathbf{0},\mathbf{B})$,

$$B = \left\{E\frac{\partial f(\boldsymbol{\alpha}^0,\mathbf{x}(0))}{\partial\alpha_k}\cdot\frac{\partial f(\boldsymbol{\alpha}^0,\mathbf{x}(0))}{\partial\alpha_l}\right\}_{k,l=1}^{p}.$$

Now assume that the stationary sequence $\varepsilon(j)$, $E\varepsilon(j) = 0$, $E\varepsilon^2(j) = 1$ in (2.4) is a difference martingale, i.e.

1. $$E\{\varepsilon(j)/\Im_{j-1}\} = 0, \tag{3.22}$$
2. $$E\{\varepsilon^2(j)/\Im_{j-1}\} = 1, \tag{3.23}$$

where $\Im_j$ is the σ-algebra generated by the sequence $\varepsilon(1),\ldots,\varepsilon(j)$. Then using Theorem 2.1, we obtain the following theorem

Theorem 3.22. *Assume that for the model (2.4) and (3.14), conditions (3.22) and (3.23) are satisfied, and the functions $f(\boldsymbol{\alpha},\mathbf{x}), g(\boldsymbol{\alpha})$ satisfy the conditions of Theorem 3.20. Then statement of Theorem 3.21 holds true.*

Consider model (2.3), in which

1. $\varepsilon(j)$ are independent identically distributed random variables with $E\varepsilon(j) = 0$, $E\varepsilon^2(j) = 1$
2. Functions $f(j,\boldsymbol{\alpha})$ and $g(\boldsymbol{\alpha})$ satisfy Conditions A1, A2, A4–A6
3. Let

$$F_n^{kl}(\boldsymbol{\alpha}) = \frac{2}{n}\sum_{i=1}^{n}\frac{\partial f(i,\boldsymbol{\alpha})}{\partial\alpha_k}\cdot\frac{\partial f(i,\boldsymbol{\alpha})}{\partial\alpha_l}.$$

Then there exists $\lim_{n\to\infty} F_n^{kl}(\boldsymbol{\alpha}) = F^{kl}(\boldsymbol{\alpha})$.

Suppose that

$$\boldsymbol{\alpha}_n \in \arg\min_{\boldsymbol{\alpha}\in J} Q_n(\boldsymbol{\alpha}) = \arg\min_{\boldsymbol{\alpha}\in J}\frac{1}{n}\sum_{j=1}^{n}[y_j - f(j,\boldsymbol{\alpha})]^2.$$

Then the theorem above implies the following statement.

Theorem 3.23. *Assume that the conditions of Theorem 2.5 and conditions 1–3 of Theorem 3.22 hold true. Then* $\boldsymbol{\eta}_n = \sqrt{n}(\boldsymbol{\alpha}_n - \boldsymbol{\alpha}^0)$ *is the solution to problems (3.17) and (3.18), where*

$$\tilde{F}(\boldsymbol{\alpha}^0) = \frac{2}{\sqrt{n}} \sum_{i=1}^{n} \nabla f(i, \boldsymbol{\alpha}^0)\varepsilon(i)$$

and the matrices $\boldsymbol{\Phi}_n(\boldsymbol{\alpha}) = \{\boldsymbol{\Phi}_n^{kl}(\boldsymbol{\alpha})\}_{k,l=1}^{p}$ *are of the form*

$$\boldsymbol{\Phi}_n^{kl} = \frac{1}{n}\sum_{i=1}^{n} \frac{\partial f(i,\boldsymbol{\alpha})}{\partial \alpha_k} \cdot \frac{\partial f(i,\boldsymbol{\alpha})}{\partial \alpha_l}\Bigg|_{\boldsymbol{\alpha}=\boldsymbol{\xi}_{1n}} + \frac{1}{n}\sum_{i=1}^{n} [y_i - f(i, \boldsymbol{\xi}_{1n})] \frac{\partial^2 f(i,\boldsymbol{\alpha})}{\partial \alpha_k \partial \alpha_l}\Bigg|_{\boldsymbol{\alpha}=\boldsymbol{\xi}_{1n}}$$

and converge weakly to random vector $\boldsymbol{\eta}$ *which is the solution to (3.19) and (3.20), where* $\boldsymbol{\Phi} = \lim_{n\to\infty} \boldsymbol{\Phi} x_n(\boldsymbol{\alpha}^0)$, *and* ς *is normally distributed* $N(\mathbf{0}, \mathbf{C})$,

$$\mathbf{C} = \lim_{n\to\infty} \frac{1}{n} \sum_{i=1}^{n} [\nabla f(i, \boldsymbol{\alpha}^0) \nabla f(i, \boldsymbol{\alpha}^0)^{\mathrm{T}}].$$

Concerning the least modules estimates for model (2.3) we note that their asymptotic normality without a priori information about the true value of the parameter is proved in Knopov and Kasitskaya (1995). Using these results, we can also prove for these estimates a theorem about the asymptotic distribution under a priory constraints on the parameter α. The main difficulty in obtaining these results consists in non-differentiability of criteria functions. At the same time, the cost function will be differentiable with respect to any direction. The proof uses the techniques proposed in Huber (1981).

In conclusion, we would like to remark that in this paragraph only discrete time parameter is considered for all models. Nevertheless, usually similar results take place for continuous time as well. More detailed discussion for this case is given in Knopov (1997b). For illustration consider the result proved in this work.

Consider

$$f(t, \mathbf{x}, y) = f(\mathbf{x}, y) \quad \text{and} \quad F_T(\mathbf{x}) = \frac{1}{T}\int_0^T f(\mathbf{x}, \xi_t)\mathrm{d}t,$$

where $\{\xi_t, t \in R\}$ is a strongly stationary ergodic stochastic process. Define the set J by

$$J = \{\mathbf{x} : g(\mathbf{x}) = (g_1(\mathbf{x}), \ldots, g_n(\mathbf{x})) \leq 0\}.$$

The following conditions are assumed to be satisfied:

A1″. The function $f(\mathbf{x}, y)$ is twice continuously differentiable in the first argument, and $E \max_{||\mathbf{x}|| \leq c} f(\mathbf{x}, \xi_0) < \infty$ for any positive c. There exists $\mathbf{x}^* \in J$ and $\gamma_0 > 0$ such that for $||\mathbf{x} - \mathbf{x}^*|| < \gamma, \gamma < \gamma_0$,

$$E\left|\frac{\partial^2 f(\mathbf{x}, \xi_0)}{\partial x_l \partial x_k}\right| \leq c.$$

A2″. The functions $g_i(\cdot), \mathbf{x} \in J$, are twice continuously differentiable, moreover, for $||\mathbf{x} - \mathbf{x}^*|| < \gamma, \gamma < \gamma_0$, we have

$$\left|\frac{\partial g_i(\mathbf{x})}{\partial x_k}\right| \leq c, \quad \left|\frac{\partial^2 g_i(\mathbf{x})}{\partial x_l \partial x_k}\right| \leq c.$$

A3″. Let $\mathbf{x}^*$ be the unique point for which $Ef(\mathbf{x}, \xi_0) > Ef(\mathbf{x}^*, \xi_0), \mathbf{x} \neq \mathbf{x}^*, \mathbf{x} \in J$, $\mathbf{x}^* \in J$ and $P\{\lim_{T\to\infty} ||\mathbf{x}_T - \mathbf{x}^*|| = 0\} = 1$.

A4″. Let N_1 be the set of indexes i for which $g_i(\mathbf{x}^*) < 0$. We assume that $\nabla g_i(\mathbf{x}^*), i \in N_1$, are linearly independent.

A5″. Functions $g_i(\mathbf{x})$ are convex.

A6″. There exists a point $\mathbf{x}_*$ such that $g(\mathbf{x}_*) < 0$.

The following theorem takes place.

Theorem 3.24. *Let Conditions* A1″–*A5″ and the conditions below hold true:*

1. *Strong mixing condition 4 of Theorem 3.4*
2. $E||\nabla f(\mathbf{x}^*, \xi_0)||^{2+\delta} < \infty, \varepsilon\delta > 2$
3. *The spectral density* $\boldsymbol{\varphi}(\lambda)$ *of the random process* $\nabla f(\mathbf{x}^*, \xi_i)$ *is non-zero at* $\lambda = 0$

Then the family of vectors $\boldsymbol{\eta}_T = \sqrt{T}(\mathbf{x}_T - \mathbf{x}^*)$ converges weakly to the random vector $\boldsymbol{\eta}$ which is the solution to the problem

$$\frac{1}{2}\mathbf{x}'\boldsymbol{\Phi}(\mathbf{x}^*)\mathbf{x} + \boldsymbol{\varsigma}\mathbf{x} \to \min,$$

$$\nabla g^{\mathrm{T}}(\mathbf{x}^*)\mathbf{x} \leq 0,$$

where $\boldsymbol{\varsigma}$ is normal random vector $N(E\nabla f(\mathbf{x}^*, \xi_0), 2\pi\boldsymbol{\varphi}(0))$, and

$$\boldsymbol{\Phi}(\mathbf{x}^*) = \left\{E\frac{\partial^2 f(\mathbf{x}^*, \xi_0)}{\partial x_l \partial x_k}\right\}_{l,k=1}^{p}.$$

Remark 3.3. If $\mathbf{x}^*$ is the inner point of J then $\boldsymbol{\eta}_T$ weakly converges to the normal distribution.

To end this subsection we would like to make a remark about the asymptotic distribution of non-parametric estimates of unknown parameters. Under the assumption that the true value of the parameter is the inner point of some set, the distribution of functionals of these estimates has been studied in Dorogovtsev (1982). The question when the true value of the parameter is not an interior point, or, in other words, the set of constraints is not closed, remains open. The problem of finding the distribution of non-parametric estimates for non-differentiable cost function has not been studied at all. Although these topics remain outside the frames of the present work, they are of essential interest.

3.4.2 *Asymptotic Distribution of Estimates for Long Memory Stochastic Systems*

In this subsection we investigate the behavior of the estimate of the parameter $\boldsymbol{\beta}$ in models with arbitrary criteria, described in Sect. 3.2. The result presented below is proved in Moldavskaya (2007).

We estimate the unknown parameter $\boldsymbol{\beta} = (\beta_1, ..., \beta_n)^{\mathrm{T}}$ in the linear regression model with continuous time

$$y(t) = \boldsymbol{\beta}^{\mathrm{T}}\mathbf{g}(t) + \eta(t), \quad 0 \leq t \leq T$$

and nonlinear constraints

$$h_j(\boldsymbol{\beta}) \leq 0, \quad j = 1, ..., r,$$

where $\mathbf{g}(t) = [g_1(t), ..., g_n(t)]^{\mathrm{T}}, \mathbf{h}(\boldsymbol{\beta}) = [h_1(\boldsymbol{\beta}), ..., h_r(\boldsymbol{\beta})]^{\mathrm{T}}$ are known functions, $\eta(t), t \in \Re$, is a measurable continuous in the mean square sense stationary process with zero mean and covariance $B_\eta(t), t \in \Re$.

Consider the conditions below.

1. Assume that $\eta(t) = G(\varepsilon(t)), t \in \Re$, is a random process, where G is an arbitrary real-valued measurable nonrandom function, and $\varepsilon(t), t \in \Re$, is a Gaussian random process with zero mean, unit variance and covariance $B_\varepsilon(t) = (1 + t^2)^{-\alpha/2}, 0 < \alpha < 1$. Assume that $EG(\varepsilon(0)) < \infty$.

If Assumption 3.3 holds, then the function G can be expanded into series of Chebyshev-Hermite polynomials in $L_2(\Re, \varphi(u)\mathrm{d}u)$, where $\varphi(u)$ is the density of $\varepsilon(i)$. Let $m \geq 1$ be the number of the first non-zero coefficient in this expansion.

2. Suppose that $g_i(t) > 0, t > 0, 1 \leq i \leq n$, are bounded on $[0, T]$. Let $\mathbf{d}(T) = \mathrm{diag}(d_1(T), \ldots, d_n(T))$, where $d_i(T) = (\int_0^T g_i^2(t)\mathrm{d}t)^{1/2}$; $\lim_{T\to\infty} T^{-1}d_i^2(T) > 0, i = 1, \ldots, n$. Define the matrices

$$\mathbf{J}_T = (J_{il,T})_{i,l=1}^n, \quad J_{il,T} = \mathbf{D}_T^{-1}\int_0^1 \mathbf{g}(tT)\mathbf{g}^{\mathrm{T}}(tT)\mathrm{d}t \cdot \mathbf{D}_T^{-1},$$

$$\mathbf{D}_T^2 = \mathrm{diag}\left(\int_0^1 g_i^2(Tt)\mathrm{d}t\right)_{i=1}^n,$$

$$\boldsymbol{\sigma}_{T,m} = \mathbf{D}_T^{-1}\left(\int_0^1\int_0^1 \frac{g(Tt)g'(Ts)}{|t-s|^{m\alpha}}\mathrm{d}t\ \mathrm{d}s\right)\mathbf{D}_T^{-1}.$$

3. Assume that $\lim_{T\to\infty}\boldsymbol{\sigma}_{T,m} = \boldsymbol{\sigma}_m$, where $\boldsymbol{\sigma}_m$ is some positive definite matrix.
4. Assume that $\lim_{T\to\infty}\mathbf{J}_T = \mathbf{J}_0$, where $\mathbf{J}_0$ is some positive definite matrix, and $\mathbf{R}_0 = \mathbf{J}_0 s_m^{-1}\mathbf{J}_0$.

5. Assume that: (1) $h_j, 1 \le j \le r$, have the first and the second order derivatives, bounded in the neighborhood of the true value $\boldsymbol{\beta}_0$; (2)$h_j(\boldsymbol{\beta}_0) = 0,\ j \in \{1,\dots,q\}, h_j(\boldsymbol{\beta}_0) < 0, j \in \{q+1,\dots,r\}$; (3) there exists $\boldsymbol{\beta}^*$ such that $\mathbf{h}(\boldsymbol{\beta}^*) < 0$; (4) vectors $\nabla h_j(\boldsymbol{\beta}_0), j \in \{1,\dots,q\}$, are linearly independent; (5) $h_j(\boldsymbol{\beta}), j \in \{1,\dots,r\}$, are convex.
6. Assume that $\boldsymbol{\beta}_T$ (the least squares estimate of the parameter $\boldsymbol{\beta}$ satisfying the condition $\mathbf{h}(\boldsymbol{\beta}) \le 0$) is consistent.
7. Diagonal matrix $\bar{\mathbf{d}}_T$ has positive elements and is such that there exists $\lim_{T\to\infty} \bar{\mathbf{h}}_T(\boldsymbol{\beta}) = \bar{\mathbf{h}}(\boldsymbol{\beta})$, where matrix $\bar{\mathbf{h}}_T(\boldsymbol{\beta})$ is determined as follows:

$$\bar{\mathbf{h}}_T(\boldsymbol{\beta}) = \bar{\mathbf{d}}_T \mathbf{h}(\boldsymbol{\beta}) \mathbf{d}_T^{-1}.$$

$\mathbf{h}(\boldsymbol{\beta})$ is the matrix which consists of the rows $\nabla h_j^{\mathrm{T}}(\boldsymbol{\beta}), j = \overline{1,r}$.

The analogous condition will be used as below in Sect. 5.2.

Condition 7 means that there exists a limit as $T \to \infty$ for rows $\bar{\mathbf{d}}_j(T)\nabla h_j^{\mathrm{T}}(\boldsymbol{\beta}_0)\mathbf{d}_T^{-1}$ of the matrix $\bar{\mathbf{h}}_T(\boldsymbol{\beta}_0)$, which is equal to $\bar{\mathbf{h}}_j(\boldsymbol{\beta}_0), j = \overline{1,r}$. Here $\bar{\mathbf{h}}_j(\boldsymbol{\beta}_0)$ is jth row of $\bar{\mathbf{h}}(\boldsymbol{\beta}_0), \bar{\mathbf{d}}_j(T)$ is the element on the main diagonal of $\bar{\mathbf{d}}_T$.

Below we formulate main results in several theorems on the asymptotic distribution of the least squares estimate in linear regression models with constraints.

Theorem 3.25. *Assume that Conditions 1, 2, 4–7 and also 3 with $m = 1$ hold. Then the random vector $\mathbf{U}_T = \mathbf{B}(T)^{-1/2}T^{-1/2}\mathbf{d}_T(\boldsymbol{\beta}_T - \boldsymbol{\beta}_0)$ converges in distribution as $T \to \infty$ to the random vector $\boldsymbol{U}$ which is the solution to the quadratic programming problem*

$$\begin{cases} \frac{1}{2}\mathbf{X}'\mathbf{R}_0\mathbf{X} - \mathbf{Q}'\mathbf{X} \to \min, \\ \bar{\mathbf{h}}_j(\boldsymbol{\beta}_0)\mathbf{X} \le 0, \quad j = 1,\dots,q, \end{cases} \tag{3.24}$$

where $\mathbf{R}_0$ is defined in (3.24) and $\boldsymbol{Q}$ is a Gaussian random vector with zero mean and covariance matrix $\mathbf{J}_0(\boldsymbol{\sigma}_1)^{-1}\mathbf{J}_0, \boldsymbol{\sigma}_1$ is defined by Conditions 2–4.

Case $m = 1$ is essential. Then the solution to the problem (3.24) is non-Gaussian, when the constraints of the problem are active, and Gaussian otherwise.

Under some additional conditions Theorem 3.25 gives the answer to the question about the limit distribution of $\mathbf{U}_T = \mathbf{B}(T)^{-m/2}T^{-1/2}\mathbf{d}_T(\boldsymbol{\beta}_T - \boldsymbol{\beta}_0)$ in the case $m \ge 2$. It is also non-Gaussian even in the case when restrictions are inactive (in case of no constraints the result is known).

It is especially important to know the asymptotic behavior of the variance of the estimate of the mean, as well as the rate of decrease of this variance for a random process with strong dependence and non-regular observations.

Consider the following model with non-regular observations. Let $x(t), t \in Z$ be a Gaussian stationary process with unknown mean $Ex(t) = m_x$, known variance $E(x - m_x)^2 = \sigma_x^2 < \infty$ and correlation function

$$R_x(t) = L(t)|t|^{-\alpha}, \quad 0 < \alpha < 1, \tag{3.25}$$

where $L(t) = L'(|t|)$, $t > 0$ is non-negative slowly varying at infinity function bounded in each bounded interval. The correlation function $R_x(t)$, $t \in Z$, satisfies the condition $\sum_{k=0}^{\infty} R_x(k) = \infty$, i.e. $x(t)$, $t \in Z$, is the random process with strong dependence.

Without loss of generality we assume that $\sigma_x^2 = 1$. We observe the random process

$$y(t) = x(t)d(t) \tag{3.26}$$

at the moments $t \in \{0, 1, \ldots, T\}$, where $d(t)$ is a Bernulli sequence with $P\{d(t) = 1\} = p > 0$, and $P\{d(t) = 0\} = q > 0$; $p + q = 1$. Assume that the values of $d(t)$ are mutually independent for $t \in \{0, 1, \ldots, T\}$, and independent of $x(t)$.

Denote by m_y, $R_y(t)$, $f_y(t)$, respectively, the expectation, the correlation function and the spectral density of the observed process $y(t)$, $t \in Z$.

Consider the estimate for m_x:

$$\hat{m}_x = (Tp)^{-1} \sum_{t=0}^{T-1} y(t). \tag{3.27}$$

The result below gives the asymptotic behavior of the variance $\text{Var}(\hat{m}_x)$ as $T \to \infty$.

Theorem 3.26. *Under the conditions of the model (3.25)–(3.27), we have*

$$\lim_{T\to\infty} R_x^{-1}(T)\text{Var}(\hat{m}_x) = 2(1-\alpha)^{-1}(2-\alpha)^{-1}.$$

3.4.3 Asymptotic Distribution of the Least Squares Estimates for Long Memory Stochastic Systems

In this subsection we present some results obtained in Ivanov and Leonenko (2001, 2002). We use the notation introduced in Sect. 3.2.

Assume that the following conditions are fulfilled.

F_1. $\liminf_{T\to\infty} d_{1T}^2(\theta)/T > 0$
F_2. $\sup_{|u|<\rho} d_{3T}(\theta + u)/d_{1T}^2(\theta)/T \le k_2 < \infty$ for some $\rho > 0$
F_3. $\sup_{|u|<\rho} d_{2T}^2(\theta + u)/d_{1T}^2(\theta) \le k_3 T^{-1}$ for some $\rho > 0$, where $k_3 < \infty$

Let us formulate the first reduction principle for least squares estimates.

Theorem 3.27. *Let conditions A, B, B′, D, E of Sect. 3.2, and F_1–F_3 hold true for $\alpha \in (0, 1/m)$, where m is the Hermitian rank of the function G. Then the limit distribution of random variables*

$$\left[B^{-m}(T) \int_0^1 g_1^2(Tu, \theta)\mathrm{d}u\right]^{1/2} (\hat{\theta}_T - \theta)$$

and

$$\frac{\int_0^T G(\xi(t))g_1(t,\theta)\mathrm{d}t}{\left[T^2B^m(T)\int_0^1 g_1^2(Tu,\theta)\mathrm{d}u\right]^{1/2}}$$

coincide as $T \to \infty$, *i.e., if the limit distribution of one of the above two families of random variables exists, then there exists the limit distribution of the other family, and they are equal.*

Note that all conditions of Theorems 3.8 and 3.27 hold true, for example, for the functions

$$g(t,\theta) = t \cdot \log(\theta + t), \quad \sqrt{t^2 + \theta t + 1}.$$

Similarly to the case treated in Theorem 3.27, consider random variables

$$\varsigma_T(\theta) = \frac{\int_0^T G(\xi(t))g_1(t,\theta)\,\mathrm{d}t}{TB^{m/2}(T)A_\theta(T)}, \tag{3.28}$$

where the function $A_\theta(T)$ is defined in (3.9).

We need some additional conditions on the derivative $g_1(t,\theta)$ of the regression function $g(t,\theta)$.

K. For $0 < ffm < 1$ the limit

$$l(\theta) = \lim_{T\to\infty} m! \int_0^1 \int_0^1 \frac{g_1(tT,\theta)g_1(sT,\theta)}{A_\theta^2(T)} \cdot \frac{\mathrm{d}t\,\mathrm{d}s}{|t-s|^{m\alpha}} \neq 0$$

exists and is finite for all $\theta \in \boldsymbol{\Theta}^c$.

K′. For $0 < \alpha m < 1$ there exists a function $\bar{g}_1(u,\theta)$, $u \in [0,1]$, $\theta \in \boldsymbol{\Theta}^c$, square integrable with respect to u and such that

$$\left|\frac{g_1(uT,\theta)}{A_\theta(T)} - \bar{g}_1(u,\theta)\right| \to 0$$

as $T \to \infty$ uniformly in $u \in [0,1]$ and $\theta \in \boldsymbol{\Theta}^c$.

Observe that condition K holds if K′ holds. In this case

$$l(\theta) = m! \int_0^1 \int_0^1 \bar{g}_1(u,\theta)\bar{g}_1(w,\theta) \cdot \frac{\mathrm{d}u\,\mathrm{d}w}{|u-w|^{m\alpha}}.$$

We formulate the second reduction principle.

Theorem 3.28. *Conditions A, B, and* B′ *hold for* $\alpha \in (0, 1/m)$, *where* m *is the Hermitian rank of the function* G.

If condition K holds true, then

1. $\mathrm{Var}[\int_0^T G(\xi(t))g_1(t,\theta)\,\mathrm{d}t] = (C_m^2/m!)T^2B^m(T)A_\theta^2(T)l(\theta)(1+o(1))$ *as* $T\to\infty$*, where* C_m *is defined in (3.28);*
2. *the limit distributions of random variables* $\varsigma_T(\theta)$ *defined in (3.28) and of random variables*

$$\varsigma_{m,T}(\theta) = \frac{C_m}{m!}\frac{\int_0^T H_m(\xi(t))g_1(t,\theta)\,\mathrm{d}t}{TB^{m/2}(T)A_\theta(T)} \tag{3.29}$$

coincide for all $\theta\in\mathbf{\Theta}^c$ *as* $T\to\infty$.

L. Let condition K′ holds and for $\alpha\in(0,1/m)$

$$\Im_m = \int_{R^m}\int_0^1\left|\int_0^1 \bar{g}_1(u,\theta)\mathrm{e}^{iu(\lambda_1+\cdots+\lambda_m)}\,\mathrm{d}u\right|^2\cdot\frac{\mathrm{d}\lambda_1\ldots\mathrm{d}\lambda_m}{|\lambda_1\ldots\lambda_m|^{1-\alpha}} < \infty$$

uniformly in $\theta\in\mathbf{\Theta}^c$.

Consider variables

$$\aleph_m(\theta) = \frac{C_m}{m!}[c(\alpha)]^{m/2}\int_{R^m}'\left[\int_0^1 \bar{g}_1(u,\theta)\mathrm{e}^{iu(\lambda_1+\cdots+\lambda_m)}\,\mathrm{d}u\right]\frac{W(\mathrm{d}\lambda_1)\ldots W(\mathrm{d}\lambda_m)}{|\lambda_1\ldots\lambda_m|^{(1-\alpha)/2}}, \tag{3.30}$$

where $\int_{R^m}'\ldots$ means the multiple stochastic integral with respect to the complex Gaussian white noise $W(\cdot)$ given in (3.7) (see, for example Major 1981) for the definition and properties of multiple stochastic integrals).

Theorem 3.29. *Let conditions A, B, B′, D, K, and L hold for* $\alpha\in(0,1/m)$*, where m is the Hermitian rank of the function G. Then random variables* $\varsigma_T(\theta)$ *and* $\varsigma_{m,T}(\theta)$ *defined in (3.28) and (3.29), respectively, converge in distribution for all* $\theta\in\mathbf{\Theta}$ *as* $T\to\infty$ *to the random variable* $\aleph_m(\theta)$ *defined in (3.30).*

From Theorems 3.27 and 3.29 we obtain the asymptotic distribution of least squares estimates for a nonlinear regression with long-range dependence.

Theorem 3.30. *If conditions of Theorems 3.27 and 3.29 are satisfied, then the random variables*

$$B^{-m/2}(T)\left[\int_0^1 g_1^2(Ts,\theta)\,\mathrm{d}s\right]^{1/2}(\hat{\theta}_T-\theta)$$

converge in distribution for all $\theta\in\mathbf{\Theta}^c$ *as* $T\to\infty$ *to the random variable* $\aleph_m(\theta)$ *defined in (3.30).*

3.5 Large Deviations of Empirical Means in Estimation and Optimization Problems

3.5.1 *Large Deviations of the Empirical Means Method for Dependent Observations*

This subsection is devoted to the stochastic optimization problem for a stationary ergodic random sequence satisfying the hypermixing condition. Assume that we have finite number of observed elements in the sequence, and instead of solving the former problem we investigate the empirical function, find its minimum points, and study their asymptotic properties. More precisely, we consider the probabilities of large deviations of minimizers and the minimal value of the empirical cost function. The conditions under which the probabilities of large deviations decrease exponentially are found.

Consider the stochastic optimization problem: minimize

$$Ef(x) = Ef(x, \xi_0), \quad x \in X, \tag{3.31}$$

where $\{\xi_i, i \in Z\}$ is a stationary in the strict sense ergodic random sequence defined on a probability space $(\boldsymbol{\Omega}, F, P)$ with values in some measurable space $(Y, \Im)$, X is a compact subset of some Banach space $\Re$ with norm $||\cdot||$, $f : X \times Y \to \Re$ is some known function continuous in the first argument and measurable in the second.

Instead of (3.31) we minimize the empirical function:

$$H_n(x) = \frac{1}{n} \sum_{k=1}^{n} f(x, \xi_k), \quad x \in X. \tag{3.32}$$

If

$$E\{\max(|f(x, \xi_0)|, x \in X)\} < \infty$$

then there exists a solution x^* to the problem (3.31). Suppose that this solution is unique.

It is known that there exists a minimum point $x_n(\omega)$ of the function (3.32). The conditions under which $x_n(\omega)$ converges to x^* with probability 1 as $n \to \infty$, were discussed in Sects. 3.1–3.3.

The purpose of the subsection is to estimate large deviations of x_n and $H_n(x_n)$.

Let us recollect some facts from functional analysis. For any $y \in Y$ the function $f(\cdot, y)$ belongs to the space $C(X)$ of continuous real functions on X. We assume that for all $y \in Y$ we have $f(\cdot, y) - Ef(\cdot) \in K$, where K is some convex compact subset from $C(X)$. Thus for any n $H_n(\cdot) - Ef(\cdot)$ is a random element defined on the probability space $(\boldsymbol{\Omega}, F, P)$ with values in K.

Definition 3.2 (Kaniovskii et al. 1995). Let $(V, ||\cdot||)$ be a normed linear space, $B(x, \rho)$ be a closed ball in V with radius ρ and centre x, $f : V \to [-\infty, +\infty]$ be some function, and $f(x_f) = \min\{f(x), x \in V\}$. A condition function ψ for f at x_f is a monotone increasing function $\psi : [0, +\infty) \to [0, +\infty]$ with $\psi(0) = 0$ such that for some $\rho > 0$ and for all $x \in B(x_f, \rho)$ we have

$$f(x) \geq f(x_f) + \psi(||x - x_f||).$$

Assume that $V_0 \subset V$, and denote by δ_{V_0} the indicator function of V_0:

$$\delta_{V_0}(x) = 0, \quad x \in V_0,$$
$$\delta_{V_0}(x) = +\infty, \quad x \notin V_0.$$

Theorem 3.31 (Kaniovskii et al. 1995). *Let $(V, ||\cdot||)$ be a normalized linear space, $V_0 \subset V$ is closed, and $f_0, g_0 : V \to \Re$ are continuous functions on V. Suppose that*

$$\varepsilon = \sup\{|f_0(x) - g_0(x)|, x \in V_0\}.$$

Define the functions $f, g : V \to (-\infty, +\infty]$ as $f = f_0 + \delta_{V_0}$, $g = g_0 + \delta_{V_0}$. Then

$$|\inf\{f(x), x \in V\} - \inf\{g(x), x \in V\}| \leq \varepsilon.$$

Let x_f be a minimum point of f:

$$f(x_f) = \inf\{f(x), x \in V\}.$$

Assume that ψ is the condition function for f at x_f with some coefficient $\rho > 0$. If ε is sufficiently small so that for all x we have $||x - x_f|| \leq \rho$ provided that $\psi(||x - x_f||) \leq 2\varepsilon$, then for any $x_g \in \arg\min\{g(x), x \in B(x_f, \rho)\}$ we have

$$\psi(||x_f - x_g||) \leq 2\varepsilon.$$

When ψ is convex and strictly increasing on $[0, \rho]$, the preceding inequality can also be expressed in the following way: if ε is small enough so that $\psi^{-1}(2\varepsilon) \leq \rho$, then for any $x_g \in \arg\min\{g(x), x \in B(x_f, \rho)\}$ we have

$$||x_f - x_g|| \leq \psi^{-1}(2\varepsilon).$$

Theorem 3.32 (Deutschel and Stroock 1989). *Let $\{\mu_\varepsilon : \varepsilon > 0\}$ be a family of probability measures on G , where G is a closed convex subset of a separable Banach space E. Assume that*

$$\Lambda(\lambda) \equiv \lim_{\varepsilon \to 0} \varepsilon \Lambda_{\mu_\varepsilon}(\lambda/\varepsilon)$$

exists for every $\lambda \in E^*$*, where* E^* *is the dual space for* E*, and for an arbitrary probability measure* μ *on* E*,*

$$\Lambda_\mu(\lambda) = \ln\left(\int_E \exp[\langle\lambda, x\rangle]\mu(\mathrm{d}x)\right),$$

where $\langle\lambda, x\rangle$ *is the corresponding duality relation. Denote*

$$\Lambda^*(q) = \sup\{\langle\lambda, q\rangle - \Lambda(\lambda), \lambda \in E^*\}, \quad q \in G.$$

Then the function Λ^* is nonnegative, lower semicontinuous and convex, and for any compact set $A \subset G$

$$\limsup_{\varepsilon\to 0} \varepsilon \ln(\mu_\varepsilon(A)) \leq -\inf\{\Lambda^*(q), q \in A\}$$

holds.

Definition 3.3 (Deutschel and Stroock 1989). Let Σ be a separable Banach space, $\{\xi_i, i \in Z\}$ be a stationary in the strict sense random sequence defined on a probability space $(\boldsymbol{\Omega}, F, P)$ with values in Σ . Let B_{mk} denote the σ -algebra over $\boldsymbol{\Omega}$ generated by random elements $\{\xi_i, m \leq i \leq k\}$. For given $l \in N$ the real random variables $\eta_1, \ldots, \eta_p, p \geq 2$ are called l -measurably separated if

$$-\infty \leq m_1 \leq k_1 < m_2 \leq k_2 < \cdots < m_p \leq k_p \leq +\infty, \quad m_j - k_{j-1} \geq l, \ j = 2, \ldots, p$$

and for each $j \in \{1, \ldots, p\}$ the random variable η_j is $B_{m_j k_j}$-measurable.

Definition 3.4 (Deutschel and Stroock 1989). A random sequence $\{\xi_i\}$ from Definition 3.3 is called a sequence with hypermixing if there exist a number $l_0 \in N \cup \{0\}$ and non-increasing functions $\alpha, \beta : \{l > l_0\} \to [1, +\infty)$ and $\gamma : \{l > l_0\} \to [0, 1]$ satisfying the conditions

$$\lim_{l\to\infty} \alpha(l) = 1, \quad \limsup_{l\to\infty} l(\beta(l) - 1) < \infty, \quad \lim_{l\to\infty} \gamma(l) = 0$$

and for which

$$||\eta_1 \ldots \eta_p||_{L^1(P)} \leq \prod_{j=1}^{p} ||\eta_j||_{L^{\alpha(l)}(P)} \tag{H-1}$$

whenever $p \geq 2, l > l_0, \eta_1, \ldots, \eta_p$ are l-measurably separated functions. Here

$$||\eta||_{L^r(P)} = \left(\int_{\boldsymbol{\Omega}} |\eta(\omega)|^r \mathrm{d}P\right)^{1/r}$$

and

$$\left| \int_{\Omega} \left(\xi(\omega) - \int_{\Omega} \xi(\omega)\, \mathrm{d}P \right) \eta(\omega)\, \mathrm{d}P \right| \leq \gamma(l) ||\xi||_{L^{\beta(l)}(P)} ||\eta||_{L^{\beta(l)}(P)} \tag{H-2}$$

for all $l > l_0, \xi, \eta \in L^1(P)$ l-measurably separated.

It is known that $C(X)^* = M(X)$ is the set of bounded signed measures on X (Danford and Schwartz 1957), and

$$\langle g, Q \rangle = \int_X g(x) Q(\mathrm{d}x)$$

for any $g \in C(X)$, $Q \in M(X)$.

Theorem 3.33. *Suppose that $\{\xi_i, i \in Z\}$ is a stationary in the strict sense ergodic random sequence satisfying the hypothesis (H-1) of hypermixing, defined on a probability space $(\boldsymbol{\Omega}, F, P)$ with values in a compact convex set $K \subset C(X)$.*

Then for any measure $Q \in M(X)$ there exists

$$\Lambda(Q) = \lim_{n\to\infty} \frac{1}{n} \ln \left(\int_{\Omega} \exp \left\{ \sum_{i=1}^{n} \int_X \xi_i(\omega)(x) Q(\mathrm{d}x) \right\} \mathrm{d}P \right)$$

and for any closed $A \subset K$

$$\limsup_{n\to\infty} \frac{1}{n} \ln \left(P \left\{ \frac{1}{n} \sum_{i=1}^{n} \xi_i \in A \right\} \right) \leq -\inf\{\Lambda^*(g), g \in A\},$$

where $\Lambda^*(g) = \sup\{\int_X g(x)Q(\mathrm{d}x) - \Lambda(Q), Q \in M(X)\}$ is the non-negative, lower semicontinuous convex function.

Proof. Consider any $Q \in M(X)$. Assume that l_0 is the number from the hypothesis (H-1). Fix $l > l_0$ and $m, n \in N$, where $l < m < n$. Then

$$n = N_n m + r_n, \quad N_n \in N,\ r_n \in N \cup \{0\},\ r_n < m.$$

We will use the following notation:

$$||g|| = \max\{|g(x)|, x \in X\}, \quad g \in C(X),$$

$$f_n = \ln \left(\int_{\Omega} \exp \left\{ \sum_{i=1}^{n} \int_X \xi_i(\omega)(x) Q(\mathrm{d}x) \right\} \mathrm{d}P \right), \quad c = \max\{||g||, g \in K\}, \tag{3.33}$$

$$v(Q, X) = \sup \left\{ \sum_{i=1}^{k} |Q(E_i)|, E_i \cap E_j = \emptyset, i \neq j, E_i \in B(X), k \in N \right\} < \infty,$$

$$Q \in M(X),$$

where the last formula is taken from Danford and Schwartz (1957).

For all ω we have

$$\sum_{i=1}^{n}\int_X \xi_i(\omega)(x)Q(\mathrm{d}x) = \sum_{j=0}^{N_n-1}\sum_{i=jm+1}^{(j+1)m-l}\int_X \xi_i(\omega)(x)Q(\mathrm{d}x) + \sum_{j=0}^{N_n-1}\sum_{i=(j+1)m-l+1}^{(j+1)m}\int_X \xi_i(\omega)(x)Q(\mathrm{d}x) + \sum_{i=N_n m+1}^{n}\int_X \xi_i(\omega)(x)Q(\mathrm{d}x). \tag{3.34}$$

Further, by (3.33) we have for each i,ω

$$\left|\int_X \xi_i(\omega)(x)Q(\mathrm{d}x)\right| \le cv(Q,X). \tag{3.35}$$

Due to (3.35) for any ω we have

$$\sum_{j=0}^{N_n-1}\sum_{i=(j+1)m-l+1}^{(j+1)m}\int_X \xi_i(\omega)(x)Q(\mathrm{d}x) \le cv(Q,X)lN_n, \tag{3.36}$$

$$\sum_{i=N_n m+1}^{n}\int_X \xi_i(\omega)(x)Q(\mathrm{d}x) \le cv(Q,X)r_n. \tag{3.37}$$

For any fixed ω denote

$$V_1 = \sum_{j=0}^{N_n-1}\sum_{i=jm+1}^{(j+1)m-l}\int_X \xi_i(\omega)(x)Q(\mathrm{d}x),$$

$$V_2 = \sum_{j=0}^{N_n-1}\sum_{i=(j+1)m-l+1}^{(j+1)m}\int_X \xi_i(\omega)(x)Q(\mathrm{d}x),$$

$$V_3 = \sum_{i=N_n m+1}^{n}\int_X \xi_i(\omega)(x)Q(\mathrm{d}x).$$

Inequalities (3.36), (3.37) imply that

$$\exp\{V_1+V_2+V_3\} \le \exp\{V_1\}\exp\{cv(Q,X)lN_n\}\exp\{cv(Q,X)r_n\}, \quad \omega\in\mathbf{\Omega}. \tag{3.38}$$

It follows from (3.38) that

$$\int_{\Omega} \exp\{V_1 + V_2 + V_3\}\, \mathrm{d}P \le \exp\{cv(Q,X)lN_n\} \exp\{cv(Q,X)r_n\} \int_{\Omega} \exp\{V_1\}\, \mathrm{d}P.$$

Due to the conditions for $\{\xi_i\}$ we obtain

$$\int_{\Omega} \prod_{j=0}^{N_n-1} \exp\left\{ \sum_{i=jm+1}^{(j+1)m-l} \int_X \xi_i(\omega)(x) Q(\mathrm{d}x) \right\} \mathrm{d}P$$

$$\le \prod_{j=0}^{N_n-1} \left(\int_{\Omega} \left(\exp\left\{ \sum_{i=jm+1}^{(j+1)m-l} \int_X \xi_i(\omega)(x) Q(\mathrm{d}x) \right\} \right)^{\alpha(l)} \mathrm{d}P \right)^{1/\alpha(l)}, \tag{3.39}$$

$$\int_{\Omega} \exp\left\{ \alpha(l) \sum_{i=jm+1}^{(j+1)m-l} \int_X \xi_i(\omega)(x) Q(\mathrm{d}x) \right\} \mathrm{d}P$$

$$= \int_{\Omega} \exp\left\{ \alpha(l) \sum_{i=1}^{m-l} \int_X \xi_i(\omega)(x) Q(\mathrm{d}x) \right\} \mathrm{d}P, \quad j = 1, \ldots, N_n - 1. \tag{3.40}$$

By (3.39) and (3.40) we have

$$\int_{\Omega} \exp\{V_1\}\, \mathrm{d}P \le \left(\int_{\Omega} \exp\left\{ \alpha(l) \sum_{i=1}^{m-l} \int_X \xi_i(\omega)(x) Q(\mathrm{d}x) \right\} \mathrm{d}P \right)^{N_n/\alpha(l)}.$$

By (3.34) we obtain

$$f_n = \ln\left(\int_{\Omega} \exp\{V_1 + V_2 + V_3\}\, \mathrm{d}P \right) \le cv(Q,X)lN_n + cv(Q,X)r_n$$

$$+ \ln\left[\left(\int_{\Omega} \exp\left\{ \alpha(l) \sum_{i=1}^{m-l} \int_X \xi_i(\omega)(x) Q(\mathrm{d}x) \right\} \mathrm{d}P \right)^{N_n/\alpha(l)} \right]$$

$$= cv(Q,X)lN_n + cv(Q,X)r_n + \frac{N_n}{\alpha(l)} \ln\left(\int_{\Omega} \exp\left\{ (\alpha(l) - 1) \right. \right.$$

$$\left. \left. \times \sum_{i=1}^{m-l} \int_X \xi_i(\omega)(x) Q(\mathrm{d}x) + \sum_{i=1}^{m-l} \int_X \xi_i(\omega)(x) Q(\mathrm{d}x) \right\} \mathrm{d}P \right)$$

$$\leq cv(Q,X)lN_n + cv(Q,X)r_n + \frac{N_n}{\alpha(l)}(\alpha(l)-1)(m-l)cv(Q,X)$$
$$+\frac{N_n}{\alpha(l)}\ln\left(\int_\Omega \exp\left\{\sum_{i=1}^{m-l}\int_X \xi_i(\omega)(x)Q(\mathrm{d}x)\right\}\,\mathrm{d}P\right)$$
$$\leq cv(Q,X)lN_n + cv(Q,X)r_n + (\alpha(l)-1)(m-l)cv(Q,X)N_n$$
$$+\frac{N_n}{\alpha(l)}\ln\left(\int_\Omega \exp\left\{\sum_{i=1}^{m}\int_X \xi_i(\omega)(x)Q(\mathrm{d}x)\right.\right.$$
$$\left.\left.-\sum_{i=m-l+1}^{m}\int_X \xi_i(\omega)(x)Q(\mathrm{d}x)\right\}\,\mathrm{d}P\right)$$
$$\leq cv(Q,X)lN_n + cv(Q,X)r_n + (\alpha(l)-1)cv(Q,X)mN_n$$
$$+\frac{N_n}{\alpha(l)}cv(Q,X)l + \frac{N_n}{\alpha(l)}\ln\left(\int_\Omega \exp\left\{\sum_{i=1}^{m}\int_X \xi_i(\omega)(x)Q(\mathrm{d}x)\right\}\,\mathrm{d}P\right)$$
$$\leq 2cv(Q,X)lN_n + cv(Q,X)r_n + (\alpha(l)-1)cv(Q,X)mN_n + \frac{N_n}{\alpha(l)}f_m. \tag{3.41}$$

Inequality (3.41) implies that

$$\frac{f_n}{n} \leq \frac{2N_n cv(Q,X)l}{N_n m} + \frac{cv(Q,X)r_n}{n} + (\alpha(l)-1)cv(Q,X) + \frac{N_n f_m}{\alpha(l)(N_n m + r_n)}$$
$$= \frac{2cv(Q,X)l}{m} + \frac{cv(Q,X)r_n}{n} + (\alpha(l)-1)cv(Q,X) + \frac{f_m}{\alpha(l)(m + r_n/N_n)}.$$

Therefore we have

$$\limsup_{n\to\infty}\frac{f_n}{n} \leq \frac{2cv(Q,X)l}{m} + (\alpha(l)-1)cv(Q,X) + \frac{1}{\alpha(l)}\frac{f_m}{m}.$$

Taking the lim inf as $m \to \infty$ in the right-hand side, we obtain

$$\limsup_{n\to\infty}\frac{f_n}{n} \leq (\alpha(l)-1)cv(Q,X) + \frac{1}{\alpha(l)}\liminf_{m\to\infty}\frac{f_m}{m}.$$

Passing to the limit as $l \to \infty$ we get

$$\limsup_{n\to\infty}\frac{f_n}{n} \leq \liminf_{m\to\infty}\frac{f_m}{m}.$$

Consequently, there exists

$$\lim_{n\to\infty}\frac{f_n}{n} = \Lambda(Q).$$

Now we can see that the statements of the theorem result from Theorem 3.32. Indeed, for

$$G = K, \quad E = C(X), \quad E^* = M(X), \quad \langle Q, g \rangle = \int_X g(x) Q(\mathrm{d}x), \quad \varepsilon = \frac{1}{n},$$

and $\mu_\varepsilon = \mu_{1/n}$, which is the probability measure on K corresponding to $(1/n) \sum_{i=1}^{n} \xi_i$, we have

$$\begin{aligned}
\lim_{\varepsilon \to 0} \varepsilon \Lambda_{\mu_\varepsilon}\left(\frac{Q}{\varepsilon}\right) &= \lim_{n \to \infty} \frac{1}{n} \ln \left(\int_K \exp\left\{ \int_X g(x) n Q(\mathrm{d}x) \right\} \mu_{1/n}(\mathrm{d}g) \right) \\
&= \lim_{n \to \infty} \frac{1}{n} \ln \left(\int_\Omega \exp\left\{ \int_X \frac{1}{n} \sum_{i=1}^{n} \xi_i(\omega)(x) n Q(\mathrm{d}x) \right\} \mathrm{d}P \right) \\
&= \lim_{n \to \infty} \frac{f_n}{n} = \Lambda(Q).
\end{aligned}$$

The proof is complete. □

Consider the problems (3.31) and (3.32). Suppose that a given sequence $\{\xi_i, i \in Z\}$ satisfies the hypothesis (H-1) of hypermixing. Then the sequence

$$\zeta_i = f(\cdot, \xi_i) - Ef(\cdot), \quad i \in Z$$

satisfies (H-1) too.

Denote

$$\begin{aligned}
A_\varepsilon &= \{z \in K : ||z|| \geq \varepsilon\}, \\
\Lambda(Q) &= \lim_{n \to \infty} \frac{1}{n} \ln \left(\int_\Omega \exp\left\{ \sum_{i=1}^{n} \int_X [f(x, \xi_i(\omega)) - Ef(x)] Q(\mathrm{d}x) \right\} \mathrm{d}P \right), \\
I(z) &= \Lambda^*(z) = \sup\left\{ \int_X z(x) Q(\mathrm{d}x) - \Lambda(Q), Q \in M(X) \right\}.
\end{aligned}$$

Theorem 3.34. *Under hypothesis (H-1) of hypermixing we have*

$$\begin{aligned}
&\limsup_{n \to \infty} \frac{1}{n} \ln P\{|\min\{Ef(x), x \in X\} - \min\{H_n(x), x \in X\}| \geq \varepsilon\} \\
&\leq -\inf\{I(z), z \in A_\varepsilon\}.
\end{aligned} \tag{3.42}$$

Assume that there exists a condition function ψ for Ef at x^* with some constant ρ. Let x_n be a minimum point of function (3.32) on the set $B(x^*, \rho)$. If ε is sufficiently small to satisfy $\psi(|x - x^*|) \le 2\varepsilon$ provided that $|x - x^*| \le \rho$, then

$$\limsup_{n\to\infty} \frac{1}{n} \ln P\{\psi(|x_n - x^*|) \ge 2\varepsilon\} \le -\inf\{I(z), z \in A_\varepsilon\}. \tag{3.43}$$

Moreover, if ψ is convex and strictly increasing on $[0, \rho]$ then

$$\limsup_{n\to\infty} \frac{1}{n} \ln P\{|x_n - x^*| \ge \psi^{-1}(2\varepsilon)\} \le -\inf\{I(z), z \in A_\varepsilon\}. \tag{3.44}$$

Proof. Theorem 3.31 implies that for each ω

$$|\min\{Ef(x), x \in X\} - \min\{H_n(x), x \in X\}| \le ||H_n - Ef||. \tag{3.45}$$

Then, for the sequence $\{\varsigma_i\}$ conditions of Theorem 3.33 are satisfied. Thus for any $\varepsilon > 0$

$$\limsup_{n\to\infty} \frac{1}{n} \ln P\{||H_n - Ef|| \ge \varepsilon\} \le -\inf\{I(z), z \in A_\varepsilon\}. \tag{3.46}$$

Inequality (3.42) followsby (3.45) and (3.46).

To proof the second part of the theorem we also use Theorem 3.31. Under conditions of the theorem we have for all ω

$$\psi(|x^* - x_n|) \le 2||H_n - Ef|| \tag{3.47}$$

or

$$|x_n - x^*| \le \psi^{-1}(2||H_n - Ef||). \tag{3.48}$$

Taking into account (3.46), inequalities (3.47) and (3.48) imply (3.43) and (3.44), respectively. Theorem is proved. □

3.5.2 *Large Deviations of Empiric Estimates for Non-Stationary Observations*

Let $\{\xi_i, i \in Z\}$ be a stationary in the strict sense ergodic random sequence on a probability space $(\boldsymbol{\Omega}, F, P)$ with values in some measurable space $(Y, \aleph)$, $X = [a; b] \subset \Re$; the function

$$\{h(i, x, y) : Z * X * Y \to \Re\}$$

is convex in the second argument and measurable in the third one.

Consider the minimization problem

$$\min_{x \in X} \left\{ f_n(x) = \frac{1}{n} \sum_{i=1}^{n} h(i, x, \xi_i) \right\}. \tag{3.49}$$

Suppose that

1. For all $i \in Z, x \in X, E|h(i, x, \xi_0)| < \infty$
2. For any $x \in X$ there exists $f(x) = \lim_{n\to\infty} Ef_n(x)$
3. There exist $\bar{x} \in X, c > 0$, such that

$$f(x) \geq f(\bar{x}) + c|x - \bar{x}| \quad \forall x \in X. \tag{3.50}$$

It follows from (3) that there exists a unique solution to the minimization problem $\min_{x \in X} f(x)$, and this solution is $\bar{x}$.

It is evident that for any n and ω the function $f_n(\cdot)$ is convex, as well as the function $Ef_n(\cdot)$ for each n.

For any function $g : \Re \to \Re$ denote

$$g'_+(x) = \lim_{\Delta \to +0} \frac{g(x + \Delta) - g(x)}{\Delta}, \tag{3.51}$$

$$g'_-(x) = \lim_{\Delta \to +0} \frac{g(x - \Delta) - g(x)}{\Delta} \tag{3.52}$$

if these limits exist.

Denote

$$g_n(x) = Ef_n(x), \quad x \in X.$$

Since the existence of limits (3.51) and (3.52) for some function follows from the convexity of this function, we obtain that the limits exist for:

1. $h(i, \cdot, y)$ for all i, y
2. $Eh(i, \cdot, \xi_0)$ for all i
3. $f_n(\cdot)$, for all n, ω
4. $g_n(\cdot)$, for all n

Lemma 3.4. *Let $u : X \times \mathbf{\Omega} \to \Re$ be a function, convex in the first argument and measurable in the second one. Suppose that for any $x \in X$ $E|u(x, \omega)| < \infty$. Denote $v(x) = Eu(x, \omega)$. Then*

$$v'_+(x) = E\{u'_+(x, \omega)\}, \quad v'_-(x) = E\{u'_-(x, \omega)\}.$$

Proof. We have

$$v'_+(x) = \lim_{\Delta \to +0} \frac{Eu(x + \Delta, \omega) - Eu(x, \omega)}{\Delta} = \lim_{\Delta \to +0} E \frac{u(x + \Delta, \omega) - u(x, \omega)}{\Delta}.$$

Due to convexity of u in x for all ω, there exist

$$u'_{+}(x,\omega) = \inf_{\Delta>0} \frac{u(x+\Delta,\omega) - u(x,\omega)}{\Delta}, \tag{3.53}$$

$$u'_{-}(x,\omega) = \inf_{\Delta>0} \frac{u(x-\Delta,\omega) - u(x,\omega)}{\Delta} \tag{3.54}$$

and the fractions on the right-hand side of (3.53) and (3.54) are monotone decreasing as $\Delta \to +0$. Then by monotone convergence theorem

$$E\frac{u(x+\Delta,\omega) - u(x,\omega)}{\Delta} \to E\{u'_{+}(x,\omega)\}, \quad \Delta \to +0.$$

Analogous statement holds for $v'_{-}(x)$. The proof is complete. □

Lemma 3.4 implies that for any $i \in Z, x \in X$,

$$(Eh)'_{+}(i,x,\xi_i) = E\{h'_{+}(i,x,\xi_i)\}, \quad (Eh)'_{-}(i,x,\xi_i) = E\{h'_{-}(i,x,\xi_i)\}$$

and for all $n \in N, x \in X$,

$$g'_{n+}(x) = E\{f'_{n+}(x)\}, \quad g'_{n-}(x) = E\{f'_{n-}(x)\}.$$

Lemma 3.5. *Suppose that Assumptions (1)–(3) are satisfied, and, in addition, assume that*

4. *Sequences* $h'_{+}(i,\bar{x},\xi_i) - E\{h'_{+}(i,\bar{x},\xi_i)\}, i \in Z$ *and* $h'_{-}(i,\bar{x},\xi_i) - E\{h'_{-}(i,\bar{x},\xi_i)\}$, $i \in Z$ *satisfy the strong mixing condition with* $\alpha(j) \le c_0/1 + j^{1+\varepsilon}, \varepsilon > 0$ *(cf. Condition 5 of Theorem 3.3).*
5. *There exists* $\delta > 2/\varepsilon$ *such that for all* i

$$E|h'_{+}(i,\bar{x},\xi_0)|^{2+\delta} < \infty, \quad E|h'_{-}(i,\bar{x},\xi_0)|^{2+\delta} < \infty.$$

6. *There exists* $c' > 0$ *such that*

$$E[h'_{+}(i,\bar{x},\xi_0)]^2 \le c', \quad E[h'_{-}(i,\bar{x},\xi_0)]^2 \le c', \quad i \in Z.$$

7. $g'_{n+}(\bar{x}) \to f'_{+}(\bar{x}), \quad g'_{n-}(\bar{x}) \to f'_{-}(\bar{x}), \quad n \to \infty.$

Then

$$P\{f'_{n+}(\bar{x}) \to f'_{+}(\bar{x}), n \to \infty\} = 1, \tag{3.55}$$

$$P\{f'_{n-}(\bar{x}) \to f'_{-}(\bar{x}), n \to \infty\} = 1. \tag{3.56}$$

Proof. Denote

$$\eta_n = f'_{n+}(\bar{x}) - E\{f'_{n+}(\bar{x})\}.$$

We have

$$E\{\eta_n^2\} = E\left[\frac{1}{n}\sum_{i=1}^{n} h'_+(i,\bar{x},\xi_i) - \frac{1}{n}\sum_{i=1}^{n} E\{h'_+(i,\bar{x},\xi_i)\}\right]^2$$

$$= E\left[\frac{1}{n^2}\sum_{i=1}^{n}\sum_{j=1}^{n}[h'_+(i,\bar{x},\xi_i) - E\{h'_+(i,\bar{x},\xi_i)\}][h'_+(j,\bar{x},\xi_j) - E\{h'_+(j,\bar{x},\xi_j)\}]\right]$$

$$= \frac{1}{n^2}\sum_{i=1}^{n}\sum_{j=1}^{n} E\zeta_i\zeta_j,$$

where $\zeta_i = h'_+(i,\bar{x},\xi_i) - E\{h'_+(i,\bar{x},\xi_i)\}, i \in Z$.

It follows from Knopov (1997b) that for all i, j

$$E\zeta_i\zeta_j \le \frac{c_1}{1 + |i-j|^{1+\varepsilon'}}, \quad \varepsilon' > 0.$$

Hence,

$$\frac{1}{n^2}\sum_{i=1}^{n}\sum_{j=1}^{n} E\zeta_i\zeta_j \le \frac{c_1}{n^2}\sum_{i=1}^{n}\sum_{j=1}^{n}\frac{1}{1+|i-j|^{1+\varepsilon'}} \le \frac{c_2}{n}.$$

Let $n = m^2$. By Borel-Cantelli lemma

$$P\left\{\lim_{m\to\infty}\eta_{m^2} = 0\right\} = 1.$$

Denote

$$\phi_m = \sup_{m^2 \le n \le (m+1)^2} |\eta_n - \eta_{m^2}|.$$

For $m^2 \le n \le (m+1)^2$ we have

$$|\eta_n| \le |\eta_{m^2}| + \phi_m,$$

$$\eta_n - \eta_{m^2} = \frac{1}{n}\sum_{i=1}^{n}\zeta_i - \frac{1}{m^2}\sum_{i=1}^{m^2}\zeta_i = \frac{1}{n}\sum_{i=m^2+1}^{n}\zeta_i + \eta_{m^2}\left(\frac{m^2}{n} - 1\right).$$

Then

$$\phi_m \le \psi_m + \sup_{m^2 \le n \le (m+1)^2} \left| \eta_{m^2} \left(\frac{m^2}{n} - 1 \right) \right|,$$

where $\psi_m = \sup_{m^2 \le n \le (m+1)^2} |1/n \sum_{i=m^2+1}^{n} \zeta_i|$.

Consider

$$E(\psi_m^2) = E \sup_{m^2 \le n \le (m+1)^2} \frac{1}{n^2} \sum_{i=m^2+1}^{n} \sum_{j=m^2+1}^{n} \zeta_i \zeta_j$$

$$\le E \frac{1}{m^4} \sum_{i=m^2+1}^{(m+1)^2} \sum_{j=m^2+1}^{(m+1)^2} |\zeta_i \zeta_j| \le \frac{c_3}{m^4} [(m+1)^2 - m^2]^2 \le \frac{c_4}{m^2}.$$

Thus

$$P\left\{ \lim_{m \to \infty} \phi_m = 0 \right\} = 1.$$

Consequently,

$$P\left\{ \lim_{n \to \infty} \eta_n = 0 \right\} = 1.$$

Now (3.55) follows from Lemma 3.4. The proof of (3.56) is completely analogous. Lemma is proved. □

Theorem 3.35. *Let Assumptions (1)–(7) be satisfied. Then with probability 1 there exists $n^* = n^*(\omega)$ such that for any $n > n^*$ problem (3.49) has a unique solution x_n, and $x_n = \bar{x}$.*

Proof. In view of (3.50),

$$f'_+(\bar{x}) \ge c, \quad f'_-(\bar{x}) \ge c.$$

Then by Lemma 3.5 with probability 1 starting from some n^* we have

$$f'_{n+}(\bar{x}) > 0, \quad f'_{n-}(\bar{x}) > 0. \tag{3.57}$$

Since the function f_n is convex, it follows from (3.57) that $\bar{x}$ is the unique minimum point f_n.

Theorem is proved. □

Definition 3.5. We say that a random sequence $\{\zeta_i, i \in Z\}$ from Definition 3.3 satisfies hypothesis (H-3) if there exist a non-negative integer l_0 and a non-increasing function $\alpha : \{l > l_0\} \to [1; +\infty)$,$\lim_{l \to \infty} \alpha(l) = 1$, such that

$$||\eta_1 \ldots \eta_q||_{L^1(P)} \le \prod_{j=1}^{q} ||\eta_j||_{L^{\alpha(l)}(P)}$$

for any $q \geq 2, l > l_0, \eta_1, \ldots, \eta_q$ l -measurably separated, where

$$||\eta||_{L^r(P)} = \left(\int_{\Omega} |\eta(\omega)|^r \, dP \right)^{1/r}.$$

Theorem 3.36. *Suppose that Assumptions (1)–(7) are satisfied together with the assumptions below:*

8. *The sequence* $\{\xi_i, i \in Z\}$ *satisfies hypothesis (H-3)*
9. *There exists* $L > 0$ *such that for all* $i \in Z, y \in Y$

$$|h'_+(i, \bar{x}, y)| \leq L, \quad |h'_-(i, \bar{x}, y)| \leq L.$$

Then

$$\limsup_{n \to \infty} \frac{1}{n} \ln(P\{A_n^c\}) \leq - \inf_{g \in F} \Lambda^*(g), \tag{3.58}$$

where $\Lambda^*(g) = \sup\{gQ(X) - \Lambda(Q), Q \in M(X)\}$,

$$\Lambda(Q) = \lim_{n \to \infty} \frac{1}{n} \ln \left(\int_{\Omega} \exp \left\{ Q(X) \sum_{i=1}^{n} \min[h'_+(i, \bar{x}, \xi_i), h'_-(i, \bar{x}, \xi_i)] \right\} dP \right),$$

$$A_n = \left\{ \omega : \arg\min_{x \in X} f_n(x) = \{\bar{x}\} \right\}, \quad A_n^c = \mathbf{\Omega} \backslash A_n, \; F = [-L; 0].$$

Proof. We have

$$\begin{aligned} P(A_n^c) &= P\{\min[f'_{n+}(\bar{x}), f'_{n-}(\bar{x})] \in F\} \\ &\leq P\left\{ \frac{1}{n} \sum_{i=1}^{n} \min[h'_+(i, \bar{x}, \xi_i), h'_-(i, \bar{x}, \xi_i)] \in F \right\}. \end{aligned} \tag{3.59}$$

Denote

$$K = \{\alpha(x) = \alpha \; \forall x \in X, \alpha \in [-L; L]\}.$$

It is evident that K is a compact convex subset of $C(X)$.

Consider the function

$$a_i = a_i(x) = \min[h'_+(i, \bar{x}, \xi_i), h'_-(i, \bar{x}, \xi_i)] \quad \forall x \in X.$$

We see that $a_i(x) \in K$ for any fixed i, ω. Define

$$F_1 = \{(\alpha(x) = \alpha) \in K : \alpha \in [-L; 0]\}.$$

Then F_1 is a closed subset of K, and

$$P\left\{\frac{1}{n}\sum_{i=1}^{n}\min[h'_{+}(i,\bar{x},\xi_i),h'_{-}(i,\bar{x},\xi_i)]\in F\right\}=P\left\{\frac{1}{n}\sum_{i=1}^{n}a_i(x)\in F_1\right\}. \tag{3.60}$$

Now we apply Theorem 3.33. By this theorem,

$$\limsup_{n\to\infty}\frac{1}{n}\ln\left(P\left\{\frac{1}{n}\sum_{i=1}^{n}a_i(x)\in F_1\right\}\right)\leq-\inf_{g\in F_1}\Lambda^*(g), \tag{3.61}$$

where $\Lambda^*(g)=\sup\{gQ(X)-\Lambda(Q),\,Q\in M(X)\}$,

$$\Lambda(Q)=\lim_{n\to\infty}\frac{1}{n}\ln\left(\int_{\Omega}\exp\left\{Q(X)\sum_{i=1}^{n}a_i\right\}\,\mathrm{d}P\right).$$

Therefore (3.59)–(3.61) imply (3.58). The proof is complete. □

3.5.3 Large Deviations in Nonlinear Regression Problems

In this subsection, we focus on large deviations for concrete nonlinear regression models, in particular, for large deviations for the least squares estimates. We present some results obtained in Ivanov (1984a, b, 1997), which we formulate them without a proof.

Let $(\Re^N, B^N)$ be a countable product of the spaces $(\Re, B)$, B is the σ-algebra of Borel subsets of the real axis $\Re$, $\mathbf{\Theta}$ is an open set in the Euclidean space $\Re^p$, $\{P_{\boldsymbol{\theta}}, \boldsymbol{\theta}\in\mathbf{\Theta}\}$ is a family of probability measures on $(\Re^N, B^N)$ corresponding to sequences of random variables $x_j=g(j,\boldsymbol{\theta})+\varepsilon_j$, $j\geq 1$, $\boldsymbol{\theta}\in\mathbf{\Theta}$, where x_j is a sequence of independent observations, ε_j is a sequence of identically distributed random variables, and $g(j,\boldsymbol{\theta})$ is a sequence of functions of the parameter $\boldsymbol{\theta}=(\theta^{(1)},\ldots,\theta^{(p)})$, which is to be estimated. Let

$$L_n(\boldsymbol{\theta})=\sum_{j=1}^{n}[x_j-g(j,\boldsymbol{\theta})]^2.$$

We say that the B^n-measurable mapping $\boldsymbol{\theta}_n:\Re^N\to\mathbf{\Theta}^c$ (where $\mathbf{\Theta}^c$ is the closure of $\mathbf{\Theta}$) for which $L_n(\boldsymbol{\theta}_n)=\inf_{\boldsymbol{\theta}\in\mathbf{\Theta}^c}L_n(\boldsymbol{\theta})$ is a least squares estimate of the parameter $\boldsymbol{\theta}\in\mathbf{\Theta}$ obtained from observations x_j, $j=1,\ldots,n$. Obviously, $\boldsymbol{\theta}_n$ is a function only of x_j, $j=1,\ldots,n$.

Let $d_n = d_n(\boldsymbol{\theta}), \boldsymbol{\theta}_n \in \boldsymbol{\Theta}$, be a diagonal matrix of order p with elements $d_{in}, i = 1, \ldots, p$ on the diagonal. We normalize $\boldsymbol{\theta}_n$ using matrix d_n. If $g(j, \boldsymbol{\theta}),\ j \geq 1$ are differentiable functions, then it is natural to take $d_{in} = (\sum_{j=1}^{n} g_i^2(j, \boldsymbol{\theta}))^{1/2}$, $g_i = \partial g / \partial \boldsymbol{\theta}_i$. Introduce the functions

$$\varphi_n(\boldsymbol{\theta}_1, \boldsymbol{\theta}_2) = \sum_{j=1}^{n} (g(j, \boldsymbol{\theta}_1) - g(j, \boldsymbol{\theta}_2))^2, \quad \boldsymbol{\theta}_1, \boldsymbol{\theta}_2 \in \boldsymbol{\Theta}^c.$$

For fixed $\boldsymbol{\theta} \in \Theta$ the function $\boldsymbol{\Psi}_n(u_1, u_2) = \varphi_n(\boldsymbol{\theta} + n^{1/2} d_n^{-1} u_1, \boldsymbol{\theta} + n^{1/2} d_n^{-1} u_2)$ is defined for $u_1, u_2 \in U_n^c(\boldsymbol{\theta})$, $U_n(\boldsymbol{\theta}) = n^{-1/2} d_n(\boldsymbol{\theta})(\boldsymbol{\Theta} \backslash \boldsymbol{\theta})$. Set $s(\tau) = \{u \in \Re^p : ||u||_p < \tau\}$. Finally, let $K \subset \boldsymbol{\Theta}$ be a compact set, and let $\sigma^2 = E\varepsilon_1^2$.

A. For any $\varepsilon > 0$ and $R > 0$ there exist $\delta > 0$ and n_0 (which might depend on K) such that for $n > n_0$

$$\sup_{\boldsymbol{\theta} \in K} \sup_{\substack{u_1, u_2 \in s^c(R) \cap U_n^c(\boldsymbol{\theta}) \\ ||u_1 - u_2||_p \leq \delta}} n^{-1} \boldsymbol{\Psi}_n(u_1, u_2) \leq \varepsilon.$$

B. For some $R > 0$ and any $r \in (0, R]$ there exist $\Delta > 0$ and $\rho > 0$ such that for $n > n_0$

$$\inf_{\boldsymbol{\theta} \in K} \inf_{u \in (s^c(R) \backslash s(r)) \cap U_n^c(\boldsymbol{\theta})} n^{-1} \boldsymbol{\Psi}_n(0, u) \geq \rho,$$

$$\inf_{\boldsymbol{\theta} \in K} \inf_{u \in U_n^c(\boldsymbol{\theta}) \backslash s(R)} n^{-1} \boldsymbol{\Psi}_n(0, u) \geq 4\sigma^2 + \Delta. \tag{3.62}$$

The following condition refines (3.62) near zero.

C. For some $R_0 > 0$ there exists a number $\kappa_0 > 0$ such that for $n > n_0$

$$\inf_{\boldsymbol{\theta} \in K} \inf_{u \in s^c(R_0) \cap U_n^c(\boldsymbol{\theta})} n^{-1} ||u||_p^{-2} \boldsymbol{\Psi}_n(0, u) \geq \kappa_0.$$

D$_1$. The set $\boldsymbol{\Theta}$ is convex. Functions $g(j, \boldsymbol{\theta}),\ j \geq 1$ are continuous on $\boldsymbol{\Theta}^c$, continuously differentiable on $\boldsymbol{\Theta}$, and for any $R > 0$ there exist $\beta_i = \beta_i(R) < \infty$, such that for $n > n_0$ and $i = 1, \ldots, p$

$$\sup_{\boldsymbol{\theta} \in K} \sup_{u \in s^c(R_0) \cap U_n^c(\boldsymbol{\theta})} d_{in}(\boldsymbol{\theta} + n^{1/2} d_n^{-1}(\boldsymbol{\theta}) u) d_{in}^{-1}(\boldsymbol{\theta}) \leq \beta_j.$$

Let

$$\varphi_{in}(\boldsymbol{\theta}_1, \boldsymbol{\theta}_2) = \sum_{j=1}^{n} (g_i(j, \boldsymbol{\theta}_1) - g_i(j, \boldsymbol{\theta}_2))^2,$$

$$\boldsymbol{\Psi}_{in}(u_1, u_2) = \varphi_{in}(\boldsymbol{\theta} + n^{1/2} d_n^{-1} u_1, \boldsymbol{\theta} + n^{1/2} d_n^{-1} u_2), \quad i = 1, \ldots, p.$$

D_2. For any $R > 0$ there exist numbers $\gamma_i = \gamma_i(R) < \infty$ such that for $n > n_0$ and $i = 1, \ldots, p$

$$\sup_{\theta \in K} \sup_{u_1, u_2 \in s(R) \cap U_n^{(\theta)}} d_{in}^{-1}(\theta) \Psi_{in}^{1/2}(u_1, u_2) \|u_1 - u_2\|_p^{-1} \leq \gamma_i .$$

If $A \subset \Re^p$ and $\tau > 0$, then $A_\tau = \bigcup_{|\rho| \leq 1} (A + \tau\rho)$ is the exterior set parallel to A.

F_s. For some $\tau_0 > 0$ such that $K_{\tau_0} \subset \Theta$, some $\alpha_i \geq 1/2$ and an integer $s \geq 3$,

$$\overline{\lim_{n \to \infty}} \sup_{\theta \in K_{\tau_0}} n^{-s(\alpha_i - 1/2) - 1} \sum_{j=1}^{n} |g_i(j, \theta)|^s < \infty, \quad i = 1, \ldots, p,$$

$$\underline{\lim_{n \to \infty}} \inf_{\theta \in K_{\tau_0}} n^{-\alpha_i} d_{in}(\theta) > 0, \quad i = 1, \ldots, p.$$

M_s. The distribution of the random variable ε_1 does not depend on θ, $E\varepsilon_1 = 0$, and $E|\varepsilon_1|^s < \infty$ for some integer $s \geq 3$.

Under the assumptions formulated above the following theorems hold true.

Theorem 3.37. *Suppose condition* M_s *holds true, and conditions A and B are satisfied for a compact* $K \subset \Theta$. *Then for any* $r > 0$

$$\sup_{\theta \in K} P_\theta\{|n^{-1/2} d_n(\theta)(\theta_n - \theta)| \geq r\} = o(n^{-(s-2)/2}).$$

Theorem 3.38. *Suppose condition* M_s *holds true, and conditions B, C,* D_1, D_2 *and* F_s *are satisfied for a compact* $K \subset \Theta$.

If $s^2 > s + p$, then there exists a constant $\kappa > 0$ such that

$$\sup_{\theta \in K} P_\theta\{|d_n(\theta)(\theta_n - \theta)| \geq \kappa(\log n)^{1/2}\} = o(n^{-(s-2)/2}).$$

Chapter 4
Determination of Accuracy of Estimation of Regression Parameters Under Inequality Constraints

This chapter is devoted to the accuracy of estimation of regression parameters under inequality constraints. In Sects. 4.2 and 4.3 we construct the truncated estimate of the matrix of m.s.e. of the estimate of multi-dimensional regression parameter. In such a construction inactive constraints are not taken into account. Another approach (which takes into account all constraints) is considered in Sects. 4.4–4.7.

4.1 Preliminary Analysis of the Problem

Consider the regression with one-dimensional parameter α^0 and one-dimensional regressors x_t, i.e., without a free term

$$y_t = \alpha^0 x_t + \varepsilon_t, \quad t = \overline{1, T}. \tag{4.1}$$

We impose the simple constraint $\alpha^0 \leq b$ on the parameter, where the value b is known. Let us estimate α^0 by the least squares method taking into account the constraints above (ICLS), and solve the minimization problem, which is a particular case of (1.6):

$$\sum_{t=1}^{T} (y_t - \alpha x_t)^2 \to \min, \quad \alpha \leq b. \tag{4.2}$$

The solution to problem (4.2) (which is ICLS estimate of α^0) is

$$\alpha_T = \begin{cases} r_T^{-2}\left(\sum\limits_{t=1}^{T} x_t y_t\right) & \text{if } r_T^{-2}\left(\sum\limits_{t=1}^{T} x_t y_t\right) \leq b, \\ b & \text{if } r_T^{-2}\left(\sum\limits_{t=1}^{T} x_t y_t\right) > b, \end{cases}$$

P.S. Knopov and A.S. Korkhin, *Regression Analysis Under A Priori Parameter Restrictions*, Springer Optimization and Its Applications 54,
DOI 10.1007/978-1-4614-0574-0_4,

where $r_T^2 = \sum_{t=1}^{T} x_t^2$. Taking into account equality (4.1), we obtain:

$$\alpha_T - \alpha^0 = \begin{cases} z_T & \text{if } z_T \le b - \alpha^0, \\ b - \alpha^0 & \text{if } z_T > b - \alpha^0, \end{cases} \tag{4.3}$$

where $z_T = r_T^{-2} \sum_{t=1}^{T} x_t \varepsilon_t$.

Assume that the random variables $\varepsilon_t, t = \overline{1,T}$ in (4.1) are independent, identically distributed with zero expectation and variance σ^2, i.e., they satisfy Assumption 2.1 in Sect. 2.2 and, moreover, their distribution functions are differentiable. Then $E\{z_T\} = 0$, $E\{z_T^2\} = \sigma^2$ and the distribution function $\Phi_T(z), z \in \mathfrak{R}^1$, of z_T is differentiable. It is easy to see that $\alpha_T^* - \alpha^0 = z_T$, where α_T^* is the least squares estimate of α^0, i.e., α_T^* is the solution to (4.2) without taking into account the constraint.

By (4.3) we obtain the distribution function of $\alpha_T - \alpha^0$:

$$F_T(z) = \begin{cases} \Phi_T(z) & \text{if } z \le b - \alpha^0, \\ 1 & \text{if } z > b - \alpha^0, \end{cases} \tag{4.4}$$

According to (4.4) the function $F_T(z)$ has a discontinuity at the point $z = b - \alpha^0$, where it changes abruptly from $\Phi_T(b - \alpha^0)$ to 1.

Consider the properties of the regression parameter estimate taking into account inequality constraints. First, we define the shift. We have

$$\begin{aligned} E\{\alpha_T - \alpha^0\} &= \int_{z\in\mathfrak{R}^1} z\mathrm{d}F_T(z) = \int_{-\infty<z\le c} z\mathrm{d}F_T(z) + cp_c \\ &= \int_{-\infty}^{c} z\varphi_T(z)\mathrm{d}z + cp_c \ne \int_{-\infty}^{c} z\varphi_T(z)\mathrm{d}z + \int_{c}^{\infty} z\varphi_T(z)\mathrm{d}z \\ &= E\{\alpha_T^* - \alpha^0\} = 0, \end{aligned} \tag{4.5}$$

where $c = b - \alpha^0 \ge 0, \varphi_T(z) = \mathrm{d}F_T(z)/\mathrm{d}z$, $p_c = \int_c^\infty z\varphi_T(z)\mathrm{d}z$.

To obtain (4.5) we use the inequality

$$cp_c = c\int_c^\infty \varphi_T(z)\mathrm{d}z < \int_c^\infty z\varphi_T(z)\mathrm{d}z, \tag{4.6}$$

which follows from the fact that $\varphi_T(z) > 0$ and in (4.6) $c \le z$.

According to (4.5), $E\{\alpha_T - \alpha^0\} \ne 0$. Thus, ICLS estimate α_T is the biased estimate (unlike to the least squares estimate α_T^*, which is unbiased). Therefore in this case, we used the mean square error (m.s.e.) instead of the variance. We have

$$\begin{aligned} E\{(\alpha_T - \alpha^0)^2\} &= \int_{z\in\mathfrak{R}^1} z^2\mathrm{d}F_T(z) = \int_{-\infty<z\le c} z^2\mathrm{d}F_T(z) + c^2 p_c \\ &= \int_{-\infty}^{c} z^2\varphi_T(z)\mathrm{d}z + c^2 p_c < \int_{-\infty}^{c} z^2\varphi_T(z)\mathrm{d}z + \int_c^\infty z^2\varphi_T(z)\mathrm{d}z = E\{(\alpha_T^* - \alpha^0)^2\}, \end{aligned} \tag{4.7}$$

since $c^2 p_c = c^2 \int_c^\infty \varphi_T(z)\mathrm{d}z < \int_c^\infty z^2 \varphi_T(z)\mathrm{d}z$. As above, we use here the inequalities $\varphi(z) > 0$ and $c \leq z$. Thus, $E\{(\alpha_T - \alpha^0)^2\} < E\{(\alpha_T^* - \alpha^0)^2\}$, i.e. m.s.e. of the ICLS estimate for the considered regression is less than the variance of LS estimate. It should be noted that for unbiased estimates, in particular, for LS estimate, variance and m.s.e. coincide.

We end up the analysis of a simple regression with inequality constraint with the remark that if we eliminate the requirement of unbiasedness we can reduce the m.s.e. of this estimate.

Since the estimate of the parameter under inequality constraints is biased, we consider in this chapter the matrix of m.s.e. as the basic characteristic of the accuracy of estimation. Moreover, special attention is paid to the calculation of the bias of the estimate.

From this simple example we can see how adding of one constraint can make the determination of accuracy more complicated. Thus it is desirable to reduce the number of constraints when we calculate the accuracy using some sample data. Therefore, in this chapter we consider two approaches for estimation of the accuracy. The first approach consists in the construction of truncated estimates by discarding inactive constraints (for a given sample). In the second approach we consider all constraints.

4.2 Accuracy of Estimation of Nonlinear Regression Parameters: Truncated Estimates

We determine the accuracy of the estimate of the nonlinear regression parameter which is obtained as a solution to problem (2.12), (2.13). The asymptotic distribution of this estimate was considered in Sects. 2.2–2.4.

We shall judge on the accuracy of obtained regression parameters estimates for finite number of observations T by the matrix $\mathbf{K}_T$ which is the estimate of the matrix $\mathbf{K} = E\{\mathbf{UU}'\}$, and the vector $\boldsymbol{\Psi}_T$, which is the estimate of the vector $\boldsymbol{\Psi} = E\{\mathbf{U}\}$, where $\mathbf{U}$ is a random variable to which $\mathbf{U}_T = \sqrt{T}(\boldsymbol{\alpha}_T - \boldsymbol{\alpha}^0)$ converges. Thus, $\mathbf{K}_T$ and $\boldsymbol{\Psi}_T$ are, respectively, the approximations for the matrix of m.s.e. of the estimate of regression parameters and of its bias.

Consider the case where $\mathbf{K}_T$ is a truncated estimate, namely, only active constraints are taken into account.

In order to determine $\mathbf{K}_T$, we prove a theorem about the asymptotic behavior of the number of active and inactive constraints.

We introduce the concept of active constraints up to a positive number ξ. Such constraint (denote its number by i) satisfies the following condition:

$$-\xi \leq g_i(\boldsymbol{\alpha}_T) \leq 0. \tag{4.8}$$

Similarly, the ith inactive constraint up to ξ satisfies

$$g_i(\boldsymbol{\alpha}_T) < -\xi. \tag{4.9}$$

The number of all possible combinations of active and inactive constraints is equal to $L = 2^m$. Clearly, L does not depend on T and ξ. We denote by p_{lT}, $l = \overline{1, L}$, the probability that for an interval of length T the lth combination of active and inactive constraints corresponds up to ξ to the solution to problem (2.12) and (2.13). Obviously, $p_{1T} + \ldots + p_{LT} = 1$.

Each combination of constraints is defined by the set of active constraints I_l which is independent of T and ξ. There always exists a number $\xi > 0$ such that

$$g_i(\boldsymbol{\alpha}^0) = 0, \quad i \in I_1^0, g_i(\boldsymbol{\alpha}^0) < -\xi, \quad i \in I_2^0. \tag{4.10}$$

Condition (4.10) can be viewed as the generalization of (2.14).

Put the number of combinations of active and inactive constraints corresponding to $\boldsymbol{\alpha} = \boldsymbol{\alpha}^0$ equal to l_0. Then $I_{l_0} = I_1^0$. If ξ is chosen in such a way that (4.10) holds true, then the concept of "active constraints" coincides with the concept of "active constraints up to ξ" at $\boldsymbol{\alpha} = \boldsymbol{\alpha}^0$.

Theorem 4.1. *Suppose that: (a) active and inactive constraints up to ξ are defined by (4.8), (4.9), where ξ is such that (4.10) holds true; (b) components of $\mathbf{g}(\boldsymbol{\alpha})$ are the continuous functions of $\boldsymbol{\alpha}$; (c) $\boldsymbol{\alpha}_T$ is a consistent estimate of $\boldsymbol{\alpha}^0$. Then we have*

$$\lim_{T\to\infty} p_{l_0T} = 1, \lim_{T\to\infty} p_{lT} = 0, \quad l = \overline{1, L}, l \neq l_0. \tag{4.11}$$

Proof. By definition of p_{lT} and the fact that α_T belongs to the admissible domain, we have

$$p_{lT} = P\{-g_i(\boldsymbol{\alpha}_T) \leq \xi, i \in I_l, -g_i(\boldsymbol{\alpha}_T) > \xi, i \notin I_l\}. \tag{4.12}$$

We can always choose ξ such that

$$0 < \xi < \beta = -\frac{1}{2} \max_{i \in I_2^0} g_i(\boldsymbol{\alpha}^0).$$

From above and (4.12) we get

$$\begin{aligned}
p_{l_0T} &= P\{g_i(\boldsymbol{\alpha}_T) < -\xi, i \in I_2^0, 0 \geq g_i(\boldsymbol{\alpha}_T) \geq -\xi, i \in I_1^0\} \\
&\geq P\{|g_i(\boldsymbol{\alpha}_T) - g_i(\boldsymbol{\alpha}^0)| < \beta, i \in I_2^0, 0 \geq g_i(\boldsymbol{\alpha}_T) \geq -\xi, i \in I_1^0\} \\
&\geq P\left\{\sum_{i \in I_2^0} |g_i(\boldsymbol{\alpha}_T) - g_i(\boldsymbol{\alpha}^0)| < \frac{\beta}{2}\right\} \\
&\quad - P\left\{\sum_{i \in I_2^0} |g_i(\boldsymbol{\alpha}_T) - g_i(\boldsymbol{\alpha}^0)| < \frac{\beta}{2}; g_i(\boldsymbol{\alpha}_T) < -\xi, i \in I_1^0\right\} = M_1 - M_2.
\end{aligned} \tag{4.13}$$

By convergence in probability of $g(\boldsymbol{\alpha}_T)$ to $g(\boldsymbol{\alpha}^0)$ for any $\delta > 0$ there exists such a value T_1 that $M_1 > 1 - \delta/2, T > T_1$. Further, by (2.14) $g_i(\boldsymbol{\alpha}^0) = 0, i \in I_1^0$, implying that for any $\delta > 0$ there exists T_2 for which the following inequality takes place:

$$M_2 \leq P\{g_i(\boldsymbol{\alpha}_T) < -\xi, i \in I_1^0\} < \delta/2, \quad T > T_2.$$

From two last inequalities and (4.13) for $T > T_0 = \max(T_1, T_2)$ we derive $p_{1_0 T} > 1 - \delta, T > T_0$, which in turn implies the statement of the theorem.

Consider such an estimate for $\mathbf{K}$:

$$\mathbf{K}_T = \sum_{l=1}^{L} \mathbf{F}_l(\mathbf{R}_T(\boldsymbol{\alpha}_T), \mathbf{G}_l(\boldsymbol{\alpha}_T), \sigma_T)\gamma_{lT}, \tag{4.14}$$

where $\mathbf{F}_l(\mathbf{R}_T(\boldsymbol{\alpha}_T)\mathbf{G}_l(\boldsymbol{\alpha}_T), \sigma_T)$ is the estimate of $\mathbf{F}_l(\mathbf{R}(\boldsymbol{\alpha}^0)\mathbf{G}_l(\boldsymbol{\alpha}^0), \sigma) = E\{\mathbf{U}_l\mathbf{U}'_l\}$, $\mathbf{G}_l(\boldsymbol{\alpha})$ is the matrix with rows $\nabla g_i'(\boldsymbol{\alpha}), i \in I_l$; σ_T^2 is some consistent estimate of σ^2 (see below).

The vector $\mathbf{U}_l$ is the solution to the following problem:

$$\frac{1}{2}\mathbf{X}'\mathbf{R}(\boldsymbol{\alpha}^0)\mathbf{X} - \mathbf{Q}'\mathbf{X} \to \min, \quad \nabla g_i'(\boldsymbol{\alpha}^0)\mathbf{X} \leq 0, \quad i \in I_l, \tag{4.15}$$

where $\mathbf{Q}$ is the normally distributed random variable with the covariance matrix $\sigma^2\mathbf{R}(\boldsymbol{\alpha}^0)$.

In (4.14) γ_{lT} is the random variable defined as follows.

If for some realization of the estimate $\boldsymbol{\alpha}_T$ we get the lth combination of active and inactive constraints up to ξ in the estimation problem (2.12), then we put $\gamma_{lT} = 1$, otherwise $\gamma_{lT} = 0$. By definition, $\sum_{l=1}^{L} \gamma_{lT} = 1$. Thus, we have $P\{\gamma_{lT} = 1\} = p_{lT}$, $P\{\gamma_{lT} = 0\} = 1 - p_{lT}$, where p_{lT} is given by (4.12).

Introduce the random variable

$$\eta_{iT} = \begin{cases} 1, & \text{if } -\xi \leq g_i(\boldsymbol{\alpha}_T) \leq 0, \\ 0, & \text{if } g_i(\boldsymbol{\alpha}_T) < -\xi. \end{cases} \tag{4.16}$$

Define $\pi_{iT} = P\{\eta_{iT} = 1\}$. Then

$$\pi_{iT} = \sum_{l \in \Theta(i)} p_{lT}, \tag{4.17}$$

where $\Theta(i)$ is the set of numbers of combinations of active and inactive constraints up to ξ, including the ith constraint as active up to ξ. In other words, if the combination number $l \in \Theta(i)$, then the constraint index $i \in I_l$.

Lemma 4.1. *Suppose that conditions of Theorem 4.1 are satisfied, l_0 is the number of the combination of active and inactive constraints which corresponds to $\boldsymbol{\alpha} = \boldsymbol{\alpha}^0$. Then*

(a) $p\lim_{T\to\infty} \gamma_{l_0T} = 1$, $p\lim_{T\to\infty} \gamma_{lT} = 0, l \neq l_0$;
(b) $p\lim_{T\to\infty} \eta_{iT} = 1, i \in I_1^0$, $p\lim_{T\to\infty} \eta_{iT} = 0, i \in I_2^0$.

Proof. According to Theorem 4.1, for any $\varepsilon > 0$ and $\delta > 0$ there exists $T_0 > 0$ for which:

$$1 - P\{|\gamma_{l_0T} - 1| < \varepsilon\} \leq 1 - P\{\gamma_{l_0T} = 1\} = 1 - p_{l_0T} \leq \delta, \quad T > T_0,$$
$$1 - P\{|\gamma_{lT}| < \varepsilon\} \leq 1 - P\{\gamma_{l_iT} = 0\} = p_{lT} \leq \delta, \quad T > T_0,\ l \neq l_0.$$

Thus (a) is proved. To prove (b) we note that if $i \in I_1^0$ then $i \in I_{l_0}$. Then $l_0 \in \Theta(i)$ for all $i \in I_1^0$, which implies, together with (4.17) and Theorem 4.1 the proof of (b). Lemma is proved.

□

According to the definition of γ_{lT} and (4.15), the estimate (4.14) is defined by taking into account not all constraints, but only active ones up to ξ. Therefore, we call the estimate (4.14) the *truncated estimate* of the matrix of m.s.e. of the regression parameter estimate.

In the model we choose

$$\sigma_T^2 = \left(T - n + \sum_{i\in I} \eta_{iT}\right)^{-1} \sum_{t=1}^{T} (y_t - f_t(\boldsymbol{\alpha}_T))^2 \tag{4.18}$$

as the estimate for the variance σ^2 of the noise in the model.

Such a choice is based on the fact that for a sample of volume T the random variable $\sum_{i\in I} \eta_{iT}$ is the number of equality constraints imposed on the parameter, or the number of additional degrees of freedom. Therefore, the denominator in formula (4.18) represents the total number of degrees of freedom.

One can check that in the case when there are no constraints or only equality constraints many known estimates can be derived from the estimate of σ^2 suggested in (4.18).

Lemma 4.2. *If Assumptions 2.1–2.3 hold true then σ_T^2 is a consistent estimate of σ^2.*

Proof. Rewrite (4.18) as follows:

$$\sigma_T^2 = \frac{T}{T - n + \sum_{i\in I} \eta_{iT}} \cdot \frac{\sum_{t=1}^{T} (y - f_t(\boldsymbol{\alpha}_T))^2}{T}.$$

According to Lemma 4.1, the first factor in this expression converges in probability to 1. According to Assumption 2.3, $\boldsymbol{\alpha}_T$ is the consistent estimate, which implies by Assumptions 2.1, 2.2 and Demidenko (1981, p. 269), the convergence of the second factor to σ^2.

□

We assume that the noise in our model is normally distributed.

Assumption 4.1. Random variables ε_t are independent and normally distributed with zero expectation and variance σ^2, and in (2.38) $\boldsymbol{\varepsilon}_T \sim N(\mathbf{O}_n, \sigma^2 \mathbf{J}_n)$.

According to Theorem 2.7 we put in expression (4.15) $\mathbf{Q} = \sigma \mathbf{N}$, where $\mathbf{N}$ is the n-dimensional random variable, normally distributed with covariance matrix $\mathbf{R}(\boldsymbol{\alpha}^0)$, $E\{\mathbf{N}\} = \mathbf{O}_n$. Denote $\mathbf{U}_l = \mathbf{S}(\mathbf{N}, \mathbf{R}(\boldsymbol{\alpha}^0), \mathbf{G}_l(\boldsymbol{\alpha}^0), \sigma)$. Then, (i,j)th element of the matrix $\mathbf{F}_l(\mathbf{R}(\boldsymbol{\alpha}^0), \mathbf{G}_l(\boldsymbol{\alpha}^0), \sigma)$ for $i, j = \overline{1, n}$ is given by the expression:

$$
\begin{aligned}
& k_{ij}^{(l)}(\mathbf{R}(\boldsymbol{\alpha}^0), \mathbf{G}_l(\boldsymbol{\alpha}^0), \sigma) \\
& = \int_{\Re^n} S_i(\mathbf{x}, \mathbf{R}(\boldsymbol{\alpha}^0), \mathbf{G}_l(\boldsymbol{\alpha}^0), \sigma) S_j(\mathbf{x}, \mathbf{R}(\boldsymbol{\alpha}^0), \mathbf{G}_l(\boldsymbol{\alpha}^0), \sigma) f(\mathbf{X}, \mathbf{R}(\boldsymbol{\alpha}^0)) \mathrm{d}\mathbf{x},
\end{aligned}
\tag{4.19}
$$

where $S_i(\mathbf{x}, \mathbf{R}(\boldsymbol{\alpha}^0), \mathbf{G}_l(\boldsymbol{\alpha}^0), \sigma)$ is the ith component of the vector $\mathbf{S}(\mathbf{N}, \mathbf{R}(\boldsymbol{\alpha}^0), \mathbf{G}_l(\boldsymbol{\alpha}^0), \sigma)$, $f(\mathbf{x}, \mathbf{R}(\boldsymbol{\alpha}^0))$ is the density of the distribution $\mathbf{N}$,

$$
f(\mathbf{x}, \mathbf{R}(\boldsymbol{\alpha}^0)) = (2\pi)^{-(1/2)n} (\det \mathbf{R}(\boldsymbol{\alpha}^0))^{-1/2} \exp\left\{ \frac{1}{2} \mathbf{x}' \mathbf{R}^{-1}(\boldsymbol{\alpha}^0) \mathbf{x} \right\}. \tag{4.20}
$$

In order to prove the consistency of $\mathbf{K}_T$ (see the expression (4.14)) we need two following lemmas.

Lemma 4.3. *Let $\boldsymbol{\alpha} \in \Re^n$, $\mathbf{Q} \in \Re^n$, $\mathbf{b} \geq \mathbf{O}_m$, $\mathbf{R}$ is the positive definite $n \times n$ matrix. If in the quadratic programming problem*

$$
\frac{1}{2} \boldsymbol{\alpha}' \mathbf{R} \boldsymbol{\alpha} - \mathbf{Q}' \boldsymbol{\alpha} \to \min, \quad \mathbf{G} \boldsymbol{\alpha} \leq \mathbf{b}, \tag{4.21}
$$

the $m \times n$-matrix $\mathbf{G}$ is of full rank, then the following inequality takes place:

$$
||\hat{\boldsymbol{\alpha}}||^2 \leq \frac{\mu_{\max}(\mathbf{R})}{\mu_{\min}(\mathbf{R})} ||\boldsymbol{\alpha}^*||^2, \tag{4.22}
$$

where $\mu_{\max}(\mathbf{R})$ and $\mu_{\min}(\mathbf{R})$ are, respectively, the maximal and minimal eigenvalues of $\mathbf{R}$, $\hat{\boldsymbol{\alpha}}$ is the solution to (4.21), $\boldsymbol{\alpha}^*$ is the solution to (4.21) without taking into account the constraints.

Proof. In view of conditions of the lemma it is appropriate to switch from the problem (4.21) to the problem

$$
\frac{1}{2} \boldsymbol{\beta}' \boldsymbol{\beta} - \mathbf{P}' \boldsymbol{\beta} \to \min, \quad \mathbf{S} \boldsymbol{\beta} \leq \mathbf{b},
$$

where $\boldsymbol{\beta} = \mathbf{H} \boldsymbol{\alpha}$, $\mathbf{P} = (\mathbf{H}^{-1})' \mathbf{Q}$, $\mathbf{S} = \mathbf{G} \mathbf{H}^{-1}$, $\mathbf{H}$ is a non-degenerate matrix such that $\mathbf{H}' \mathbf{H} = \mathbf{R}$. Denote by $\hat{\boldsymbol{\beta}}$ its solution under constraints, and by $\boldsymbol{\beta}^*$ its solution without constraints. Let $\boldsymbol{\beta}^1$ be a point belonging to the admissible convex set $\mathrm{M} = \{\boldsymbol{\beta} : \mathbf{S} \boldsymbol{\beta} \leq \mathbf{b}, \boldsymbol{\beta} \in \Re^n\}$. Clearly, $\hat{\boldsymbol{\beta}}$ is the projection of $\boldsymbol{\beta}^*$ on M.

Similarly to the proof of Theorem 2.12 we use the properties of a convex set, namely, that the distance from any point a, not belonging to the set M, to the projection of this point to the boundary of M is less than the distance from a to an arbitrary point belonging to M. Therefore, we have $||\hat{\boldsymbol{\beta}} - \boldsymbol{\beta}^1|| \leq ||\boldsymbol{\beta}^* - \boldsymbol{\beta}^1||$.

It follows from the inequality above with $\boldsymbol{\beta}^1 = \mathbf{O}_n$, that $||\mathbf{H}\hat{\boldsymbol{\alpha}}||^2 \leq ||\mathbf{H}\boldsymbol{\alpha}^*||^2$, implying $\hat{\boldsymbol{\alpha}}'\mathbf{R}\hat{\boldsymbol{\alpha}} \leq (\boldsymbol{\alpha}^*)'\mathbf{R}\boldsymbol{\alpha}^*$. Then

$$\mu_{\min}(\mathbf{R})||\hat{\boldsymbol{\alpha}}||^2 \leq \hat{\boldsymbol{\alpha}}'\mathbf{R}\hat{\boldsymbol{\alpha}} \leq (\boldsymbol{\alpha}^*)'\mathbf{R}\boldsymbol{\alpha}^* \leq \mu_{\max}(\mathbf{R})||\boldsymbol{\alpha}^*||^2$$

and the statement of the lemma follows.

Lemma 4.4. *Let $\mathbf{S}(\mathbf{N}, \mathbf{R}, \mathbf{G}, \mathbf{b}, \sigma)$ be the vector-function of the variables $\sigma > 0$, $\mathbf{N} \in \Re^n$, $n \times n$ matrix $\mathbf{R}$, $m \times n$ matrix $\mathbf{G}$, $\mathbf{b} \in \Re^m$. Assume that $\mathbf{S}(\mathbf{N}, \mathbf{R}, \mathbf{G}, \mathbf{b}, \sigma)$ is the solution to problem (4.21), where $\mathbf{Q} = \sigma\mathbf{N}$, and $\mathbf{N}$ is a random variable, normally distributed with the covariance matrix $\mathbf{R}$, $E\{\mathbf{N}\} = \mathbf{O}_n$. Assume that for $\mathbf{R}$, $\mathbf{G}$ and $\mathbf{b}$ the conditions of Lemma 4.3 are fulfilled. Then for $i, j = \overline{1, n}$, the function*

$$k_{ij}(\mathbf{R}, \mathbf{G}, \mathbf{b}, \sigma) = \int_{\Re^n} s_i(\mathbf{x}, \mathbf{R}, \mathbf{G}, \mathbf{b}, \sigma) s_j(\mathbf{x}, \mathbf{R}, \mathbf{G}, \mathbf{b}, \sigma) f(\mathbf{x}, \mathbf{R}) d\mathbf{x} \tag{4.23}$$

is continuous in $\mathbf{R}, \mathbf{G}, \mathbf{b}$ and σ provided that $\mathbf{b} \geq \mathbf{O}_m$ and

$$|k_{ij}(\mathbf{R}, \mathbf{G}, \mathbf{b}, \sigma)| < \infty, \quad \forall \mathbf{R}, \forall \mathbf{G}, \forall \mathbf{b}, \forall \sigma. \tag{4.24}$$

In (4.23) $k_{ij}(\mathbf{R}, \mathbf{G}, \mathbf{b}, \sigma), i, j = \overline{1, n}$ is an element of the matrix $\mathbf{K} = E\{\mathbf{S}(\mathbf{N}, \mathbf{R}, \mathbf{G}, \mathbf{b}, \sigma)\, \mathbf{S}'(\mathbf{N}, \mathbf{R}, \mathbf{G}, \mathbf{b}, \sigma)\}$; $f(\mathbf{x}, \mathbf{R})$ is the density of distribution $\mathbf{N}$, see (4.20) with $\mathbf{R}(\boldsymbol{\alpha}^0) = \mathbf{R}$.

Proof. As results from (4.21), condition (4.24) is fulfilled. Indeed, in the absence of constraints in (4.21) the solution to this problem is a normally distributed centered n-dimensional random variable with finite second moments. Imposing of the constraint $\mathbf{G}\boldsymbol{\alpha} \leq \mathbf{b}$ does not influence the finiteness of the moments, since when we calculate the second moments of the estimate we integrate over the subset $\mathrm{M} \subset \Re^n$, $\mathrm{M} = \{\boldsymbol{\alpha} : \mathbf{G}\boldsymbol{\alpha} \leq \mathbf{b}, \boldsymbol{\alpha} \in \Re^n\}$. Moreover, the mass on the boundary of M is equal to $\int_{\Re^n \setminus \mathrm{M}} \mathbf{f}(\mathbf{x}, \mathbf{R}) d\mathbf{x}$.

1. First we prove the continuity of (4.23) on $\mathbf{R}$. Here by continuity of the function $k_{ij}(\mathbf{R}, \mathbf{G}, \mathbf{b}, \sigma)$ with respect to the quadratic matrix $\mathbf{R}$ of order n we understand the continuity of the function with respect to the vector of dimension n^2.

For any matrices $\mathbf{R}_1, \mathbf{R}_2$ we have

$$|k_{ij}(\mathbf{R}_1, \mathbf{G}, \mathbf{b}, \sigma) - k_{ij}(\mathbf{R}_2, \mathbf{G}, \mathbf{b}, \sigma)|$$
$$\leq \left| \int_{\Re^n} [\, s_i(\mathbf{x}, \mathbf{R}_1, \mathbf{G}, \mathbf{b}, \sigma) s_j(\mathbf{x}, \mathbf{R}_1, \mathbf{G}, \mathbf{b}, \sigma) \right.$$

$$- s_i(\mathbf{x}, \mathbf{R}_2, \mathbf{G}, \mathbf{b}, \sigma)s_j(\mathbf{x}, \mathbf{R}_2, \mathbf{G}, \mathbf{b}, \sigma)] f(\mathbf{x}, \mathbf{R}_2)\mathrm{d}\mathbf{x}\Big|$$

$$+ \left| \int_{\Re^n} s_i(\mathbf{x}, \mathbf{R}_1, \mathbf{G}, \mathbf{b}, \sigma)s_j(\mathbf{x}, \mathbf{R}_1, \mathbf{G}, \mathbf{b}, \sigma)[f(\mathbf{x}, \mathbf{R}_1) - (\mathbf{x}, \mathbf{R}_2)]\mathrm{d}\mathbf{x} \right|. \quad (4.25)$$

Let

$$||\mathbf{R}_1 - \mathbf{R}_2|| < \delta. \quad (4.26)$$

Put $\varphi(\mathbf{R}, \sigma, \boldsymbol{\alpha}) = (1/2)\boldsymbol{\alpha}'\mathbf{R}\boldsymbol{\alpha} - \mathbf{Q}'\boldsymbol{\alpha} = (1/2)\boldsymbol{\alpha}'\mathbf{R}\boldsymbol{\alpha} - \sigma\mathbf{N}'\boldsymbol{\alpha}$. We have from (4.21)

$$|\varphi(\mathbf{R}_1, \sigma, \boldsymbol{\alpha} - \varphi(\mathbf{R}_2, \sigma, \boldsymbol{\alpha}) \le \delta||\boldsymbol{\alpha}||^2. \quad (4.27)$$

Using the property of strongly convex functions (see Karmanov 1975, p. 36) we obtain:

$$||\mathbf{S}(\mathbf{N}, \mathbf{R}_2, \mathbf{G}, \mathbf{b}, \sigma) - \mathbf{S}(\mathbf{N}, \mathbf{R}_1, \mathbf{G}, \mathbf{b}, \sigma)||^2$$
$$\le \frac{2}{\mu}[\varphi(\mathbf{R}_1, \mathbf{S}(\mathbf{N}, \mathbf{R}_2, \mathbf{G}, \mathbf{b}, \sigma)) - \varphi(\mathbf{R}_1, \mathbf{S}(\mathbf{N}, \mathbf{R}_1, \mathbf{G}, \mathbf{b}, \sigma))],$$

where $\mu > 0$ is some constant.

Taking into account (4.27) and the fact that $\varphi(\mathbf{R}_2, \mathbf{S}(\mathbf{N}, \mathbf{R}_1, \mathbf{G}, \mathbf{b}, \sigma)) \ge \varphi(\mathbf{R}_2, \mathbf{S}(\mathbf{N}, \mathbf{R}_2, \mathbf{G}, \mathbf{b}, \sigma))$, we obtain

$$\varphi(\mathbf{R}_1, \mathbf{S}(\mathbf{N}, \mathbf{R}_2, \mathbf{G}, \mathbf{b}, \sigma)) - \varphi(\mathbf{R}_1, \mathbf{S}(\mathbf{N}, \mathbf{R}_1, \mathbf{G}, \mathbf{b}, \sigma))$$
$$\le |\varphi(\mathbf{R}_1, \mathbf{S}(\mathbf{N}, \mathbf{R}_2, \mathbf{G}, \mathbf{b}, \sigma)) - \varphi(\mathbf{R}_2, \mathbf{S}(\mathbf{N}, \mathbf{R}_2, \mathbf{G}, \mathbf{b}, \sigma))|$$
$$+ |\varphi(\mathbf{R}_2, \mathbf{S}(\mathbf{N}, \mathbf{R}_1, \mathbf{G}, \mathbf{b}, \sigma)) - \varphi(\mathbf{R}_1, \mathbf{S}(\mathbf{N}, \mathbf{R}_1, \mathbf{G}, \mathbf{b}, \sigma))|$$
$$\le \delta(||\mathbf{S}(\mathbf{N}, \mathbf{R}_2, \mathbf{G}, \mathbf{b}, \sigma)||^2 + ||\mathbf{S}(\mathbf{N}, \mathbf{R}_1, \mathbf{G}, \mathbf{b}, \sigma)||^2).$$

Using the last two equalities and (4.27) we obtain after transformation

$$||\mathbf{S}(\mathbf{N}, \mathbf{R}_2, \mathbf{G}, \mathbf{b}, \sigma) - \mathbf{S}(\mathbf{N}, \mathbf{R}_1, \mathbf{G}, \mathbf{b}, \sigma)||^2$$
$$\le \frac{2\delta}{\mu}(||\mathbf{S}(\mathbf{N}, \mathbf{R}_2, \mathbf{G}, \mathbf{b}, \sigma)||^2 + ||\mathbf{S}(\mathbf{N}, \mathbf{R}_1, \mathbf{G}, \mathbf{b}, \sigma)||^2). \quad (4.28)$$

The solution to problem (4.21) without taking into account the constraints is $\boldsymbol{\alpha}^* = \mathbf{R}^{-1}\boldsymbol{\sigma}\mathbf{N}$. Then its solution with constraints satisfies, according to Lemma 4.3, the inequality

$$||\mathbf{S}(\mathbf{N}, \mathbf{R}, \mathbf{G}, \mathbf{b}, \sigma)|| \le c||\mathbf{N}||, \quad (4.29)$$

where $c > 0$ is some value independent of $\mathbf{N}$. From (4.28) and (4.29) it follows that

$$||\mathbf{S}(\mathbf{N}, \mathbf{R}_2, \mathbf{G}, \mathbf{b}, \sigma) - \mathbf{S}(\mathbf{N}, \mathbf{R}_1, \mathbf{G}, \mathbf{b}, \sigma)||^2 \le (4/\mu)\delta c^2||\mathbf{N}||^2. \quad (4.30)$$

Denote the first term in the right-hand side of (4.25) by M_1, and the second one by M_2. From (4.29) and (4.30) after some transformation we get

$$M_1 \leq \int_{\Re^n} |s_i(\mathbf{x}, \mathbf{R}_1, \mathbf{G}, \mathbf{b}, \sigma) - s_i(\mathbf{x}, \mathbf{R}_2, \mathbf{G}, \mathbf{b}, \sigma)| \cdot |s_j(\mathbf{x}, \mathbf{R}_2, \mathbf{G}, \mathbf{b}, \sigma)|$$
$$\times f(\mathbf{x}, \mathbf{R}_2)\mathrm{d}\mathbf{x} + \int_{\Re^n} |s_j(\mathbf{x}, \mathbf{R}_1, \mathbf{G}, \mathbf{b}, \sigma) - s_j(\mathbf{x}, \mathbf{R}_2, \mathbf{G}, \mathbf{b}, \sigma)|$$
$$\times |s_j(\mathbf{x}, \mathbf{R}_2, \mathbf{G}, \mathbf{b}, \sigma)| f(\mathbf{x}, \mathbf{R}_2)\mathrm{d}\mathbf{x} \leq 4\sqrt{\frac{\delta}{\mu}c^2} \int_{\Re^n} ||\mathbf{x}||^2 f(\mathbf{x}, \mathbf{R}_2)\mathrm{d}\mathbf{x} \leq \varepsilon. \tag{4.31}$$

Let us estimate M_2. The functions $s_i(\mathbf{x}, \mathbf{R}, \mathbf{G}, \mathbf{b}, \sigma)$, $i = \overline{1, n}$, are continuous in $\mathbf{x}$, see Lemma 2.3. Consequently, for any sphere S_r centered at zero and with radius r

$$\int_{S_r} |s_i(\mathbf{x}, \mathbf{R}, \mathbf{G}, \mathbf{b}, \sigma) s_j(\mathbf{x}, \mathbf{R}, \mathbf{G}, \mathbf{b}, \sigma)|\mathrm{d}\mathbf{x} < \infty, \quad i, j = \overline{1, n}. \tag{4.32}$$

We have

$$M_2 \leq M_3 + M_4 = \int_{S_r} |s_i(\mathbf{x}, \mathbf{R}_1, \mathbf{G}, \mathbf{b}, \sigma) s_j(\mathbf{x}, \mathbf{R}_1, \mathbf{G}, \mathbf{b}, \sigma)|$$
$$\times |f(\mathbf{x}, \mathbf{R}_1) - f(\mathbf{x}, \mathbf{R}_2)|\mathrm{d}\mathbf{x} + \left| \int_{\Re^n \setminus S_r} s_i(\mathbf{x}, \mathbf{R}_1, \mathbf{G}, \mathbf{b}, \sigma) s_j(\mathbf{x}, \mathbf{R}_1, \mathbf{G}, \mathbf{b}, \sigma) \right.$$
$$\left. \times [f(\mathbf{x}, \mathbf{R}_1) - f(\mathbf{x}, \mathbf{R}_2)]\mathrm{d}\mathbf{x} \right|.$$

For a given number $\varepsilon > 0$ we select r so large that

$$M_4 \leq \left| \int_{\Re^n \setminus S_r} s_i(\mathbf{x}, \mathbf{R}_1, \mathbf{G}, \mathbf{b}, \sigma) s_j(\mathbf{x}, \mathbf{R}_1, \mathbf{G}, \mathbf{b}, \sigma) f(\mathbf{x}, \mathbf{R}_1)\mathrm{d}\mathbf{x} \right|$$
$$+ \left| \int_{\Re^n \setminus S_r} s_i(\mathbf{x}, \mathbf{R}_1, \mathbf{G}, \mathbf{b}, \sigma) s_j(\mathbf{x}, \mathbf{R}_1, \mathbf{G}, \mathbf{b}, \sigma) f(\mathbf{x}, \mathbf{R}_2)\mathrm{d}\mathbf{x} \right| \leq 2\varepsilon.$$

It is always possible because of (4.31) and the finiteness of $k_{ij}(\mathbf{R}, \mathbf{G}, \mathbf{b}, \sigma)$ for all $\mathbf{R}$. Thus $M_4 \leq 2\varepsilon$. Further, since the integral (4.32) is finite, we have

$$M_3 \leq \max_{\mathbf{x} \in S_r} |f(\mathbf{x}, \mathbf{R}_1) - f(\mathbf{x}, \mathbf{R}_2)| \int_{S_r} |s_i(\mathbf{x}, \mathbf{R}_1, \mathbf{G}, \mathbf{b}, \sigma) s_j(\mathbf{x}, \mathbf{R}_1, \mathbf{G}, \mathbf{b}, \sigma)|\mathrm{d}\mathbf{x}.$$

It is easy to see that for fixed $r = r(\varepsilon)$ it is possible to select $\mathbf{R}_2$ such that

$$\max_{\mathbf{x} \in S_r} |f(\mathbf{x}, \mathbf{R}_1) - f(\mathbf{x}, \mathbf{R}_2)| < \frac{\varepsilon}{\int_{S_r} |s_i(\mathbf{x}, \mathbf{R}_1, \mathbf{G}, \mathbf{b}, \sigma) s_j(\mathbf{x}, \mathbf{R}_1, \mathbf{G}, \mathbf{b}, \sigma)|\mathrm{d}\mathbf{x}}.$$

Summarizing all obtained estimates, we finally get for $||\mathbf{R}_1 - \mathbf{R}_2|| < \delta$

$$|k_{ij}(\mathbf{R}_1, \mathbf{G}, \mathbf{b}, \sigma) - k_{ij}(\mathbf{R}_2, \mathbf{G}, \mathbf{b}, \sigma) \leq 4\varepsilon.$$

Thus, the function $k_{ij}(\mathbf{R}, \mathbf{G}, \mathbf{b}, \sigma)$ is continuous in $\mathbf{R}$.

Let us prove the continuity of $k_{ij}(\mathbf{R}, \mathbf{G}, \mathbf{b}, \sigma)$ with respect to σ. Let $|\sigma_1 - \sigma_2| \leq \Delta\sigma$. After transformations we have from (4.23)

$$|k_{ij}(\mathbf{R}, \mathbf{G}, \mathbf{b}, \sigma_1) - k_{ij}(\mathbf{R}, \mathbf{G}, \mathbf{b}, \sigma_2)| \leq \int_{\mathfrak{R}^n} |s_i(\mathbf{x}, \mathbf{R}, \mathbf{G}, \mathbf{b}, \sigma_1) - s_i(\mathbf{x}, \mathbf{R}, \mathbf{G}, \mathbf{b}, \sigma_2)|$$
$$\times |s_j(\mathbf{x}, \mathbf{R}, \mathbf{G}, \mathbf{b}, \sigma_1)| f(\mathbf{x}, \mathbf{R}) d\mathbf{x} + \int_{\mathfrak{R}^n} |s_j(\mathbf{x}, \mathbf{R}, \mathbf{G}, \mathbf{b}, \sigma_1) - s_j(\mathbf{x}, \mathbf{R}, \mathbf{G}, \mathbf{b}, \sigma_2)|$$
$$\times |s_i(\mathbf{x}, \mathbf{R}, \mathbf{G}, \mathbf{b}, \sigma_2)| f(\mathbf{x}, \mathbf{R}) d\mathbf{x}. \tag{4.33}$$

It follows from (4.21) that

$$|\varphi(\mathbf{R}, \sigma_1, \boldsymbol{\alpha}) - \varphi(\mathbf{R}, \sigma_2, \boldsymbol{\alpha})| \leq \Delta\sigma ||\mathbf{N}|| ||\boldsymbol{\alpha}||. \tag{4.34}$$

By strong convexity, (4.29) and (4.34), we get

$$||\mathbf{S}(\mathbf{N}, \mathbf{R}, \mathbf{B}, \mathbf{b}, \sigma_2) - \mathbf{S}(\mathbf{N}, \mathbf{R}, \mathbf{B}, \mathbf{b}, \sigma_1)||^2 \leq \frac{2}{\mu} [\varphi(\mathbf{R}, \sigma_1, \mathbf{S}(\mathbf{N}, \mathbf{R}, \mathbf{B}, \mathbf{b}, \sigma_2))$$
$$-\varphi(\mathbf{R}, \sigma_1, \mathbf{S}(\mathbf{N}, \mathbf{R}, \mathbf{B}, \mathbf{b}, \sigma_1))]$$
$$\leq \frac{2\Delta\sigma}{\mu} ||\mathbf{N}|| \cdot [||\mathbf{S}(\mathbf{N}, \mathbf{R}, \mathbf{B}, \mathbf{b}, \sigma_2)|| + ||\mathbf{S}(\mathbf{N}, \mathbf{R}, \mathbf{B}, \mathbf{b}, \sigma_1)||] \leq \frac{4c\Delta}{\mu} ||\mathbf{N}||^2.$$

From above, using (4.29), (4.33) we obtain for $|\sigma_1 - \sigma_2| < \Delta\sigma$

$$|k_{ij}(\mathbf{R}, \mathbf{G}, \mathbf{b}, \sigma_1) - k_{ij}(\mathbf{R}, \mathbf{G}, \mathbf{b}, \sigma_2)| \leq 4c \sqrt{\frac{\Delta c}{\mu}} \int_{\mathfrak{R}^n} ||\mathbf{x}||^2 f(\mathbf{x}, \mathbf{R}) d\mathbf{x} \leq \varepsilon_1,$$

i.e. $k_{ij}(\mathbf{R}, \mathbf{G}, \mathbf{b}, \sigma)$ is continuous in σ.

Now we prove the continuity of $k_{ij}(\mathbf{R}, \mathbf{G}, \mathbf{b}, \sigma)$ with respect to $\mathbf{G}$. Let $\mathbf{G}_2 - \mathbf{G}_1 = \Delta\mathbf{G}$. From (4.23) we get the expression analogous to (4.33):

$$|k_{ij}(\mathbf{R}, \mathbf{G}_2, \mathbf{b}, \sigma) - k_{ij}(\mathbf{R}, \mathbf{G}_2, \mathbf{b}, \sigma)| \leq \int_{\mathfrak{R}^n} |s_i(\mathbf{x}, \mathbf{R}, \mathbf{G}_2, \mathbf{b}, \sigma) - s_i(\mathbf{x}, \mathbf{R}, \mathbf{G}_2, \mathbf{b}, \sigma)|$$
$$\times |s_j(\mathbf{x}, \mathbf{R}, \mathbf{G}_2, \mathbf{b}, \sigma)| f(\mathbf{x}, \mathbf{R}) d\mathbf{x} + \int_{\mathfrak{R}^n} |s_j(\mathbf{x}, \mathbf{R}, \mathbf{G}_2, \mathbf{b}, \sigma) - s_j(\mathbf{x}, \mathbf{R}, \mathbf{G}_1, \mathbf{b}, \sigma)|$$
$$\times |s_i(\mathbf{x}, \mathbf{R}, \mathbf{G}_1, \mathbf{b}, \sigma)| f(\mathbf{x}, \mathbf{R}) d\mathbf{x}. \tag{4.35}$$

By the necessary and sufficient conditions for the existence of an extremum (4.21) we obtain

$$\begin{aligned}&||\mathbf{S}(\mathbf{N},\mathbf{R},\mathbf{G}_1,\mathbf{b},\sigma)-\mathbf{S}(\mathbf{N},\mathbf{R},\mathbf{G}_2,\mathbf{b},\sigma)||^2\\&\le||\mathbf{R}^{-1}\mathbf{G}_2'||^2||\lambda(\mathbf{N},\mathbf{R},\mathbf{G}_2,\mathbf{b},\sigma)-\lambda(\mathbf{N},\mathbf{R},\mathbf{G}_1,\mathbf{b},\sigma)||^2\\&+||\mathbf{R}^{-1}||^2||\Delta\mathbf{G}'||^2||\lambda(\mathbf{N},\mathbf{R},\mathbf{G}_1,\mathbf{b},\sigma)||^2,\end{aligned} \tag{4.36}$$

where $\lambda(\mathbf{N},\mathbf{R},\mathbf{G}_2,\mathbf{b},\sigma)$ is the Lagrange multiplier for problem (4.21) with $\mathbf{G}=\mathbf{G}_i$, $i=1,2$, which is obtained in the solution to the dual problem

$$\Psi(\mathbf{R},\mathbf{G},\mathbf{b},\boldsymbol{\lambda})=\frac{1}{2}\boldsymbol{\lambda}'\mathbf{G}\mathbf{R}^{-1}\mathbf{G}'\boldsymbol{\lambda}+\boldsymbol{\lambda}'(\mathbf{b}-\mathbf{G}\mathbf{R}^{-1}\sigma\mathbf{N})\to\min,\quad \boldsymbol{\lambda}\ge\mathbf{O}_m. \tag{4.37}$$

Thus,

$$\begin{aligned}&\Psi(\mathbf{R},\mathbf{G}_1,\mathbf{b},\boldsymbol{\lambda})-\Psi(\mathbf{R},\mathbf{G}_2,\mathbf{b},\boldsymbol{\lambda})\\&\le\frac{1}{2}||\mathbf{G}_1\mathbf{R}^{-1}\mathbf{G}_1'-\mathbf{G}_2\mathbf{R}^{-1}\mathbf{G}_2'||\cdot||\boldsymbol{\lambda}||^2+||\sigma\mathbf{R}^{-1}||\cdot||\Delta\mathbf{G}||\cdot||\mathbf{N}||\cdot||\boldsymbol{\lambda}||.\end{aligned}$$

Taking into account this inequality and strong convexity, we get

$$\begin{aligned}&||\boldsymbol{\lambda}(\mathbf{N},\mathbf{R},\mathbf{G}_2,\mathbf{b},\sigma)-\boldsymbol{\lambda}(\mathbf{N},\mathbf{R},\mathbf{G}_1,\mathbf{b},\sigma)||^2\le\frac{2}{\mu_1}[||\mathbf{G}_1\mathbf{R}^{-1}\mathbf{G}'_1-\mathbf{G}_2\mathbf{R}^{-1}\mathbf{G}'_2||\\&\times(||\boldsymbol{\lambda}(\mathbf{N},\mathbf{R},\mathbf{G}_2,\mathbf{b},\sigma)||^2+||\boldsymbol{\lambda}(\mathbf{N},\mathbf{R},\mathbf{G}_1,\mathbf{b},\sigma)||^2)+||\sigma\mathbf{R}^{-1}||\cdot||\Delta\mathbf{G}||\cdot||\mathbf{N}||\\&\times(||\boldsymbol{\lambda}(\mathbf{N},\mathbf{R},\mathbf{G}_2,\mathbf{b},\sigma)||+||\boldsymbol{\lambda}(\mathbf{N},\mathbf{R},\mathbf{G}_1,\mathbf{b},\sigma).\end{aligned}$$

Using Lemma 4.3, we find similarly to (4.29) that

$$||\boldsymbol{\lambda}(\mathbf{N},\mathbf{R},\mathbf{G},\mathbf{b},\sigma)||\le c_2+c_1||\mathbf{N}||, \tag{4.38}$$

where $c_j>0$, $j=1,2$ are some values independent of $\mathbf{N}$.

From two last inequalities, (4.36), and the finiteness of norms of matrices $\mathbf{R}$, $\mathbf{G}_i$, $i=1,2$, we obtain

$$\begin{aligned}&||\mathbf{S}(\mathbf{N},\mathbf{R},\mathbf{G}_2,\mathbf{b},\sigma)-S(\mathbf{N},\mathbf{R},\mathbf{G}_1,\mathbf{b},\sigma)||^2\\&\le(c_1'||\Delta\mathbf{G}'||+c_2'||\Delta\mathbf{G}||+c_3'||\Delta\mathbf{G}'||||\Delta\mathbf{G}||+c_4'||\Delta\mathbf{G}'||^2)\left(\sum_{j=0}^{2}a_j||\mathbf{N}||^j\right),\end{aligned} \tag{4.39}$$

where $c_1'>0$, $i=\overline{1,4}$, are some constants, a_j, $j=0,1,2$, are the functions of constants c_1, μ_1 and $c_i''>0$, $i=\overline{1,4}$. Here $||\mathbf{G}_1\mathbf{R}^{-1}||\le c_1''$, $||\mathbf{R}^{-1}\mathbf{G}'_1||\le c_2''$, $||\mathbf{R}^{-1}||\le c_3''$, $||\sigma||\le c_4''$.

It follows from (4.29), (4.35) and (4.39) that

$$|k_{ij}(\mathbf{R},\mathbf{G}_2,\mathbf{b},\sigma)-k_{ij}(\mathbf{R},\mathbf{G}_1,\mathbf{b},\sigma|\leq 2(d_1||\Delta\mathbf{G}'||^2+d_2||\Delta\mathbf{G}|| \\ +d_3||\Delta\mathbf{G}'||\cdot||\Delta\mathbf{G}||+d_4||\Delta\mathbf{G}'||)^{1/2}\left(\sum_{j=0}^{2}a_j\int_{\mathfrak{R}^n}|\mathbf{x}|^j f(\mathbf{x},\mathbf{R})d\mathbf{x}\right),$$

where d_i, $i=\overline{1,4}$ are functions of constants introduced above: $c_i'>0$, $i=\overline{1,4}$; c_1, c_2, μ_1 and $c_i''>0$, $i=\overline{1,4}$.

Since $\int_{\mathfrak{R}^n}||\mathbf{x}||^j f(\mathbf{x},\mathbf{R})d\mathbf{x}=\text{const}$, $j=0, 1, 2$, we derive from the last inequality that $|k_{ij}(\mathbf{R},\mathbf{G}_2,\mathbf{b},\sigma)-k_{ij}(\mathbf{R},\mathbf{G}_1,\mathbf{b},\sigma)|\to 0$ as $||\Delta\mathbf{G}||=||\mathbf{G}_2-\mathbf{G}_1||\to 0$, which proves the continuity of $k_{ij}(\mathbf{R},\mathbf{G},\mathbf{b},\sigma)$ in $\mathbf{G}$.

Finally we prove the continuity of $k_{ij}(\mathbf{R},\mathbf{G},\mathbf{b},\sigma)$ with respect to $\mathbf{b}$. Let $||\mathbf{b}_1-\mathbf{b}_2||\leq\Delta\mathbf{b}$. Similar to (4.33), (4.35) we get, in this case,

$$|k_{ij}(\mathbf{R},\mathbf{G},\mathbf{b}_1,\sigma)-k_{ij}(\mathbf{R},\mathbf{G},\mathbf{b}_2,\sigma)|\leq\int_{\mathfrak{R}^n}|s_i(\mathbf{x},\mathbf{R},\mathbf{G},\mathbf{b}_1,\sigma)-s_i(\mathbf{x},\mathbf{R},\mathbf{G},\mathbf{b}_2,\sigma)| \\ \times|s_j(\mathbf{x},\mathbf{R},\mathbf{G},\mathbf{b}_1,\sigma)|f(\mathbf{x},\mathbf{R})d\mathbf{x}+\int_{\mathfrak{R}^n}|s_j(\mathbf{x},\mathbf{R},\mathbf{G},\mathbf{b}_1,\sigma)-s_j(\mathbf{x},\mathbf{R},\mathbf{G},\mathbf{b}_2,\sigma)| \\ \times|s_i(\mathbf{x},\mathbf{R},\mathbf{G},\mathbf{b}_2,\sigma)|f(\mathbf{x},\mathbf{R})d\mathbf{x}. \quad (4.40)$$

From the necessary and sufficient conditions for the existence of a minimum in (4.21) we get

$$||\mathbf{S}(\mathbf{N},\mathbf{R},\mathbf{G},\mathbf{b}_1,\sigma)-\mathbf{S}(\mathbf{N},\mathbf{R},\mathbf{G},\mathbf{b}_2,\sigma)|| \\ \leq||\mathbf{R}^{-1}\mathbf{G}'||\cdot||\boldsymbol{\lambda}(\mathbf{N},\mathbf{R},\mathbf{G},\mathbf{b}_1,\sigma)-\boldsymbol{\lambda}(\mathbf{N},\mathbf{R},\mathbf{G},\mathbf{b}_2,\sigma)||. \quad (4.41)$$

Let us estimate the square of the norm of the difference of Lagrange multipliers. Considering problem (4.37) which is dual to (4.21), we obtain

$$\Psi(\mathbf{R},\mathbf{G},\mathbf{b}_1,\boldsymbol{\lambda})-\Psi(\mathbf{R},\mathbf{G},\mathbf{b}_2,\boldsymbol{\lambda})\leq\boldsymbol{\lambda}'\Delta\mathbf{b}. \quad (4.42)$$

By conditions of the lemma, the function $\Psi(\mathbf{R},\mathbf{G},\mathbf{b},\boldsymbol{\lambda})$ is strongly convex. Hence, it satisfies (see Karmanov 1975, p. 54) the inequality

$$||\Delta\boldsymbol{\lambda}||^2=||\boldsymbol{\lambda}(\mathbf{N},\mathbf{R},\mathbf{G},\mathbf{b}_1,\sigma)-\boldsymbol{\lambda}(\mathbf{N},\mathbf{R},\mathbf{G},\mathbf{b}_2,\sigma)||^2 \\ \leq\frac{2}{\mu_2}(\Psi(\mathbf{R},\mathbf{G},\mathbf{b}_1,\boldsymbol{\lambda}(\mathbf{b}_2))-\Psi(\mathbf{R},\mathbf{G},\mathbf{b}_1,\boldsymbol{\lambda}(\mathbf{b}_1))), \quad (4.43)$$

where $\mu_2>0$ is some constant, $\boldsymbol{\lambda}(\mathbf{b}_i)=\boldsymbol{\lambda}(\mathbf{N},\mathbf{R},\mathbf{G},\mathbf{b}_i,\sigma)$, $i=1,2$; $\Delta\boldsymbol{\lambda}=\boldsymbol{\lambda}(\mathbf{b}_2)-\boldsymbol{\lambda}(\mathbf{b}_1)$.

Taking into account (4.42), we get

$$\begin{aligned}
&\Psi(\mathbf{R},\mathbf{G},\mathbf{b}_1,\boldsymbol{\lambda}(\mathbf{b}_2)) - \Psi(\mathbf{R},\mathbf{G},\mathbf{b}_1,\boldsymbol{\lambda}(\mathbf{b}_1))\\
&\le \Psi(\mathbf{R},\mathbf{G},\mathbf{b}_1,\boldsymbol{\lambda}(\mathbf{b}_2)) - \Psi(\mathbf{R},\mathbf{G},\mathbf{b}_2,\boldsymbol{\lambda}(\mathbf{b}_2))\\
&\quad + \Psi(\mathbf{R},\mathbf{G},\mathbf{b}_2,\boldsymbol{\lambda}(\mathbf{b}_1)) - \Psi(\mathbf{R},\mathbf{G},\mathbf{b}_1,\boldsymbol{\lambda}(\mathbf{b}_1))\\
&\le (\boldsymbol{\lambda}'(\mathbf{b}_1) - \boldsymbol{\lambda}'(\mathbf{b}_2))\Delta\mathbf{b} = -\Delta\boldsymbol{\lambda}'\Delta\mathbf{b} \ge 0,
\end{aligned} \tag{4.44}$$

where we used that $\Psi(\mathbf{R},\mathbf{G},\mathbf{b}_2,\boldsymbol{\lambda}(\mathbf{b}_2)) \le \Psi(\mathbf{R},\mathbf{G},\mathbf{b}_2,\boldsymbol{\lambda}(\mathbf{b}_1))$.

From (4.43) and (4.44) we obtain

$$||\boldsymbol{\lambda}(\mathbf{N},\mathbf{R},\mathbf{B},\mathbf{b}_1,\sigma) - \boldsymbol{\lambda}(\mathbf{N},\mathbf{R},\mathbf{B},\mathbf{b}_2,\sigma)|| \le \frac{2}{\mu_2}||\Delta\mathbf{b}||.$$

Then from (4.41)

$$||\mathbf{S}(\mathbf{N},\mathbf{R},\mathbf{G},\mathbf{b}_1,\sigma) - \mathbf{S}(\mathbf{N},\mathbf{R},\mathbf{G},\mathbf{b}_2,\sigma)|| \le \frac{2}{\mu_2}||\mathbf{R}^{-1}\mathbf{G}'|| \cdot ||\Delta\mathbf{b}||.$$

From this inequality and (4.29), (4.40) it follows that

$$\begin{aligned}
&|k_{ij}(\mathbf{R},\mathbf{G},\mathbf{b}_1,\sigma) - k_{ij}(\mathbf{R},\mathbf{G},\mathbf{b}_2,\sigma)|\\
&\le \left(\frac{4c}{\mu_2}||\mathbf{R}^{-1}\mathbf{G}'|| \cdot ||\Delta\mathbf{b}||\right)\int_{\Re^n} ||\mathbf{x}||f(\mathbf{x},\mathbf{R})d\mathbf{x}.
\end{aligned}$$

Therefore,

$$\begin{aligned}
\int_{\Re^n} ||\mathbf{x}||f(\mathbf{x},\mathbf{R})d\mathbf{x} &= \int_{||\mathbf{x}||\le 1} ||\mathbf{x}||f(\mathbf{x},\mathbf{R})d\mathbf{x} + \int_{||\mathbf{x}||>1} ||\mathbf{x}||f(\mathbf{x},\mathbf{R})d\mathbf{x}\\
&\le \int_{||\mathbf{x}||\le 1} ||\mathbf{x}||f(\mathbf{x},\mathbf{R})d\mathbf{x} + \int_{||\mathbf{x}||>1} ||\mathbf{x}||^2 f(\mathbf{x},\mathbf{R})d\mathbf{x}\\
&\le \int_{||\mathbf{x}||\le 1} ||\mathbf{x}||f(\mathbf{x},\mathbf{R})d\mathbf{x} + \int_{\Re^n} ||\mathbf{x}||^2 f(\mathbf{x},\mathbf{R})d\mathbf{x} = \text{const}.
\end{aligned}$$

From the last two inequalities we have $|k_{ij}(\mathbf{R},\mathbf{G},\mathbf{b}_2,\sigma) - k_{ij}(\mathbf{R},\mathbf{G},\mathbf{b}_1,\sigma)| \to 0$ as $||\mathbf{b}_1 - \mathbf{b}_2|| \le \Delta\mathbf{b} \to 0$, which proves continuity of $k_{ij}(\mathbf{R},\mathbf{G},\mathbf{b},\sigma)$ in $\mathbf{b}$.

Lemma is proved.

One can see that the solution $\mathbf{U}_l = \mathbf{S}(\mathbf{N},\mathbf{R}(\boldsymbol{\alpha}^0),\mathbf{G}_l(\boldsymbol{\alpha}^0),\sigma)$ to the problem (4.15) satisfies the conditions of Lemma 4.4. Therefore, this lemma can be applied to elements of the matrix $\mathbf{K}_l(\mathbf{R}(\boldsymbol{\alpha}^0),\mathbf{G}_l(\boldsymbol{\alpha}^0),\sigma)$, given by (4.19), since the elements of matrices $\mathbf{R}(\boldsymbol{\alpha}^0)$ and $\mathbf{G}_l(\boldsymbol{\alpha}^0)$ are continuous in $\boldsymbol{\alpha}^0$.

Replacing $\boldsymbol{\alpha}^0$ and σ, respectively, with $\boldsymbol{\alpha}_T$ and σ_T in the expression for the matrix $\mathbf{K}_l(\mathbf{R}(\boldsymbol{\alpha}^0), \mathbf{G}_l(\boldsymbol{\alpha}^0), \sigma)$, we obtain its estimate $\mathbf{K}_l(\mathbf{R}_T(\boldsymbol{\alpha}_T), \mathbf{G}_l(\boldsymbol{\alpha}_T), \sigma_T)$ where

$$k_{ij}^{(l)}(\mathbf{R}_T(\boldsymbol{\alpha}_T), \mathbf{G}_l(\boldsymbol{\alpha}_T), \sigma_T), \quad i, j = \overline{1, n}. \tag{4.45}$$

By consistency of $\boldsymbol{\alpha}_T$ and continuity of $\nabla g_i(\boldsymbol{\alpha})$ (see Assumption 2.2A) we get $p \lim_{T\to\infty} \mathbf{G}_l(\boldsymbol{\alpha}_T) = \mathbf{G}_l(\boldsymbol{\alpha}^0)$; and, according to (2.29), $p \lim_{T\to\infty} \mathbf{R}_T(\boldsymbol{\alpha}_T) = \mathbf{R}(\boldsymbol{\alpha}^0)$.

By convergence in probability of $I_l = \{i\}, l = i + 1$ (according to Lemma 4.2), of the matrices $\mathbf{G}_l(\boldsymbol{\alpha}_T)$, $\mathbf{R}(\boldsymbol{\alpha}_T)$, and Lemma 2.11 together with (4.19) (4.45), we obtain

$$p \lim_{T\to\infty} \mathbf{K}_l(\mathbf{R}_T(\boldsymbol{\alpha}_T), \mathbf{G}_l(\boldsymbol{\alpha}_T), \sigma_T) = \mathbf{K}_l(\mathbf{R}, \mathbf{G}_l, \sigma). \tag{4.46}$$

By Lemma 4.1,

$$p \lim_{T\to\infty} \gamma_{l_0 T} = 1, \quad p \lim_{T\to\infty} \gamma_{lT} = 0, \quad l \neq l_0. \tag{4.47}$$

Since $J_{l_0} = I_1^0$, (2.74) and (4.15) imply that $\mathbf{U}_{l_0} = \mathbf{U}$, which proves $\mathbf{K} = \mathbf{K}_{l_0}(\mathbf{R}, \mathbf{G}_{l_0}, \sigma)$. Then, from (4.14), (4.46) and (4.47) we deduce that

$$\begin{aligned} p \lim_{T\to\infty} \mathbf{K}_T &= p \lim_{T\to\infty} \mathbf{K}_{l_0}(\mathbf{R}_T(\boldsymbol{\alpha}_T), \mathbf{G}_l(\boldsymbol{\alpha}_T), \sigma_T) \\ &= \mathbf{K}_{l_0}(\mathbf{R}(\boldsymbol{\alpha}^0), \mathbf{G}_{l_0}(\boldsymbol{\alpha}^0), \sigma) = \mathbf{K}, \end{aligned} \tag{4.48}$$

i.e. the matrix $\hat{\mathbf{K}}$, defined by (4.14), is the consistent estimate of the matrix $\mathbf{K}$.

We proved

Theorem 4.2. *If the Assumptions 2.2–2.7 and 4.1 are satisfied, then the matrix* $\mathbf{K}_T$, *defined by (4.14), converges in probability to the matrix* $\mathbf{K} = E\{\mathbf{U}\mathbf{U}'\}$, *where* $\mathbf{U}$ *is the solution to (2.74).*

Denote the matrix of m.s.e. of the estimates of regression parameters by $\mathbf{K}_T^0 = E\{\mathbf{U}_T\mathbf{U}'_T\}$.

Corollary 4.1. *Let the conditions of Theorem 4.2 be satisfied.Then*

$$p \lim_{T\to\infty} ||\mathbf{K}_T^0 - \mathbf{K}_T|| = 0. \tag{4.49}$$

Proof. We have $||\mathbf{K}_T^0 - \mathbf{K}_T|| \leq ||\mathbf{K}_T^0 - \mathbf{K}|| + ||\mathbf{K} - \mathbf{K}_T||$. It follows by Theorem 2.7 (see Theorems (VI) and (VII) in Rao 1965, Section 2.4) that $\mathbf{K}_T^0 \to \mathbf{K} = E\{\mathbf{U}\mathbf{U}'\}$ as $T \to \infty$, i.e. for arbitrary $\varepsilon > 0$ there exists T_0 such that

$$||\mathbf{K}_T^0 - \mathbf{K}|| < \frac{\varepsilon}{2}, \quad T > T_0.$$

According to Theorem 4.2 for arbitrary values $\varepsilon > 0$ and $\delta > 0$

$$P\left\{||\mathbf{K}_T - \mathbf{K}|| < \frac{\varepsilon}{2}\right\} > 1 - \delta, \quad T > T_1.$$

From these three inequalities we obtain for $T > \max(T_1, T_0)$

$$P\{||\mathbf{K}_T^0 - \mathbf{K}_T|| < \varepsilon\} \geq P\{||\mathbf{K}_T^0 - \mathbf{K}|| + ||\mathbf{K}_T - \mathbf{K}|| < \varepsilon\}$$
$$\geq P\left\{||\mathbf{K}_T^0 - \mathbf{K}|| < \frac{\varepsilon}{2}\right\} - P\left\{||\mathbf{K}_T - \mathbf{K}|| \geq \frac{\varepsilon}{2}\right\} = P\left\{||\mathbf{K}_T - \mathbf{K}|| < \frac{\varepsilon}{2}\right\} > 1 - \delta,$$

This completes the proof of the corollary. ■

Corollary 4.1 explains the motivation why $\mathbf{K}_T$ is used as the approximate matrix of m.s.e. of the estimate of the parameter $E\{T(\boldsymbol{\alpha}_T - \boldsymbol{\alpha}^0)(\boldsymbol{\alpha}_T - \boldsymbol{\alpha}^0)'\}$.

Consider the bias of regression parameters estimates. For the lth combination of active constraints we have $E\{\mathbf{U}_l\} = \mathbf{F}_l(\mathbf{R}(\boldsymbol{\alpha}^0), \mathbf{G}_l(\boldsymbol{\alpha}^0), \sigma)$, where $\mathbf{U}_l$ is the solution to (4.15).

Put

$$\boldsymbol{\Psi}_T = \sum_{l=1}^{L} \mathbf{F}_l(\mathbf{R}_T(\boldsymbol{\alpha}_T), \mathbf{G}_l(\boldsymbol{\alpha}_T), \sigma_T)\gamma_{lT}. \tag{4.50}$$

We set $\gamma_{lT} = 1$, if we get the lth combination of active and inactive constraints in the solution to the estimation problem with T observation, and $\gamma_{lT} = 0$ otherwise. Then for the ith component $E\{\mathbf{U}_l\}$ we have

$$\varphi_{li}(\mathbf{R}(\boldsymbol{\alpha}^0), \mathbf{G}_l(\boldsymbol{\alpha}^0), \sigma) = \int_{\mathfrak{R}^n} s_{li}(\mathbf{x}, \mathbf{R}(\boldsymbol{\alpha}^0), \mathbf{G}_l(\boldsymbol{\alpha}^0), \sigma) f(\mathbf{x}, \mathbf{R}(\boldsymbol{\alpha}^0))d\mathbf{x}, \tag{4.51}$$

where $s_{li}(\mathbf{x}, \mathbf{R}(\boldsymbol{\alpha}^0), \mathbf{G}_l(\boldsymbol{\alpha}^0), \sigma)$ is the ith component of the solution to (4.15), $f(\mathbf{x}, \mathbf{R}(\boldsymbol{\alpha}^0))$ is the distribution density of $\mathbf{N}$, see (4.20).

Lemma 4.5. *Let* $\mathbf{S}(\mathbf{N}, \mathbf{R}, \mathbf{G}, \mathbf{b}, \sigma)$ *be the solution to (4.21). Then for* $i = \overline{1, n}$ *the function*

$$\varphi_i(\mathbf{R}, \mathbf{G}, \mathbf{b}, \sigma) = \int_{\mathfrak{R}^n} s_i(\mathbf{x}, \mathbf{R}, \mathbf{B}, \mathbf{b}, \sigma) f(\mathbf{x}, \mathbf{R})d\mathbf{x} \tag{4.52}$$

is continuous with respect to $\mathbf{R}$, $\mathbf{G}$, $\mathbf{b}$ and σ provided that $|\varphi_i(\mathbf{R}, \mathbf{G}, \mathbf{b}, \sigma)| < \infty$, $\forall \mathbf{R}, \forall \mathbf{G}, \forall \mathbf{b}, \forall \sigma$. In (4.52), $\varphi_i(\mathbf{R}, \mathbf{G}, \mathbf{b}, \sigma)$ is the ith component of $E\mathbf{S}\{(\mathbf{N}, \mathbf{R}, \mathbf{G}, \mathbf{b}, \sigma)\}$.

The proof is completely analogous to the proof of Lemma 4.4 and therefore is omitted.

Theorem 4.3. *If the Assumptions 2.2–2.7 and 4.1 are satisfied, then* $p\lim_{T\to\infty} \boldsymbol{\Psi}_T = E\{\mathbf{U}\}$, *where* $\mathbf{U}$ *is the solution to (2.74).*

Proof. We have $p\lim_{T\to\infty} \sigma_T^2 = \sigma^2$, $p\lim_{T\to\infty} \mathbf{R}_T(\boldsymbol{\alpha}_T) = \mathbf{R}(\boldsymbol{\alpha}^0)$, which implies together with Lemma 4.4

$$p\lim_{T\to\infty} \mathbf{F}_l(\mathbf{R}_T(\boldsymbol{\alpha}_T), \mathbf{G}_l(\boldsymbol{\alpha}_T), \sigma_T) = \mathbf{F}_l(\mathbf{R}(\boldsymbol{\alpha}^0), \mathbf{G}_l(\boldsymbol{\alpha}^0), \sigma).$$

Then, according to (4.47), we derive from (4.50)

$$p\lim_{T\to\infty} \boldsymbol{\Psi}_T = \mathbf{F}_{l_0}(\mathbf{R}(\boldsymbol{\alpha}^0), \mathbf{G}_{l_0}(\boldsymbol{\alpha}^0), \sigma) = E\{\mathbf{U}_{l_0}\}.$$

Since $\mathbf{U}_{l_0} = \mathbf{U}$, we have $E\{\mathbf{U}\} = E\{\mathbf{U}_{l_0}\}$. Theorem is proved. □

Corollary 4.2. *Let the conditions of Theorem 4.3 be satisfied. Then*

$$p\lim_{T\to\infty} ||\boldsymbol{\Psi}_T^0 - \boldsymbol{\Psi}_T|| = 0.$$

The proof is similar to the proof of Corollary 4.1.

4.3 Determination of the Truncated Sample Matrix of m.s.e. of the Estimate of Parameters in Nonlinear Regression

In this section we investigate truncated sample estimates of parameters of nonlinear regression.

Denote by $\hat{\mathbf{K}}_T$ the truncated sample estimate of the matrix of m.s.e. of the estimate of a multidimensional regression parameter. According to Sect. 4.2, to calculate $\hat{\mathbf{K}}_T$ it is sufficient to consider in (2.12) only active constraints up to $\xi > 0$. The number ξ measures the inaccuracy with which the left-hand side of the constraints is assumed to be equal to 0.

Assume that the sample consists of T observations of the pairs $(\mathbf{x}_t, y_t)$, and that Assumption 4.1 holds true. According to Theorems 2.7 and 2.8, the value of T should be sufficiently large. Denote by $\boldsymbol{\alpha}_T = \hat{\boldsymbol{\alpha}}_T$ and $\sigma_T = \hat{\sigma}_T$ the sample LS estimates of $\boldsymbol{\alpha}^0$ and σ, taking into account the constraints (2.12). From (4.14)

$$\hat{\mathbf{K}}_T = \mathbf{K}_l(\mathbf{R}_T(\hat{\boldsymbol{\alpha}}_T), \mathbf{G}_l(\hat{\boldsymbol{\alpha}}_T), \hat{\sigma}_T), \tag{4.53}$$

where $l = l(\hat{\boldsymbol{\alpha}}_T)$ is the number of the combination of active constraints up to ξ, corresponding to the sample estimate $\hat{\boldsymbol{\alpha}}_T$. For this value we have in (4.14) $\gamma_{lT} = 1$.

According to (4.18), $\hat{\sigma}_T$ in (4.53) is determined by

$$\hat{\sigma}_T^2 = \left(T - n + \sum_{i\in I} \hat{\eta}_{iT}\right)^{-1} \sum_{t=1}^{T} (y_t - f_t(\hat{\boldsymbol{\alpha}}_T))^2, \tag{4.54}$$

where

$$\hat{\eta}_{iT} = \begin{cases} 1, & \text{if } -\xi \leq \mathbf{g}_i(\hat{\boldsymbol{\alpha}}_T) \leq 0, \\ 0, & \text{if } \mathbf{g}_i(\hat{\boldsymbol{\alpha}}_T) < -\xi. \end{cases}$$

Let us replace in Assumption 4.1 the variance σ^2 of $\boldsymbol{\varepsilon}_T$ by its sample estimate (4.54), and in (2.38) the parameter $\boldsymbol{\alpha}^0$ by its sample estimate. We get

$$\hat{\mathbf{Q}}_T = \tilde{\mathbf{D}}'_T(\hat{\boldsymbol{\alpha}}_T)\boldsymbol{\varepsilon}_T \sim N(\mathbf{O}_n, \hat{\sigma}_T^2 \mathbf{R}_T(\hat{\boldsymbol{\alpha}}_T)). \tag{4.55}$$

According to Sect. 4.2,

$$\hat{\mathbf{K}}_T = E\{\hat{\mathbf{U}}_T \hat{\mathbf{U}}'_T | \hat{\boldsymbol{\alpha}}_T, \hat{\sigma}_T\}, \tag{4.56}$$

where $\hat{\mathbf{U}}_T$ is the solution to the problem

$$\frac{1}{2}\mathbf{X}'\mathbf{R}_T(\hat{\boldsymbol{\alpha}}_T)\mathbf{X} - \hat{\mathbf{Q}}'_T\mathbf{X} \to \min, \quad \nabla g'_i(\hat{\boldsymbol{\alpha}}_T)\mathbf{X} \leq 0, \quad i \in I(\hat{\boldsymbol{\alpha}}_T), \tag{4.57}$$

obtained from (4.15) by replacing $\boldsymbol{\alpha}^0$ with $\hat{\boldsymbol{\alpha}}_T$, $\mathbf{Q}$ with $\hat{\mathbf{Q}}_T$ (see (4.55)), and σ with $\hat{\sigma}_T$, when $I(\hat{\boldsymbol{\alpha}}_T) = \{i : -\xi \leq g_i(\hat{\boldsymbol{\alpha}}_T) \leq 0, \quad i \in I\}$.

In the general case, the matrix $\hat{\mathbf{K}}_T$ can be calculated by the Monte-Carlo method. For this it is necessary to generate iteratively the random vector $\hat{\mathbf{Q}}_T$, and then to solve the quadratic programming problem (4.57). If the number of constraints in (4.57) is less than or equal to three, it is possible to use the method for calculation $\hat{\mathbf{K}}_T$ considered in Sect. 4.6.

4.4 Accuracy of Parameter Estimation in Linear Regression with Constraints and without a Trend

Consider the problem

$$S_T(\boldsymbol{\alpha}) = \frac{1}{2}\sum_{t=1}^{T}(y_t - \mathbf{x}'_t\boldsymbol{\alpha})^2, \quad g_i(\boldsymbol{\alpha}) = \mathbf{g}'_i\boldsymbol{\alpha} - b_i \leq 0, \quad i = \overline{1,m} \tag{4.58}$$

of parameter estimation by the LS method in the linear regression with liner inequality constraints and without a trend. This problem is a special case of the more general situation, see (2.114), when the admissible domain is convex and the regressor has a trend.

Asymptotic properties of the estimates naturally follow from the results of Sect. 2.5.

Under the assumptions made in Sect. 2.5, $\mathbf{E}_T = \sqrt{T}\mathbf{J}_n$, $\bar{\mathbf{E}}_T = \sqrt{T}\mathbf{J}_m$, and the functions $g_i(\boldsymbol{\alpha})$, $i = \overline{1,m}$ are linear, see (4.58). Therefore, instead of Assumptions 2.9–2.12 we consider Assumption 2.9 and Assumption 4.2 given below, which is a special case of Assumption 2.10.

Assumption 4.2. The matrix $\mathbf{G}$ with lines $\mathbf{g}'_i$, $i = \overline{1,m}$ is of full rank.

Assume that the random variables ε_t are identically distributed, i.e., that Assumption 2.1 (which is a particular case of Assumption 2.8) is satisfied.

When Assumptions 2.1 and 4.1 about the normality of the noise hold true, the result of Sect. 4.3 on the truncated estimate of the matrix of m.s.e. of the estimate of the nonlinear regression parameter is transferred to case of linear regression, if we put $f_t(\alpha) = x_t'\alpha$. In particular, the truncated estimate can be defined (4.14) with $\mathbf{R}_T(\hat{\alpha}_T) = \mathbf{R}_T$, $\nabla g_i'(\hat{\alpha}_T) = \mathbf{g}'_i$.

The estimate of the matrix of m.s.e. considered below takes into account all constraints and does not require that the errors are normally distributed, i.e., it does not require Assumption 4.1 to be satisfied. Thus, such an estimate is the generalization of the truncated estimate.

The advantage of the truncated estimate is that for its construction only part of constraints is required. It considerably simplifies the calculations based on the Monte-Carlo method which demands considerable computing resources.

To show the consistency of the estimate below we prove several auxiliary statements.

4.4.1 *Auxiliary Results*

Let us estimate the dependence of the solution to quadratic programming problem (4.21) on the matrix $\mathbf{R}$ and the vector $\mathbf{b}$. Investigations of this dependence will naturally complement the parametric properties of the solution to the quadratic programming problem discussed in Lemma 2.5, and further on in Sect. 2.2.

Denote the solutions to (4.21) by

$$\alpha_1 = \alpha(\mathbf{R}_1, \mathbf{b}_1), \quad \alpha_2 = \alpha(\mathbf{R}_2, \mathbf{b}_2), \tag{4.59}$$

where $\mathbf{R}_1$ and $\mathbf{R}_2$ are the arbitrary positive definite matrices, $\mathbf{R}_2 = \mathbf{R}_1 + \Delta\mathbf{R}$, $\mathbf{b}_1$ and $\mathbf{b}_2$ are arbitrary vectors, $\mathbf{b}_2 = \mathbf{b}_1 + \Delta\mathbf{b}$.

From necessary and sufficient conditions for the existence of the minimum in (4.21) we have

$$\mathbf{R}\alpha - \mathbf{Q} + \mathbf{G}'\mathbf{\Lambda} = \mathbf{O}_n, \quad \mathbf{\Lambda} \geq \mathbf{O}_m, \tag{4.60}$$

where $\mathbf{\Lambda} \in \mathfrak{R}^m$ is the Lagrange multiplier for problem (4.21). From the first equality in (4.60) we obtain

$$\alpha_1 - \alpha_2 = -\mathbf{R}_1^{-1}(\mathbf{R}_1 - \mathbf{R}_2)\alpha_2 - \mathbf{R}_1^{-1}\mathbf{G}'(\mathbf{\Lambda}_1 - \mathbf{\Lambda}_2), \tag{4.61}$$

where $\mathbf{\Lambda}_1$ и $\mathbf{\Lambda}_2$ are Lagrange multipliers in problem (4.21), respectively, for pairs $\mathbf{R} = \mathbf{R}_1$, $\mathbf{b} = \mathbf{b}_1$, and $\mathbf{R} = \mathbf{R}_2$, $\mathbf{b} = \mathbf{b}_2$. The problems, dual to (4.21) with $\mathbf{R} = \mathbf{R}_i$, $\mathbf{b} = \mathbf{b}_i$, $i = 1, 2$, are:

$$\varphi_i = \frac{1}{2}\mathbf{\Lambda}'\mathbf{M}_i\mathbf{\Lambda} + \mathbf{V}'_i\mathbf{\Lambda} \to \min, \quad -\mathbf{\Lambda} \leq \mathbf{O}_m, \; i = 1, 2, \tag{4.62}$$

where

$$\mathbf{M}_i = \mathbf{G}\mathbf{R}_i^{-1}\mathbf{G}', \mathbf{V}_i = \mathbf{b}_i - \mathbf{G}\mathbf{R}_i^{-1}\mathbf{Q}. \tag{4.63}$$

According to Lemma 4.3 the solution $\mathbf{\Lambda}_i$ to (4.62) satisfies the condition

$$||\mathbf{\Lambda}_i||^2 \leq (\mu_{\max}(\mathbf{M}_i)/\mu_{\min}(\mathbf{M}_i)||\mathbf{\Lambda}_i^*||^2, \tag{4.64}$$

where $\mathbf{\Lambda}_i^*$ is the solution to problem (4.62) without constraints:

$$\mathbf{\Lambda}_i^* = -\mathbf{M}_i^{-1}\mathbf{V}_i. \tag{4.65}$$

We have by (4.62)

$$(\varphi_1 - \varphi_2) \leq \frac{1}{2}||\mathbf{M}_1 - \mathbf{M}_2|| \cdot ||\mathbf{\Lambda}||^2 + ||\mathbf{V}_1 - \mathbf{V}_2|| \cdot ||\mathbf{\Lambda}||. \tag{4.66}$$

Using strong convexity, (see Karmanov 1975, p. 36) for $\varphi_i, i = 1, 2$ we derive from (4.66)

$$||\mathbf{\Lambda}_1 - \mathbf{\Lambda}_2||^2 \leq \frac{2}{\nu}[||\mathbf{M}_1 - \mathbf{M}_2|| \cdot (||\mathbf{\Lambda}_2||^2 + ||\mathbf{\Lambda}_1||^2) + ||\mathbf{V}_1 - \mathbf{V}_2|| \cdot (||\mathbf{\Lambda}_2|| + ||\mathbf{\Lambda}_1||)], \tag{4.67}$$

where $\nu > 0$.

After transformation we obtain

$$||\mathbf{M}_1 - \mathbf{M}_2|| \geq ||\mathbf{R}_1^{-1}|| \cdot ||\mathbf{R}_2^{-1}|| \cdot ||\mathbf{G}\mathbf{G}'|| \cdot ||\Delta\mathbf{R}||,$$
$$||\mathbf{V}_1 - \mathbf{V}_2|| \leq ||\Delta\mathbf{b}|| + ||\mathbf{R}_1^{-1}|| \cdot ||\mathbf{R}_2^{-1}|| \cdot ||\mathbf{G}|| \cdot ||\mathbf{Q}|| \cdot ||\Delta\mathbf{R}||.$$

Put

$$||\Delta\mathbf{b}|| = \Delta_1, \quad ||\Delta\mathbf{R}|| = \Delta_2. \tag{4.68}$$

Then

$$||\mathbf{M}_1 - \mathbf{M}_2|| \leq c_1\Delta_2, \quad ||\mathbf{V}_1 - \mathbf{V}_2|| \leq \Delta_1 + c_2||\mathbf{Q}||\Delta_2, \tag{4.69}$$

where according to Assumptions 2.9 and 4.2

$$c_1 = ||\mathbf{R}_1^{-1}|| \cdot ||\mathbf{R}_2^{-1}|| \cdot ||\mathbf{G}\mathbf{G}'|| > 0, \quad c_2 = ||\mathbf{R}_1^{-1}|| \cdot ||\mathbf{R}_2^{-1}|| \cdot ||\mathbf{G}|| > 0.$$

To get (4.69) we used the following inequalities for eigenvalues of a matrix (Demidenko 1981, p. 289):

$$\mu_{\max}(\mathbf{G}\mathbf{R}\mathbf{G}') \leq \mu_{\max}(\mathbf{R})\mu_{\max}(\mathbf{G}\mathbf{G}'),$$
$$\mu_{\min}(\mathbf{G}\mathbf{R}\mathbf{G}') \geq \mu_{\min}(\mathbf{R})\mu_{\min}(\mathbf{G}\mathbf{G}'), \tag{4.70}$$

where $\mathbf{R}$ is a symmetric $(n \times n)$ matrix, $\mathbf{G}$ is a rectangular $(m \times n)$ matrix.

According to (4.64) and (4.63), (4.65), (4.68), (4.70) we have

$$||\mathbf{\Lambda}_i|| \leq \left(\frac{\mu_{\max}(\mathbf{R}_i)}{\mu_{\min}(\mathbf{R}_i)}\frac{\mu_{\max}(\mathbf{G}\mathbf{G}')}{\mu_{\min}(\mathbf{G}\mathbf{G}')}\right)^{1/2}\frac{\mu_{\max}(\mathbf{R}_i)}{\mu_{\min}(\mathbf{G}\mathbf{G}')}(c_{i2}||\mathbf{Q}||+c_{i1}\Delta_1+c_{i0}), \quad i = 1, 2, \tag{4.71}$$

where

$$c_{i2} = ||\mathbf{R}_i^{-1}|| \cdot ||\mathbf{G}||, \qquad c_{i1} = \begin{cases} 0, & i = 1, \\ 1, & i = 2, \end{cases} \qquad c_{i0} = ||\mathbf{b}_1||, \quad i = 1, 2. \tag{4.72}$$

Then it follows from expressions (4.67) and (4.69) that

$$||\mathbf{\Lambda}_1 - \mathbf{\Lambda}_2||^2 \leq c_{\lambda 2}||\mathbf{Q}||^2 + c_{\lambda 1}||\mathbf{Q}|| + c_{\lambda 0}, \tag{4.73}$$

where the coefficients $c_{\lambda i}$, $i = 0, 1, 2$, do not depend on $\mathbf{Q}$ and are the polynomials of Δ_1 and Δ_2 (the values Δ_1 и Δ_2 are defined in (4.68)):

$$c_{\lambda i} = \phi(\Delta_1, \Delta_2), \quad i = 0, 1, 2. \tag{4.74}$$

Coefficients of polynomials $\phi(\Delta_1, \Delta_2)$, $i = 0, 1, 2$, are non-negative functions of

$$\rho_j = ||\mathbf{R}_j^{-1}||, \; j = 1, 2 \quad \text{and} \quad \rho_{3j} = \left(\frac{\mu_{\max}(\mathbf{R}_i)}{\mu_{\min}(\mathbf{R}_i)} \frac{\mu_{\max}(\mathbf{GG}')}{\mu_{\min}(\mathbf{GG}')} \right)^{1/2} \frac{\mu_{\max}(\mathbf{R}_i)}{\mu_{\min}(\mathbf{GG}')}, \; j = 1, 2.$$

Moreover, for bounded coefficients ρ_j, $j = 1, 2$, ρ_{3j}, $j = 1, 2$

$$c_{\lambda i} = \phi(0, 0) = 0, \quad i = 0, 1, 2. \tag{4.75}$$

From (2.9) and (2.13), we have

$$||\boldsymbol{\alpha}_i||^2 \leq c_{3i}^2 ||\mathbf{Q}||^2, \quad c_{3i} = (\mu_{\max}(\mathbf{R}_i)/\mu_{\min}^3(\mathbf{R}_i))^{1/2}, \quad i = 1, 2. \tag{4.76}$$

From (4.61) we obtain

$$||\boldsymbol{\alpha}_1 - \boldsymbol{\alpha}_2|| \leq ||\mathbf{R}_1|| \cdot ||\boldsymbol{\alpha}_2|| \cdot ||\Delta \mathbf{R}|| + ||\mathbf{R}_1^{-1}|| \cdot ||\mathbf{G}'|| \cdot ||\mathbf{\Lambda}_1 - \mathbf{\Lambda}_2||. \tag{4.77}$$

From here and expressions (4.73)–(4.76) we derive the estimate

$$||\boldsymbol{\alpha}_1 - \boldsymbol{\alpha}_2|| \leq c_{\alpha 3}||\mathbf{Q}|| + (c_{\alpha 2}||\mathbf{Q}||^2 + c_{\alpha 1}||\mathbf{Q}|| + c_{\alpha 0})^{1/2}, \tag{4.78}$$

where

$$\begin{aligned} c_{\alpha 3} &= c_{32}||\mathbf{R}_1||\Delta_2, \quad c_{\alpha 2} = (\rho_1||\mathbf{G}'||)^{1/2} c_{\lambda 2}, \\ c_{\alpha 1} &= (\rho_1||\mathbf{G}'||)^{1/2} c_{\lambda 1}, \quad c_{\alpha 0} = (\rho_1||\mathbf{G}'||)^{1/2} c_{\lambda 0}. \end{aligned} \tag{4.79}$$

Consider the case when in (4.59)

$$\mathbf{R}_1 = \mathbf{R}, \quad \mathbf{R}_2 = \mathbf{R}_T, \quad \mathbf{b}_1 = \mathbf{b}, \quad \mathbf{b}_2 = \mathbf{b}_T^{(1)}. \tag{4.80}$$

where $\mathbf{b}$ is a fixed vector, $\mathbf{b}_T^{(1)}$ is a random vector, $\mathbf{R}$ is a matrix with fixed elements, and

$$\lim_{T\to\infty} \mathbf{R}_T = \mathbf{R}, \quad p\lim_{T\to\infty} \mathbf{b}_T^{(1)} = \mathbf{b}^{(1)}, \quad ||\mathbf{R}_T|| < \infty, \forall T. \tag{4.81}$$

Thus,

$$\mu_{\max}(\mathbf{R}_T) \to \mu_{\max}(\mathbf{R}), \quad \mu_{min}(\mathbf{R}_T) \to \mu_{min}(\mathbf{R}) > 0 \quad \text{as } T \to \infty.$$

Then, for coefficients of polynomials in (4.74), (4.76) we have:

$$\rho_1 = \text{const}, \quad \lim_{T\to\infty} \rho_2 = ||\mathbf{R}^{-1}|| = \rho_1, \quad \rho_{31} = \text{const},$$

$$\lim_{T\to\infty} \rho_{32} = \rho_{31}, \quad \lim_{T\to\infty} c_{32} = c_{31}, \quad c_{31} = \text{const}. \tag{4.82}$$

According to (4.81) and (4.68):

$$p\lim_{T\to\infty} \Delta_1 = p\lim_{T\to\infty} ||\mathbf{b} - \mathbf{b}_{1T}|| = 0, \quad \lim_{T\to\infty} \Delta_2 = \lim_{T\to\infty} ||\mathbf{R} - \mathbf{R}_T|| = 0, \tag{4.83}$$

where $\Delta_1 = \mathbf{b} - \mathbf{b}_{1T}$, $\Delta_2 = \mathbf{R} - \mathbf{R}_T$.

By (4.82) the coefficients of functions $\phi(\Delta_1, \Delta_2)$, $i = 0, 1, 2$, converge to finite values. Then, according to (4.75), we have

$$p\lim_{T\to\infty} c_{\lambda i} = 0, \quad i = 0, 1, 2,$$

which implies together with (4.79)

$$p\lim_{T\to\infty} c_{\alpha i} = 0, \quad i = 0, 1, 2, 3. \tag{4.84}$$

We have proved the following lemma.

Lemma 4.6. *Consider two quadratic programming problems*

1. $$(1/2)\boldsymbol{\alpha}'\mathbf{R}\boldsymbol{\alpha} - \mathbf{Q}'\boldsymbol{\alpha} \to \min, \mathbf{G}\boldsymbol{\alpha} \le \mathbf{b},$$

2. $$(1/2)\boldsymbol{\alpha}' R_T\boldsymbol{\alpha} - Q'\boldsymbol{\alpha} \to \min, \mathbf{G}\boldsymbol{\alpha} \le \mathbf{b}_T^{(1)}, \tag{4.85}$$

where $\boldsymbol{\alpha} \in \Re^n$, the matrices $\mathbf{R}$ and $\mathbf{R}_T$ satisfy Assumption 2.9, the matrix $\mathbf{G}$ **satisfies** Assumption 4.2 and, moreover, conditions (4.81) hold true. Denote the solutions to the first and the second problems, respectively, by $\boldsymbol{\alpha}_1$ and $\boldsymbol{\alpha}_2$.

Then, $\boldsymbol{\alpha}_1$ and $\boldsymbol{\alpha}_2$ satisfy inequalities (4.76), (4.78). In these inequalities the coefficients in the right-hand side of (4.76) are $c_{31} = \text{const}$, $\lim_{T\to\infty} c_{32} = c_{31}$; by (4.84) the coefficients in the right-hand side of (4.78) have zero limits.

Consider two problems:

$$\frac{1}{2}\mathbf{y}'\mathbf{R}_T\mathbf{y} - \mathbf{q}'_T\mathbf{y} \to \min, \quad \mathbf{G}_1\mathbf{y} \le \mathbf{b}_T^{(1)}, \tag{4.86}$$

$$\frac{1}{2}\mathbf{y}'\mathbf{R}\mathbf{y} - \mathbf{q}'\mathbf{y} \to \min, \quad \mathbf{G}_1\mathbf{y} \le \mathbf{b}^{(1)}, \tag{4.87}$$

where $\mathbf{q}_T = \mathbf{Q}_T/\sigma$, $q = \mathbf{Q}/\sigma$ (the vector $\mathbf{Q}_T$ is defined in (2.123)), $m_1 \times n$ matrix $\mathbf{G}_1$ is of full rank (see Assumption 2.12). We denote by $\mathbf{s}(\mathbf{R}_T, \mathbf{b}_T^{(1)}, \mathbf{q}_T)$ and $\mathbf{s}(\mathbf{R}, \mathbf{b}^{(1)}, \mathbf{q})$, respectively, the solutions to (4.86) and (4.87). Put $\mathbf{k} = E\{\mathbf{s}(\mathbf{R}, \mathbf{b}^{(1)}, \mathbf{q})(\mathbf{R}, \mathbf{b}^{(1)}, \mathbf{q})'\}$. Elements of the $(n \times n)$ matrix $\mathbf{k}$ are of the form

$$k_{ij}(\mathbf{R}, \mathbf{b}^{(1)}) = \int_{\Re^n} s_i(\mathbf{R}, \mathbf{b}^{(1)}, \mathbf{x}) s_j(\mathbf{R}, \mathbf{b}^{(1)}, \mathbf{x}) \mathrm{d}F(\mathbf{x}), \quad i, j = \overline{1, n}, \tag{4.88}$$

where $s_i(\mathbf{R}, \mathbf{b}^{(1)}, \mathbf{x})$ is the ith component of $\mathbf{u} = \mathbf{s}(\mathbf{R}, \mathbf{b}^{(1)}, \mathbf{q})$ at $\mathbf{q} = \mathbf{x}$, $F(\mathbf{x})$ is the distribution function of $\mathbf{q} \sim N(\mathbf{O}_n, \mathbf{R})$.

Let us introduce the $(n \times n)$ matrix $\boldsymbol{\kappa}_T$ with elements

$$k_{ij}^T(\mathbf{R}_T, \mathbf{b}_{1T}) = \int_{\Re^n} s_i(\mathbf{R}_T, \mathbf{b}_T^{(1)}, \mathbf{x},) s_j(\mathbf{R}_T, \mathbf{b}_T^{(1)}, \mathbf{x}) \mathrm{d}F_T(\mathbf{x}), \tag{4.89}$$

where $s_i(\mathbf{R}_T, \mathbf{b}_T^{(1)}, \mathbf{x})$ is the ith component of $\mathbf{s}(\mathbf{R}_T, \mathbf{b}_T^{(1)}, \mathbf{q}_T)$ as $\mathbf{q}_T = \mathbf{x}$, $F_T(\mathbf{x})$ is the distribution function of $\mathbf{q}_T$.

Lemma 4.7. *Suppose that in (4.86), (4.87)*

1. $\mathbf{q}_T \in \Re^n$ *is the random variable with distribution function* $F_T(\mathbf{x})$ *and* $E\{\mathbf{q}_T\mathbf{q}'_T\} = \mathbf{R}_T$*;*
2. *Matrices* $\mathbf{R}_T$ *and* $\mathbf{R}$ *are of full rank;*
3. $\lim_{T\to\infty} F_T(\mathbf{x}) = F(\mathbf{x})$, *where* $F(\mathbf{x})$ *is the distribution function of* $\mathbf{q} \sim N(\mathbf{O}_n, \mathbf{R})$,
4. $\mathbf{b}_{1T} \in \Re^{m_1}$ *is the random variable, and* $p\lim_{T\to\infty} \mathbf{b}_T^{(1)} = \mathbf{b}^{(1)}$.

Then, $p\lim_{T\to\infty} k_{ij}^T(\mathbf{R}_T, \mathbf{b}_T^{(1)}) = k_{ij}(\mathbf{R}, \mathbf{b}^{(1)}), i, j = \overline{1, n}$.

Proof. To prove the lemma it is enough to consider (i, j)th elements of matrices $\boldsymbol{\kappa}_T$ and $\mathbf{k}$. We proceed in the same way as in Lemma 4.4. After transformations we obtain from (4.88) and (4.89) with probability 1

$$|k_{ij}^T(\mathbf{R}_T, \mathbf{b}_T^{(1)}) - k_{ij}(\mathbf{R}, \mathbf{b}^{(1)})| \le N_1 + |N_2|, \tag{4.90}$$

where

$$N_1 = \int_{\Re^n} |s_i(\mathbf{R}_T, \mathbf{b}_T^{(1)}, \mathbf{x}) s_j(\mathbf{R}_T, \mathbf{b}_T^{(1)}, \mathbf{x}) - s_i(\mathbf{R}, \mathbf{b}^{(1)}, \mathbf{x}) s_j(\mathbf{R}, \mathbf{b}^{(1)}, \mathbf{x})| \mathrm{d}F_T(\mathbf{x}), \tag{4.91}$$

$$N_2 = \int_{\Re^n} s_i(\mathbf{R}, \mathbf{b}^{(1)}, \mathbf{x}) s_j(\mathbf{R}, \mathbf{b}^{(1)}, \mathbf{x}) (\mathrm{d}F_T(\mathbf{x}) - \mathrm{d}F(\mathbf{x})). \tag{4.92}$$

Let us estimate the parts in the right-hand side in (4.91) and (4.92).

Estimation of N_1. From (4.91) we have with probability 1

$$\begin{aligned} N_1 &\leq N_{11} + N_{12} \\ &= \int_{\Re^n} |s_i(\mathbf{R}_T, \mathbf{b}_T^{(1)}, \mathbf{x}) - s_i(\mathbf{R}, \mathbf{b}^{(1)}, \mathbf{x})| |s_j(\mathbf{R}_T, \mathbf{b}_T^{(1)}, \mathbf{x})| \mathrm{d}F_T(\mathbf{x}) \\ &+ \int_{\Re^n} |s_j(\mathbf{R}_T, \mathbf{b}_T^{(1)}, \mathbf{x}) - s_j(\mathbf{R}, \mathbf{b}^{(1)}, \mathbf{x})| |s_i(\mathbf{R}_T, \mathbf{b}^{(1)}, \mathbf{x})| \mathrm{d}F_T(\mathbf{x}). \end{aligned} \tag{4.93}$$

Put in (4.85) $\mathbf{G} = \mathbf{G}_1, \mathbf{b} = \mathbf{b}^{(1)}$. Then we can apply Lemma 4.6 to estimate the integrals (4.93). Using inequalities (4.76), (4.78), we see that

$$|s_i(\mathbf{R}_T, \mathbf{b}_T^{(1)}, \mathbf{x}) - s_i(\mathbf{R}, \mathbf{b}^{(1)}, \mathbf{x})| \leq c_{\alpha 3} ||\mathbf{x}|| + (c_{\alpha 2} ||\mathbf{x}||^2 + c_{\alpha 1} ||\mathbf{x}|| + c_{\alpha 0})^{1/2},$$
$$|s_i(\mathbf{R}_T, \mathbf{b}_T^{(1)}, \mathbf{x})| \leq c_{32} ||\mathbf{x}||.$$

From above and (4.93) we have with probability 1

$$\begin{aligned} N_{11} &\leq \int_{\Re^n} c_{\alpha 3} c_{32} ||\mathbf{x}||^2 \mathrm{d}F_T(\mathbf{x}) \\ &+ \int_{\Re^n} [c_{\alpha 2} ||\mathbf{x}||^2 + c_{\alpha 1} ||\mathbf{x}|| + c_{\alpha 0}]^{1/2} c_{32} ||\mathbf{x}|| \mathrm{d}F_T(\mathbf{x}) = N_{11}^{(1)} + N_{11}^{(2)}, \end{aligned}$$

where $N_{11}^{(1)} = \int_{\Re^n} c_{\alpha 3} c_{32} ||\mathbf{x}||^2 \mathrm{d}F_T(\mathbf{x})$,

$$N_{11}^{(2)} = \int_{\Re^n} [c_{\alpha 2} ||\mathbf{x}||^2 + c_{\alpha 1} ||\mathbf{x}|| + c_{\alpha 0}]^{1/2} c_{32} ||\mathbf{x}|| \mathrm{d}F_T(\mathbf{x}).$$

The integral $\int_{\Re^n} ||\mathbf{x}||^2 \mathrm{d}F_T(\mathbf{x})$ is the sum of components of variances of $\mathbf{q}_T$, and equals to the trace of the matrix $\mathbf{R}_T$. According to Assumption 2.9, this value converges to the trace $\mathrm{tr}(\mathbf{R})$ of the matrix $\mathbf{R}$. Therefore, we have

$$\lim_{T \to \infty} \int_{\Re^n} ||\mathbf{x}||^2 \mathrm{d}F_T(\mathbf{x}) = \lim_{T \to \infty} \mathrm{tr}(\mathbf{R}_T) = \mathrm{tr}(\mathbf{R}) = \int_{\Re^n} ||\mathbf{x}||^2 \mathrm{d}F(\mathbf{x}). \tag{4.94}$$

According to Lemma 4.6 and (4.84) we have, respectively, $\lim_{T\to\infty} c_{32} = \text{const}$, and $p\lim_{T\to\infty} c_{\alpha 3} = 0$. Then $N_{11}^{(1)}$ is bounded from above by the value converging in probability to 0. Hence,

$$p\lim_{T\to\infty} N_{11}^{(1)} = 0. \tag{4.95}$$

According to Hölder inequality, we have with probability 1

$$N_{11}^{(2)} \le c_{32}\left[\int_{\Re^n} (c_{\alpha 2}||\mathbf{x}||^2 + c_{\alpha 1}||\mathbf{x}|| + c_{\alpha 0})\mathrm{d}F_T(\mathbf{x})\right]^{1/2}\left[\int_{\Re^n} ||\mathbf{x}||^2\,\mathrm{d}F_T(\mathbf{x})\right]^{1/2}. \tag{4.96}$$

Further,

$$\begin{aligned}\int_{\Re^n} ||\mathbf{x}||\,\mathbf{d}F_T(\mathbf{x}) &\le \int_{||x||\le 1} ||\mathbf{x}||\,\mathbf{d}F_T(\mathbf{x}) + \int_{||x||>1} ||\mathbf{x}||^2\,\mathbf{d}F_T(\mathbf{x})\\ &= \int_{||x||\le 1} (||\mathbf{x}|| - ||\mathbf{x}||^2)\mathbf{d}F_T(\mathbf{x}) + \int_{\Re^n} ||\mathbf{x}||^2\,\mathbf{d}F_T(\mathbf{x}) \le c_4,\ c_4 > 0,\end{aligned} \tag{4.97}$$

Put

$$\psi(\mathbf{x}) = \begin{cases} |\mathbf{x}| - |\mathbf{x}|^2, & |\mathbf{x}| \le 1,\\ 0, & |\mathbf{x}| \ge 1.\end{cases}$$

Then we have $\int_{||x||\le 1} (||\mathbf{x}|| - ||\mathbf{x}||^2)\mathbf{d}F_T(\mathbf{x}) = \int_{\Re^n} \psi(\mathbf{x})\mathbf{d}F_T(\mathbf{x})$. Since the function $\psi(\mathbf{x})$ is continuous and bounded on $\Re^n$, we have by Helli-Brey theorem (see Rao 1965, Sect. 2c4)

$$\int_{\Re^n} \psi(\mathbf{x})\mathbf{d}F_T(\mathbf{x}) \to \int_{\Re^n} \psi(x)\mathbf{d}F(\mathbf{x}) = \int_{||\mathbf{x}||\le 1} (||\mathbf{x}|| - ||\mathbf{x}||^2)\mathbf{d}F(\mathbf{x}) \quad \text{as } T\to\infty.$$

Thus, we obtain that the right-hand side of (4.96) converges in probability to zero. Consequently, $p\lim_{T\to\infty} N_{11}^{(2)} = 0$. Hence, taking into account (4.95),

$$p\lim_{T\to\infty} N_{11} = 0.$$

Similarly we get $p\lim_{T\to\infty} N_{12} = 0$.

Thus,

$$p\lim_{T\to\infty} N_1 = 0. \tag{4.98}$$

Estimation of N_2. Denote $\Phi_{ij}(\mathbf{R}, \mathbf{b}^{(1)}, \mathbf{x}) = s_i(\mathbf{R}, \mathbf{b}^{(1)}, \mathbf{x})s_j(\mathbf{R}, \mathbf{b}^{(1)}, \mathbf{x})$. Then for any sphere S_r centered at zero and with radius r we have

$$
\begin{aligned}
N_2 &= \int_{\Re^n} \Phi_{ij}(\mathbf{R}, \mathbf{b}^{(1)}, \mathbf{x}) \mathrm{d}F_T(\mathbf{x}) - \int_{\Re^n} \Phi_{ij}(\mathbf{R}, \mathbf{b}^{(1)}, \mathbf{x}) \mathrm{d}F(\mathbf{x}) \\
&= \int_{S_r} \Phi_{ij}(\mathbf{R}, \mathbf{b}^{(1)}, \mathbf{x}) \mathrm{d}F_T(\mathbf{x}) - \int_{S_r} \Phi_{ij}(\mathbf{R}, \mathbf{b}^{(1)}, \mathbf{x}) \mathrm{d}F(\mathbf{x}) \\
&+ \int_{\Re^n \backslash S_r} \Phi_{ij}(\mathbf{R}, \mathbf{b}^{(1)}, \mathbf{x}) \mathrm{d}F_T(\mathbf{x}) - \int_{\Re^n \backslash S_r} \Phi_{ij}(\mathbf{R}, \mathbf{b}^{(1)}, \mathbf{x}) \mathrm{d}F(\mathbf{x}).
\end{aligned} \tag{4.99}
$$

We can always choose values of r and T large enough to make the right-hand side of (4.99) arbitrarily small. Thus, we have $\lim_{T\to\infty} N_2 = 0$. From this expression, limit (4.98) and inequality (4.90) we obtain that the left-hand side of (4.90) converges in probability to 0. Lemma is proved. □

Denote by $s_i(\mathbf{R}, \mathbf{b}, \mathbf{Q})$ the ith component of the solution $\mathbf{s}(\mathbf{R}, \mathbf{b}, \mathbf{Q})$ to (4.21). Let $\mathbf{Q}$ in expression (4.21) be a random variable with distribution function $\Psi(\mathbf{x})$, $\mathbf{x} \in \Re^n$. The following statement takes place.

Lemma 4.8. *Assume that in (4.21) the* $(n \times n)$ *matrix* $\mathbf{R}$ *is positive definite. Then the solution* $\mathbf{s}(\mathbf{R}, \mathbf{b}, \mathbf{Q}) \in \Re^n$ *to (4.21) is continuous with respect to* $\mathbf{R}$ *and* $\mathbf{b}$.

Proof. We prove first the continuity of $\mathbf{s}(\mathbf{R}, \mathbf{b}, \mathbf{Q})$ with respect to $\mathbf{R}$ for fixed $\mathbf{Q}$ and $\mathbf{b}$. Put $\mathbf{R}_1 = \mathbf{R}$, $\mathbf{R}_2 = \mathbf{R} + \Delta\mathbf{R}$, where $\Delta\mathbf{R}$ is the non-negative definite matrix. Since $\mathbf{R}_1$ and $\mathbf{R}_2$ are positive definite there always exist non-degenerate matrices $\mathbf{H}_i, i = 1, 2$, such that $\mathbf{R}_i = \mathbf{H}'_i \mathbf{H}_i, i = 1, 2$. Since $\Delta\mathbf{R}$ is non-negative definite it follows that $\Delta\mathbf{R} = \Delta\mathbf{H}'\Delta\mathbf{H}$, where $\Delta\mathbf{H}$ is some square matrix. From these expressions we obtain

$$
\Delta\mathbf{R} \to \mathbf{O}_{nn} \quad \text{implying } \Delta\mathbf{H} \to \mathbf{O}_{nn} \text{ and } \mathbf{H}_2 \to \mathbf{H}_1. \tag{4.100}
$$

The solutions to (4.21) without constraints for $\mathbf{R} = \mathbf{R}_i$, $i = 1, 2$ are, respectively, $\boldsymbol{\alpha}_1^* = \mathbf{R}^{-1}\mathbf{Q}$ and $\boldsymbol{\alpha}_2^* = (\mathbf{R} + \Delta\mathbf{R})^{-1}\mathbf{Q}$. Clearly,

$$
\boldsymbol{\alpha}_2^* \to \boldsymbol{\alpha}_1^* \quad \text{as } \Delta\mathbf{R} \to \mathbf{O}_{nn}, \tag{4.101}
$$

i.e. the solution $\boldsymbol{\alpha}^* = \mathbf{R}^{-1}\mathbf{Q}$ to (4.21) without constraints is continuous in $\mathbf{R}$.

Put $\boldsymbol{\beta}_i^* = \mathbf{H}_{i\alpha_i^*}$, $\boldsymbol{\beta}_i = \mathbf{H}_i\mathbf{s}(\mathbf{R}_i, \mathbf{b}, \mathbf{Q})$, $i = 1, 2$. According to the proof of Theorem 2.12, $\boldsymbol{\beta}_i$ is the projection of $\boldsymbol{\beta}_i^*$ onto the convex set, determined by the constraints of the problem (4.21). Then (see Poljak 1983, p. 116)

$$
||\boldsymbol{\beta}_1 - \boldsymbol{\beta}_2|| \leq ||\boldsymbol{\beta}_1^* - \boldsymbol{\beta}_2^*||. \tag{4.102}
$$

Then by (4.100) and (4.101) we have

$$
\boldsymbol{\beta}_1^* - \boldsymbol{\beta}_2^* = (\mathbf{H}_1 - \mathbf{H}_2)\boldsymbol{\alpha}_1^* + \mathbf{H}_2(\boldsymbol{\alpha}_1^* - \boldsymbol{\alpha}_2^*) \to \mathbf{O}_n \quad \text{as } \Delta\mathbf{R} \to \mathbf{O}_{nn},
$$

and according to (4.102)

$$
||\boldsymbol{\beta}_1 - \boldsymbol{\beta}_2|| \to 0 \quad \text{as } \Delta\mathbf{R} \to \mathbf{O}_{nn}. \tag{4.103}
$$

Analogously, for $\boldsymbol{\beta}_1^* - \boldsymbol{\beta}_2^*$ we have by (4.103)

$$\boldsymbol{\beta}_1 - \boldsymbol{\beta}_2 = (\mathbf{H}_1 - \mathbf{H}_2)\mathbf{s}(\mathbf{R}_1, \mathbf{b}, \mathbf{Q}) + \mathbf{H}_2(\mathbf{s}(\mathbf{R}_1, \mathbf{b}, \mathbf{Q}) - \mathbf{s}(\mathbf{R}_2, \mathbf{b}, \mathbf{Q})) \to \mathbf{O}_n.$$

Taking into account (4.100) and that $\mathbf{H}_1$ is non-degenerate, we have $\mathbf{s}(\mathbf{R}_2, \mathbf{b}, \mathbf{Q}) \to \mathbf{s}(\mathbf{R}_1, \mathbf{b}, \mathbf{Q})$ as $\Delta\mathbf{R} \to \mathbf{O}_{nn}$. Thus, $\mathbf{s}(\mathbf{R}, \mathbf{b}, \mathbf{Q})$ is continuous in $\mathbf{R}$.

Now we prove the continuity of $\mathbf{s}(\mathbf{R}, \mathbf{b}, \mathbf{Q})$ with respect to $\mathbf{b}$. The necessary and sufficient conditions for the existence of the minimum in (4.21) are

$$\mathbf{R}\boldsymbol{\alpha}(\mathbf{b}) - \mathbf{Q} + \sum_{i \in I} \lambda_i(\mathbf{b})\mathbf{g}_i = \mathbf{O}_n, \quad \lambda_i(\mathbf{b})(\mathbf{g}'_i\boldsymbol{\alpha}(\mathbf{b}) - b_i) = 0, \quad \lambda_i(\mathbf{b}) \geq 0,\ i \in I,$$

where $\lambda_i(\mathbf{b})$, $i \in I$ are Lagrange multipliers, $\mathbf{g}'_i$ is the ith row of the matrix $\mathbf{G}$, $i \in I$; b_i, $i \in I$ are the components of $\mathbf{b}$. Taking $\mathbf{b}_1$ and $\mathbf{b}_2 = \mathbf{b}_1 + \Delta\mathbf{b}$ in the right-hand side of (4.21) we get $\boldsymbol{\alpha}(\mathbf{b}_2) = \boldsymbol{\alpha}(\mathbf{b}_1) + \Delta\boldsymbol{\alpha}$, $\lambda_i(\mathbf{b}_2) = \lambda_i(\mathbf{b}_1) + \Delta\lambda_i$, $i \in I$. Inserting $\boldsymbol{\alpha}(\mathbf{b}_2)$ and $\boldsymbol{\alpha}(\mathbf{b}_1)$ in the first condition for the minimum, we obtain

$$\Delta\boldsymbol{\alpha}'\mathbf{R}\boldsymbol{\alpha} + \sum_{i \in I} \Delta\lambda_i \Delta\boldsymbol{\alpha}'\mathbf{g}_i = 0.$$

Now define the lower bound for the second term in this equality. There are four possible cases.

1. $\lambda_i(\mathbf{b}_1) > 0,\ \lambda_i(\mathbf{b}_2) > 0$. Then by the second equation in the conditions for the existence of the minimum we see that $\mathbf{g}'_i\boldsymbol{\alpha}(\mathbf{b}_1) = b_{1i}$, $\mathbf{g}'_i\boldsymbol{\alpha}(\mathbf{b}_2) = b_{2i}$ where b_{1i}, b_{2i} are the ith components of vectors $\mathbf{b}_1$ and $\mathbf{b}_2$, respectively. Hence, $\mathbf{g}'_i\Delta\boldsymbol{\alpha} = \Delta b_i = b_{2i} - b_{1i}$ and $\Delta\lambda_i\Delta\boldsymbol{\alpha}'\mathbf{g}_i = \Delta\lambda_i\Delta b_i$.
2. $\lambda_i(\mathbf{b}_1) = 0,\ \lambda_i(\mathbf{b}_2) > 0$. By the second equation in the conditions for the existence of the minimum we have $\mathbf{g}'_i\boldsymbol{\alpha}(\mathbf{b}_1) \leq b_{1i}$, $\mathbf{g}'_i\boldsymbol{\alpha}(\mathbf{b}_2) = b_{2i}$. Hence, $\mathbf{g}'_i\Delta\boldsymbol{\alpha} \geq \Delta b_i$. Since $\Delta\lambda_i > 0$, we get $\Delta\lambda_i\Delta\boldsymbol{\alpha}'\mathbf{g}_i \geq \Delta\lambda_i\Delta b_i$.
3. $\lambda_i(\mathbf{b}_1) > 0,\ \lambda_i(\mathbf{b}_2) = 0$. Then $\mathbf{g}'_i\boldsymbol{\alpha}(\mathbf{b}_1) = b_{1i}$, $\mathbf{g}'_i\boldsymbol{\alpha}(\mathbf{b}_2) \leq b_{2i}$ and, therefore, $\mathbf{g}'_i\Delta\boldsymbol{\alpha} \leq \Delta b_i$. Since $\Delta\lambda_i < 0$ we have $\Delta\lambda_i\Delta\boldsymbol{\alpha}'\mathbf{g}_i \geq \Delta\lambda_i\Delta b_i$.
4. $\lambda_i(\mathbf{b}_1) = 0,\ \lambda_i(\mathbf{b}_2) = 0$, implying that $\mathbf{g}'_i\boldsymbol{\alpha}(\mathbf{b}_1) \leq b_{1i}$, $\mathbf{g}'_i\boldsymbol{\alpha}(\mathbf{b}_2) \leq b_{2i}$ and $\Delta\lambda_i\Delta\boldsymbol{\alpha}'\mathbf{g}_i = \Delta\lambda_i\Delta b_i = 0$.

Thus, $\sum_{i \in I}\Delta\lambda_i\Delta\boldsymbol{\alpha}'\mathbf{g}_i \geq \sum_{i \in I}\Delta\lambda_i\Delta b_i$. From this expression and the conditions for the existence of the minimum

$$\Delta\boldsymbol{\alpha}'\mathbf{R}\Delta\boldsymbol{\alpha} \leq -\sum_{i \in I}\Delta\lambda_i\Delta b_i \leq ||\Delta\boldsymbol{\lambda}|| \cdot ||\Delta\mathbf{b}||.$$

By (4.43) and (4.44) we have

$$||\Delta\boldsymbol{\lambda}|| \leq \frac{2}{\mu_2}||\Delta\mathbf{b}||, \quad \mu_2 = \text{const}.$$

From the last two inequalities $\Delta\alpha'\mathbf{R}\Delta\alpha \leq (2/\mu_2)||\Delta\mathbf{b}||^2$. Hence, $||\Delta\mathbf{b}|| \to 0$ implying $\Delta\alpha'\mathbf{R}\Delta\alpha \to 0$. Since the matrix $\mathbf{R}$ is positive definite, this limit implies $\Delta\alpha \to \mathbf{O}_n$. Thus statement of the lemma on continuity of the solution to (4.21) with respect to $\mathbf{b}$ is proved. □

Lemma 4.9. *Let the conditions of Lemma 4.8 be satisfied, and assume that the distribution* $\mathbf{Q}$ *has the first and the second moments. Then the functions*

$$k_{ij}(\mathbf{R},\mathbf{b}) = \int_{\Re^n} s_i(\mathbf{R},\mathbf{b},\mathbf{x})s_j(\mathbf{R},\mathbf{b},\mathbf{x})\mathrm{d}\Psi(\mathbf{x}), \quad i,j=\overline{1,n} \tag{4.104}$$

are continuous with respect to $\mathbf{R}$ and $\mathbf{b}$.

Proof. If constraints in the problem (4.21) are absent, then its solution is $\alpha^* = \mathbf{s}(\mathbf{R},\mathbf{b},\mathbf{Q}) = \mathbf{R}^{-1}\mathbf{Q}$. By conditions of the lemma, the matrix of second moments $E\{\alpha^*(\alpha^*)'\} = \mathbf{R}^{-1}E\{\mathbf{Q}\mathbf{Q}'\}\mathbf{R}^{-1}$ exists, implying the existence of the integral (4.104). Adding to the problem (4.21) constraints $\mathbf{G}\alpha \leq \mathbf{b}$ does not affect the existence of the integral (4.104), because in this case we integrate over the set $\Omega \subset \Re^n$, $\Omega = \{\alpha : \mathbf{G}\alpha \leq \mathbf{b}, \alpha \in \Re^n\}$. Moreover, the mass concentrated on the boundary is $\int_{\Re^n\setminus\Omega}\mathrm{d}\Psi(\mathbf{x})$. Thus, under constraints $\mathbf{G}\alpha \leq \mathbf{b}$ the integral (4.104) exists.

Putting $\mathbf{Q} = \mathbf{x}$ in (4.21) we obtain from (4.22)

$$|s_i(\mathbf{R},\mathbf{b},\mathbf{x})s_j(\mathbf{R},\mathbf{b},\mathbf{x})| \leq \frac{\mu_{\max}(\mathbf{R})}{\mu_{\min}(\mathbf{R})}||\mathbf{R}^{-1}||^2||\mathbf{x}||^2 \ \forall\mathbf{R}, \quad \forall\mathbf{b} \geq \mathbf{O}_m.$$

According to the conditions of the lemma, the function $||\mathbf{x}||^2$ is integrable with respect to $\Psi(\mathbf{x})$. Hence, $|s_i(\mathbf{R},\mathbf{b},\mathbf{x})s_j(\mathbf{R},\mathbf{b},\mathbf{x})|$ is bounded from above by a function, integrable with respect to $\Psi(\mathbf{x})$ for all elements $\mathbf{R}$ and components $\mathbf{b}$. Moreover, by Lemma 4.8, the function $s(\mathbf{R},\mathbf{b},\mathbf{x})$ is continuous in $\mathbf{R}$ and $\mathbf{b}$. Thus, the function $s_i(\mathbf{R},\mathbf{b},\mathbf{x})s_j(\mathbf{R},\mathbf{b},\mathbf{x})$ satisfies the conditions of the continuity theorem, implying the statement of the lemma. □

4.4.2 Main Results

Let $\mathbf{E}_T = \sqrt{T}\mathbf{J}_n$, $\bar{\mathbf{E}}_T = \sqrt{T}\mathbf{J}_m$. The corollary below follows from Theorem 2.13.

Corollary 4.3. *If Assumptions 2.1, 2.9 and 4.2 hold true, and the constraints are of the form (4.58), the random variable* $\mathbf{U}_T = \sqrt{T}(\alpha_T - \alpha^0)$ *converges in distribution as* $T \to \infty$ *to the random variable* $\mathbf{U}$ *which is the solution to the problem*

$$\frac{1}{2}\mathbf{X}'\mathbf{R}\mathbf{X} - \mathbf{Q}'\mathbf{X} \to \min, \quad \mathbf{G}_1\mathbf{X} \leq \mathbf{O}_{m_1}. \tag{4.105}$$

Let us transform the initial estimation problem to the problem which allows to get a consistent estimate of the matrix of m.s.e. of the parameter estimate and in which all constraints in (4.58) are taken into account. Note that the truncated estimate takes into account only active constraints.

We transform the problem (4.58) as follows:

$$\frac{1}{2}\boldsymbol{\alpha}'\mathbf{X}'_T\mathbf{X}_T\boldsymbol{\alpha} - \boldsymbol{\alpha}'\mathbf{X}'_T\mathbf{y}_T \rightarrow \min, \quad \mathbf{G}\boldsymbol{\alpha} \leq \mathbf{b}, \tag{4.106}$$

where $\mathbf{X}_T$ is $(T \times n)$ matrix with tth row $\mathbf{x}'_t$, $\mathbf{y}_T = [y_1, y_2, ..., y_T]'$, $\mathbf{G}$ is the $(m \times n)$ matrix with rows $\mathbf{g}'_i$, $i = \overline{1, m}$, $\mathbf{b}$ is the vector consisting of the constraints in the right-hand side of (4.58).

Let us transform problem (4.106). We add in the cost function the term $(1/2)(\boldsymbol{\alpha}^0)'\mathbf{X}'_T\mathbf{X}_T\boldsymbol{\alpha}^0 + (\boldsymbol{\alpha}^0)'\mathbf{X}'_T\boldsymbol{\varepsilon}_T$ and observe that $\mathbf{y}_T = \mathbf{X}_T\boldsymbol{\alpha}^0 + \boldsymbol{\varepsilon}_T$ ($\boldsymbol{\varepsilon}_T = [\varepsilon_1\varepsilon_2 \ldots \varepsilon_T]'$ is the noise in the regression). Then, we get

$$\frac{1}{2}\mathbf{Y}'\mathbf{R}_T\mathbf{Y} - \mathbf{Y}'\mathbf{Q}_T \rightarrow \min, \quad \mathbf{GY} \leq \sqrt{T}(\mathbf{b} - \mathbf{G}\boldsymbol{\alpha}^0), \tag{4.107}$$

where

$$\mathbf{R}_T = T^{-1}\mathbf{X}'_T\mathbf{X}_T, \quad \mathbf{Y} = \sqrt{T}(\boldsymbol{\alpha} - \boldsymbol{\alpha}^0),$$
$$\mathbf{Q}_T = (\sqrt{T})^{-1}\mathbf{X}'_T\boldsymbol{\varepsilon}_T = (\sqrt{T})^{-1}\sum_{t=1}^{T}\boldsymbol{\varepsilon}_t\mathbf{x}_t. \tag{4.108}$$

Put in (4.107) $\mathbf{Y} = \sigma\mathbf{y}$, $\mathbf{Q}_T = \sigma\mathbf{q}_T$, where σ^2 is the variance of $\varepsilon_t, t = \overline{1, T}$. Then (4.107) transforms to

$$\frac{1}{2}\mathbf{y}'\mathbf{R}_T\mathbf{y} - \mathbf{y}'\mathbf{q}_T \rightarrow \min, \quad \mathbf{G}_1\mathbf{y} \leq \mathbf{B}_T^{(1)}, \quad \mathbf{G}_2\mathbf{y} \leq \mathbf{B}_T^{(2)}, \tag{4.109}$$

where $\mathbf{G}_j$ is the $(m_j \times n)$ matrix with rows $\mathbf{g}'_i, i \in I_j^0, j = 1, 2$ ($m_j = |I_j^0|$), and

$$\mathbf{B}_T^{(i)} = \sqrt{T}(\boldsymbol{\beta}^{(i)} - \mathbf{G}_i\boldsymbol{\alpha}^0)/\sigma, \quad i = 1, 2, \tag{4.110}$$

where $\mathbf{b}^{(i)} \in \Re^{m_i}$ is the vector with components $b_j, j \in I_i^0$ (see (4.58)), with $\mathbf{b}^{(1)} = \mathbf{G}_1\boldsymbol{\alpha}^0$. Thus $\mathbf{b}_T^{(1)} = \mathbf{O}_{m_1}$, and components $\mathbf{b}_T^{(2)}$ are non-negative.

Denote by $\mathbf{u}_T$ the solution to (4.109). By (4.109), $\mathbf{u}_T$ is a function of $\mathbf{R}_T$, $\mathbf{B}_T^{(1)}$, $\mathbf{B}_T^{(2)}$ and $\mathbf{q}_T$: $\mathbf{u}_T = \mathbf{S}(\mathbf{R}_T, \mathbf{B}_T^{(1)}, \mathbf{B}_T^{(2)}, \mathbf{q}_T)$. Let $\mathbf{U}_T = \sigma\mathbf{u}_T$. Then

$$\mathbf{K}_T^0 = E\{\mathbf{U}_T\mathbf{U}'_T\} = \sigma^2\mathbf{k}_T^0, \quad \mathbf{k}_T^0 = E\{\mathbf{u}_T\mathbf{u}'_T\}. \tag{4.111}$$

Corollary 4.3 implies that $\mathbf{u}_T$ converges in distribution to the random variable $\mathbf{u} = \mathbf{s}(\mathbf{R}, \mathbf{b}^{(1)}, \mathbf{q})$ which is the solution to

$$\frac{1}{2}\mathbf{y}'\mathbf{R}\mathbf{y} - \mathbf{q}'\mathbf{y} \rightarrow \min, \quad \mathbf{G}_1\mathbf{y} \leq \mathbf{b}^{(1)} = \mathbf{O}_{m_1}, \tag{4.112}$$

where $\mathbf{q} \sim N(\mathbf{O}_n, \mathbf{R})$ is the random variable, to which $\mathbf{q}_T$ converges in distribution as $T \to \infty$. Moreover, $\mathbf{U} = \sigma\mathbf{u}$, where $\mathbf{U} = \mathbf{S}(\mathbf{R}, \mathbf{b}^{(1)}, \mathbf{q})$ is the solution to problem (4.105). Then we have

$$\mathbf{K} = E\{\mathbf{U}\mathbf{U}'\} = \sigma^2\mathbf{k}, \quad \mathbf{k} = E\{\mathbf{u}\mathbf{u}'\}, \tag{4.113}$$

where $\mathbf{k} = E\{\mathbf{s}(\mathbf{R}, \mathbf{b}^{(1)}, \mathbf{q})\mathbf{s}'(\mathbf{R}, \mathbf{b}^{(1)}, \mathbf{q})\}$.

The matrix $\mathbf{k}_T^0$ depends on unknown values $\boldsymbol{\alpha}^0$ and σ^2 from the expressions for $\mathbf{B}_T^{(i)}, i = 1, 2$, in (4.110). Let us replace $\boldsymbol{\alpha}^0$ and σ^2 by their estimates $\boldsymbol{\alpha}_T$ and σ_T (σ_T is some consistent estimate of σ). Then right-hand side parts in (4.109) can be written as $\tilde{\mathbf{B}}_T^{(i)} = \sqrt{T}(\mathbf{b}^{(i)} - \mathbf{G}_i\boldsymbol{\alpha}_T)/\sigma_T, i = 1, 2$ (see (4.110)). Since $\mathbf{b}^{(1)} = \mathbf{G}_1\boldsymbol{\alpha}^0$, we get by Corollary 4.3.

$$p\lim_{T\to\infty} \tilde{\mathbf{B}}_T^{(1)} = -p\lim_{T\to\infty} \mathbf{G}_1\mathbf{U}_T/\sigma_T = -\mathbf{G}_1\mathbf{U}/\sigma \neq \mathbf{b}^{(1)} = \mathbf{O}_{m_1}.$$

Consequently, $\mathbf{u}_T$ does not converge to $\mathbf{u}$ in distribution. In order to get the consistent estimate of the matrix of m.s.e. we use the concept of active constraints up to order $\xi > 0$, introduced in Sect. 4.2.

Let us replace the right-hand side parts of the constraints in (4.109) with the values

$$b_{iT} = \frac{\sqrt{T}(b_i - \mathbf{g}'_i\boldsymbol{\alpha}_T)}{\sigma_T}(1 - \eta_{iT}), \quad i \in I. \tag{4.114}$$

In (4.114) σ_T^2 is the estimate of the variance σ^2 of the noise:

$$\sigma_T^2 = \left(T - n + \sum_{i\in I}\eta_{iT}\right)^{-1}\sum_{t=1}^{T}(y_t - \mathbf{x}'_t\boldsymbol{\alpha}_T)^2. \tag{4.115}$$

The expression (4.115) is a particular case of the estimate (4.18) for the regression function $f_t(\boldsymbol{\alpha}_T) = \mathbf{x}'_t\boldsymbol{\alpha}_T$. According to Lemma 4.2, the estimate σ_T^2 is consistent.

Using (4.114), we pass from the problem (4.109) to

$$\frac{1}{2}\mathbf{y}'\mathbf{R}_T\mathbf{y} - \mathbf{y}'\mathbf{q}_T \rightarrow \min, \quad \mathbf{G}_1\mathbf{y} \leq \mathbf{b}_T^{(1)}, \quad \mathbf{G}_2\mathbf{y} \leq \mathbf{b}_T^{(2)}.$$

Here, $\mathbf{b}_T^{(j)} \in \Re^{m_j}$ is the vector with components b_{iT} given by (4.114), $i \in I_j^0$, $j = 1, 2$.

The solution to the problem above is of the form $\mathbf{S}(\mathbf{R}_T, \mathbf{b}_T^{(1)}, \mathbf{b}_T^{(2)}, \mathbf{q}_T)$. By consistency of $\boldsymbol{\alpha}_T$ (see Theorem 2.12) and σ_T^2 (see Lemma 4.2), we have

$$p \lim_{T\to\infty} \mathbf{b}_T^{(1)} = \mathbf{b}^{(1)} = \mathbf{O}_{m_1}, \quad \mathbf{b}_T^{(2)} \to \infty \quad \text{as } T \to \infty. \tag{4.116}$$

Consider matrix $\hat{\mathbf{k}}_T$ with elements

$$K_{ij}^T(\mathbf{R}_T, \mathbf{b}_T^{(1)}, \mathbf{b}_T^{(2)}) = \int_{\Re^n} S_i(\mathbf{R}_T, \mathbf{b}_T^{(1)}, \mathbf{b}_T^{(2)}, \mathbf{x}) S_j\left(\mathbf{R}_T, \mathbf{b}_T^{(1)}, \mathbf{b}_T^{(2)}, \mathbf{x}\right) \mathrm{d}F_T(\mathbf{x}), \tag{4.117}$$

as the estimate of the matrix $\mathbf{k}_T^0$ in (4.111). Here $S_i(\mathbf{R}_T, \mathbf{b}_T^{(1)}, \mathbf{b}_T^{(2)}, \mathbf{x})$ is the ith component of $\mathbf{S}(\mathbf{R}_T, \mathbf{b}_T^{(1)}, \mathbf{b}_T^{(2)}, \mathbf{x})$, $F_T(\mathbf{x})$ is the distribution function of $\mathbf{q}_T = \mathbf{Q}_T/\sigma$. The estimate $\mathbf{K}_T^0$ of the matrix of m.s.e. of regression parameter estimate is, according to (4.111),

$$\mathbf{K}_T = \sigma_T^2 \mathbf{k}_T, \tag{4.118}$$

where $\mathbf{k}_T = [K_{ij}^T(\mathbf{R}_T, \mathbf{b}_T^{(1)}, \mathbf{b}_T^{(2)})], i, j = \overline{1, n}$.

Theorem 4.4. *If Assumptions 2.1, 2.9, 4.2 hold true, then* $p \lim_{T\to\infty} \mathbf{k}_T = \mathbf{k}$, *where the matrix* $\mathbf{k}$ *is defined in (4.113).*

Proof. Consider the problem:

$$\frac{1}{2}\mathbf{y}'\mathbf{R}_T\mathbf{y} - \mathbf{y}'\mathbf{q}_T \to \min, \quad \mathbf{G}_1\mathbf{y} \le \mathbf{b}_T^{(1)},$$

and denote its solution by $\mathbf{s}(\mathbf{R}_\mathrm{T}, \mathbf{b}_T^{(1)}, \mathbf{q}_T)$. Put

$$k_{ij}^T(\mathbf{R}_\mathrm{T}, \mathbf{b}_T^{(1)}) = \int_{\Re^n} s_i(\mathbf{R}_\mathrm{T}, \mathbf{b}_T^{(1)}, \mathbf{x}) s_j(\mathbf{R}_\mathrm{T}, \mathbf{b}_T^{(1)}, \mathbf{x}) \mathrm{d}F_T(\mathbf{x}), \tag{4.119}$$

where $s_i(\mathbf{R}_\mathrm{T}, \mathbf{b}_T^{(1)}, \mathbf{x})$ is the ith component of $\mathbf{s}(\mathbf{R}_\mathrm{T}, \mathbf{b}_T^{(1)}, \mathbf{x})$.

Consider elements $k_{ij}(\mathbf{R}, \mathbf{b}^{(1)})$ of the matrix $\mathbf{k}$, determined by (4.88), where $F(\mathbf{x})$ is the distribution function. Equation (2.131) implies that $\mathbf{q}_T = \mathbf{Q}_T/\sigma$ converges in distribution to $\mathbf{q} = \mathbf{Q}/\sigma$, i.e. $\lim_{T\to\infty} F_T(\mathbf{x}) = F(\mathbf{x})$. Then by Assumption 2.9, (4.116) and Lemma 4.7 we obtain

$$p \lim_{T\to\infty} k_{ij}^T(\mathbf{R}_T, \mathbf{b}_T^{(1)}) = k_{ij}(\mathbf{R}, \mathbf{b}^{(1)}), \quad i, j = \overline{1, n}. \tag{4.120}$$

Let us show that $k_{ij}^T(\mathbf{R}_T, \mathbf{b}_T^{(1)})$ and $K_{ij}^T(\mathbf{R}_T, \mathbf{b}_T^{(1)}, \mathbf{b}_T^{(2)})$ (see (4.117)) have the same limit as $T \to \infty$. Introduce the set

$$\omega(\mathbf{b}_{1T}, \mathbf{b}_{2T}) = \{\mathbf{x} : \mathbf{s}(R_T, \mathbf{b}_T^{(1)}, \mathbf{x}) \le \mathbf{b}_T^{(2)}, \quad \mathbf{x} \in \Re^n\}.$$

It is easy to verify that $\omega(\mathbf{b}_T^{(1)}, \mathbf{b}_{1T}^{(2)}) \subseteq \omega(\mathbf{b}_T^{(1)}, \mathbf{b}_{2T}^{(2)})$ provided that $\mathbf{b}_{1T}^{(2)} \le \mathbf{b}_{2T}^{(2)}$.

According to (4.116), for arbitrary values $\varepsilon > 0$, $\delta > 0$, $M > 0$ and $\eta > 0$ there exists such $T_0 > 0$ that

$$P\{||\mathbf{b}_T^{(1)}|| \leq \varepsilon\} > 1 - \delta, \quad P\{||\mathbf{b}_T^{(2)}|| \geq M\} > 1 - \eta, \quad T > T_0. \tag{4.121}$$

Put

$$\varphi_{ij}(\mathbf{R}_T, \mathbf{b}_T^{(1)}, \mathbf{x}) = s_i(\mathbf{R}_T, \mathbf{b}_T^{(1)}, \mathbf{x}) s_j(\mathbf{R}_T, \mathbf{b}_T^{(1)}, \mathbf{x}),$$
$$\Phi_{ij}(\mathbf{R}_T, \mathbf{b}_T^{(1)}, \mathbf{b}_T^{(2)}, \mathbf{x}) = S_i(\mathbf{R}_T, \mathbf{b}_T^{(1)}, \mathbf{b}_T^{(2)}, \mathbf{x}) S_j(\mathbf{R}_T, \mathbf{b}_T^{(1)}, \mathbf{b}_T^{(2)}, \mathbf{x}).$$

Then we get from (4.107) with $\tilde{\mathbf{G}}_T = \mathbf{G}$, $\overline{\mathbf{E}}_T = \sqrt{T}\mathbf{J}_n$ and (4.119), that

$$\begin{aligned}
|K_{ij}^T(\mathbf{R}_T, \mathbf{b}_T^{(1)}, \mathbf{b}_T^{(2)}) - k_{ij}^T(\mathbf{R}_T, \mathbf{b}_T^{(1)})| &= \Bigg| \int_{\omega(\mathbf{b}_{1T}, \mathbf{b}_{2T})} \phi_{ij}(\mathbf{R}_T, \mathbf{b}_T^{(1)}, \mathbf{x}) \mathrm{d}F_T(\mathbf{x}) \\
&+ \int_{\Re^n \setminus \omega(\mathbf{b}_T^{(1)}, \mathbf{b}_T^{(2)})} \phi_{ij}(\mathbf{R}_T, \mathbf{b}_T^{(1)}, \mathbf{x}) \mathrm{d}F_T(\mathbf{x}) - \int_{\omega(\mathbf{b}_{1T}, \mathbf{b}_{2T})} \Phi_{ij}(\mathbf{R}_T, \mathbf{b}_T^{(1)}, \mathbf{b}_T^{(2)}, \mathbf{x}) \mathrm{d}F_T(\mathbf{x}) \\
&- \int_{\Re^n \setminus \omega(\mathbf{b}_T^{(1)}, \mathbf{b}_T^{(2)})} \Phi_{ij}(\mathbf{R}_T, \mathbf{b}_T^{(1)}, \mathbf{b}_T^{(2)}, \mathbf{x}) \mathrm{d}F_T(\mathbf{x}) \Bigg| \\
&= \int_{\Re^n \setminus \omega(\mathbf{b}_T^{(1)}, \mathbf{b}_T^{(2)})} |\phi_{ij}(\mathbf{R}_T, \mathbf{b}_T^{(1)}, \mathbf{x}) - \Phi_{ij}(\mathbf{R}_\mathrm{T}, \mathbf{b}_T^{(1)}, \mathbf{b}_T^{(21)}, \mathbf{x})| \mathrm{d}F_T(\mathbf{x}) = L_T(\mathbf{R}_T, \mathbf{b}_T^{(1)}, \mathbf{b}_T^{(2)})
\end{aligned} \tag{4.122}$$

since $s_i(\mathbf{R}_T, \mathbf{b}_T^{(1)}, \mathbf{x}) = S_i(\mathbf{R}_T, \mathbf{b}_T^{(1)}, \mathbf{b}_T^{(2)}, \mathbf{x})$, $i = \overline{1, n}$ for $\mathbf{x} \in \omega(\mathbf{b}_T^{(1)}, \mathbf{R}_T^{(2)})$.

For fixed $\mathbf{x}$ we have $S_i(\mathbf{R}_T, \mathbf{b}_T^{(1)}, \mathbf{b}_T^{(2)}, \mathbf{x}) \to s_i(\mathbf{R}_T, \mathbf{b}_T^{(1)}, \mathbf{x})$ as $T \to \infty$, since $\mathbf{b}_T^{(2)} \to \infty$. Thus, for any $\gamma > 0$ there exist $M > 0$ and $N > 0$ such that the following inequality holds true:

$$\max_{||\mathbf{b}_T^{(1)}|| \leq \varepsilon, ||\mathbf{b}_T^{(2)}|| \geq M} \int_{\Re^n \setminus \omega(\mathbf{b}_T^{(1)}, \mathbf{b}_T^{(2)})} |\varphi_{ij}(\mathbf{R}_T, \mathbf{b}_T^{(1)}, \mathbf{x}) - \Phi_{ij}(\mathbf{R}_T, \mathbf{b}_T^{(1)}, \mathbf{b}_T^{(2)}, \mathbf{x})| \mathrm{d}F_T(\mathbf{x}) \leq \gamma,$$

where $\varepsilon > 0$ is known. Then by (4.121) we obtain

$$\begin{aligned}
P\{L_T(\mathbf{R}_T, \mathbf{b}_T^{(1)}, \mathbf{b}_T^{(2)}) \leq \gamma\} &\geq P\{||\mathbf{b}_T^{(1)}|| \leq \varepsilon, ||\mathbf{b}_T^{(2)}|| \geq M\} \\
&= P\{||\mathbf{b}_T^{(1)}|| \leq \varepsilon\} - P\{||\mathbf{b}_T^{(1)}|| \leq \varepsilon, ||\mathbf{b}_T^{(2)}|| < M\} \\
&\geq P\{||\mathbf{b}_T^{(1)}|| \leq \varepsilon\} - P\{||\mathbf{b}_T^{(2)}|| < M\} \geq 1 - \varepsilon - \eta, \quad T > T_0.
\end{aligned}$$

The above inequality and (4.122) imply that

$$P\{|K_{ij}^T(\mathbf{R}_T, \mathbf{b}_T^{(1)}, \mathbf{b}_T^{(2)}) - k_{ij}^T(\mathbf{R}_T, \mathbf{b}_T^{(1)})| \leq \gamma\} \geq 1 - \varepsilon - \eta, \quad T > T_0.$$

Thus,

$$p \lim_{T\to\infty} |K_{ij}^T(\mathbf{R}_T, \mathbf{b}_T^{(1)}, \mathbf{b}_T^{(2)}) - k_{ij}^T(\mathbf{R}_T, \mathbf{b}_T^{(1)})| = 0. \quad (4.123)$$

By (4.120), $k_{ij}^T(\mathbf{R}_T, \mathbf{b}_T^{(1)})$ converges in probability (and, hence, in distribution) to $k_{ij}(\mathbf{R}, \mathbf{b}^{(1)})$. Moreover, the limit distribution function $k_{ij}^T(\mathbf{R}_T, \mathbf{b}_T^{(1)})$ has a jump equal to 1 at $k_{ij}(\mathbf{R}, \mathbf{b}^{(1)})$. Then, according to (4.123), $K_{ij}^T(\mathbf{R}_T, \mathbf{b}_T^{(1)}, \mathbf{b}_T^{(2)})$ converges in distribution to $k_{ij}(\mathbf{R}, \mathbf{b}^{(1)})$, since $k_{ij}(\mathbf{R}, \mathbf{b}^{(1)}) = \text{const}$. Theorem is proved. □

Theorem 4.5. *If Assumptions 2.1, 2.9, 4.2 are satisfied, then matrix $\mathbf{K}_T$, defined in the (4.118), is the consistent estimate of the matrix of m.s.e. of the estimate of the regression parameter in regression with inequality constraints (4.58).*

The proof follows from Lemma 4.2 and Theorem 4.4.

Now we turn to the calculation of the sample estimate of the matrix of m.s.e. of the estimate $\hat{\mathbf{K}}_T$. According to obtained results, $\hat{\mathbf{K}}_T$ is the conditional expectation, $\hat{\mathbf{K}}_T = E\{|\hat{\mathbf{U}}_T\hat{\mathbf{U}}'_T|\hat{\boldsymbol{\alpha}}_T, \hat{\sigma}_T\}$, where $\hat{\mathbf{U}}_T$ is the solution to the problem

$$\frac{1}{2}\mathbf{Y}'\mathbf{R}_T\mathbf{Y} - \mathbf{Q}'_T\mathbf{Y} \to \min, \quad \mathbf{G}\mathbf{Y} \le \hat{\mathbf{b}}_T, \quad \mathbf{Q}_T = (\sqrt{T})^{-1}\sum_{t=1}^{T} \varepsilon_t \mathbf{x}_t, \quad (4.124)$$

and $\hat{b}_{iT} = [(\sqrt{T}(b_i - \mathbf{g}'_i\hat{\boldsymbol{\alpha}}_T)/\hat{\sigma}_T)(1 - \hat{\eta}_{iT})]$, $i \in I$. Here fixed values $\hat{\boldsymbol{\alpha}}_T$ and $\hat{\sigma}_T$ are the sample estimates of $\boldsymbol{\alpha}^0$ and σ, respectively, $\hat{\eta}_{iT}$ is the value of η_{iT} for a concrete sample. Hence, under known distribution of ε_t for all t (not necessarily normal) the matrix $\hat{\mathbf{K}}_T$ can be calculated for arbitrary number of regressors by the Monte-Carlo method, solving iteratively (4.124).

It is also interesting how the multiplier $(1 - \eta_{iT})$ inserted in the right-hand side of the ith constraint in (4.114), affects the accuracy of calculations of $\mathbf{k}_T^0$.This multiplier is necessary for the consistency of $\mathbf{k}_T$ and, therefore, of $\mathbf{K}_T$. For this purpose, we consider another estimate of $\mathbf{k}_T^0$:

$$\mathbf{k}_T^* = E\{\mathbf{S}(\mathbf{R}_T, \mathbf{W}_T, \mathbf{q}_T)\mathbf{S}'(\mathbf{R}_T, \mathbf{W}_T, \mathbf{q}_T)\},$$

where $\mathbf{W}_T = \sqrt{T}(\mathbf{b} - \mathbf{G}\boldsymbol{\alpha}_T)/\sigma_T = [w_{iT}]$, $i \in I$, $w_{iT} = \sqrt{T}(b_i - \mathbf{g}'_i\boldsymbol{\alpha}_T)/\sigma_T$.

The estimate $\mathbf{k}_T^*$ differs from the estimate $\mathbf{k}_T$ in that $\mathbf{k}_T$ depends on $l_{iT} = w_{iT}(1 - \eta_{iT})$, while k_T^* depends on w_{iT}, $i \in I$. Therefore, we compare the vectors $\mathbf{W}_T$ and $\mathbf{b}_T$, where $\mathbf{b}_T = (\sqrt{T}/\sigma_T)((\mathbf{b} - \mathbf{G}\boldsymbol{\alpha}_T)(\mathbf{J}_m - \mathbf{H}_T))$, $\mathbf{H}_T = \text{diag}(\eta_{iT})$, $i = \overline{1, m}$.

Taking into account that $\hat{\mathbf{b}}_T$ in (4.124) is the sample estimate of $\mathbf{b}_T$, we show that

$$||\mathbf{W}_T - \mathbf{b}_T|| \le \xi\sqrt{nT}. \quad (4.125)$$

There are two possible cases.

1. The ith constraint is active up to ξ : $-\xi \leq g_i(\boldsymbol{\alpha}_T) \leq 0$. Let $-g_i(\boldsymbol{\alpha}_T) \leq \xi$. This case includes situations when the ith constraint may be active or inactive. We have

$$l_{iT} - w_{iT} = \sqrt{T} g_i(\boldsymbol{\alpha}_T)\eta_{iT} = \begin{cases} 0 & \text{if} - \xi \leq g_i(\boldsymbol{\alpha}_T) \leq 0, \\ g_i(\boldsymbol{\alpha}_T)\sqrt{T} & \text{otherwise.} \end{cases}$$

Then $l_{iT} - w_{iT} \leq |l_{iT} - w_{iT}| \leq \xi\sqrt{\mathrm{T}}$, since $\boldsymbol{\eta}_{iT} = 1$. If all constraints are active up to ξ, (4.125) holds true.

2. The ith constraint is inactive up to ξ : $-g_i(\boldsymbol{\alpha}_T) > \xi$. This case includes only usual inactive constraints, therefore we have $\gamma_{iT} = 0$. Then $l_{iT} - w_{iT} = 0$. If all constraints are inactive up to ξ, we have $||\mathbf{W}_T - \mathbf{b}_T|| = 0$, i.e., (4.125) is true. If the part of constraints is inactive up to ξ, (4.125) also holds true. Denote by $\hat{\mathbf{k}}_T^*$ and $\hat{\mathbf{k}}_T$, respectively, the sample estimates of $\hat{\mathbf{k}}_T^*$ and $\hat{\mathbf{k}}_T$ under fixed $\boldsymbol{\alpha}_T = \hat{\boldsymbol{\alpha}}_T$, $\sigma_T = \hat{\sigma}_T$ and $\eta_{iT} = \hat{\eta}_{iT}$. By Lemma 4.9, $||\hat{\mathbf{k}}_T - \hat{\mathbf{k}}_T^*|| \leq \delta$, where δ is small provided that the value in the right-hand side of (4.125) is small. This situation takes place when the value ξ is sufficiently small, for example, represents the accuracy of computer calculations.

4.5 Determination of Accuracy of Estimation of Linear Regression Parameters in Regression with Trend

In this section we use Assumptions 2.9–2.12 concerning the regressor and the constraints, given in Sect. 2.5. We assume that the noise in the model has a normal distribution, i.e., satisfies Assumption 4.1 (see Sect. 4.2).

As in Sect. 4.4 we estimate the matrix of m.s.e. of the estimate of the regression parameter $\mathbf{K}_T^0$ by (4.118), where the estimate of the noise dispersion σ_T^2 is given by (4.115).

Lemma 4.10. *If Assumptions 2.8–2.12 are satisfied, then* $\boldsymbol{\sigma}_T^2$ *is the consistent estimate of* $\boldsymbol{\sigma}^2$.

Proof. We have from (4.115)

$$\sigma_T^2 = \frac{\mathbf{U}'_T\mathbf{R}_T\mathbf{U}_T - 2\mathbf{U}'_T\mathbf{Q}_T}{T - n + \sum_{i\in I}\eta_{iT}} + \frac{\sum_{t=1}^T (y_t - x_t'\boldsymbol{\alpha}^0)^2}{T} \cdot \frac{T}{T - n + \sum_{i\in I}\eta_{iT}}, \tag{4.126}$$

where the random variable $\mathbf{Q}_T$ is defined in (4.108), $\mathbf{U}_T = \mathbf{E}_T(\boldsymbol{\alpha}_T - \boldsymbol{\alpha}^0)$, $\mathbf{E}_T = \mathrm{diag}(\sqrt{\rho_{11}^T}, \sqrt{\rho_{22}^T}, ..., \sqrt{\rho_{nn}^T})$, and ρ_{ii}^T are the elements on the main diagonal of $\mathbf{P}_T = \sum_{t=1}^T \mathbf{x}_t\mathbf{x}'_t$.

Assumptions 2.8, 2.9 imply the convergence (2.131). According to Theorem 2.13, under Assumptions 2.8–2.12, $\mathbf{U}_T$ converges in distribution as $T \to \infty$. Consequently, the numerator of the first term in (4.126) has a limit distribution. Its denominator tends to infinity, since the sum $\sum_{i\in I} \eta_{iT}$ converges in probability to some finite number m_1 according to Lemma 4.1. Then the first term in (4.126) converges in probability to zero. By Assumption 2.8 and Lemma 4.1, the second term converges in probability to σ^2 by the law of large numbers. Lemma is proved. □

In the case when the regression has a trend the matrix k_T in (4.118) is defined as

$$\mathbf{k}_T = [\mathbf{K}_{ij}^T(\mathbf{R}_T, \tilde{\mathbf{G}}_{T1}, \tilde{\mathbf{G}}_{T2}, \mathbf{b}_T^{(1)}, \mathbf{b}_T^{(2)})], \quad i,j = \overline{1,n}. \tag{4.127}$$

In (4.127) the matrices $\tilde{\mathbf{G}}_{Tk}$, $k = 1, 2$ are determined in terms of the notation used in (2.140), vectors $\mathbf{b}_T^{(j)} \in \Re^{m_j}$ have the components b_{iT}, $i \in I_j^0$, $j = 1, 2$,

$$b_{iT} = \frac{\bar{e}_{iT}(b_i - \mathbf{g}'_i \boldsymbol{\alpha}_T)}{\sigma_T}(1 - \eta_{iT}), \quad i \in I. \tag{4.128}$$

In (4.128) $\bar{e}_{iT}$ is an element on the main diagonal of the matrix $\overline{\mathbf{E}}_T$, determined by Assumption 2.12, and (i, j)th elements of the matrix $\mathbf{k}_T$ are defined by

$$K_{ij}^T(\mathbf{R}_T, \tilde{\mathbf{G}}_{T1}, \tilde{\mathbf{G}}_{T2}, \mathbf{b}_T^{(1)}, \mathbf{b}_T^{(2)}) = \int_{\Re^n} S_i(\mathbf{R}_T, \tilde{\mathbf{G}}_{T1}, \tilde{\mathbf{G}}_{T2}, \mathbf{b}_T^{(1)}, \mathbf{b}_T^{(2)}, \mathbf{x}) \times S_j(\mathbf{R}_T, \tilde{\mathbf{G}}_{T1}, \tilde{\mathbf{G}}_{T2}, \mathbf{b}_T^{(1)}, \mathbf{b}_T^{(2)}, \mathbf{x}) f(\mathbf{x}, \mathbf{R}_T) d\mathbf{x}. \tag{4.129}$$

Here $S_i(\mathbf{R}_T, \tilde{\mathbf{G}}_{T1}, \tilde{\mathbf{G}}_{T2}, \mathbf{b}_T^{(1)}, \mathbf{b}_T^{(2)}, \mathbf{x})$ is the ith component of $\mathbf{S}(\mathbf{R}_T, \tilde{\mathbf{G}}_{T1}, \tilde{\mathbf{G}}_{T2}, \mathbf{b}_T^{(1)}, \mathbf{b}_T^{(2)}, \mathbf{x})$ which is the solution to the problem

$$\frac{1}{2}\mathbf{z}'\mathbf{R}_T\mathbf{z} - \mathbf{z}'\mathbf{q}_T \to \min, \quad \tilde{\mathbf{G}}_{T1}\mathbf{z} \le \mathbf{b}_T^{(1)}, \quad \tilde{\mathbf{G}}_{T2}\mathbf{z} \le \mathbf{b}_T^{(2)}, \tag{4.130}$$

where (see (4.108) and Assumption 4.1)

$$\mathbf{q}_T = \mathbf{Q}_T/\sigma = \mathbf{E}_T^{-1}\sum_{t=1}^T \varepsilon_t \mathbf{x}_t/\sigma \sim N(\mathbf{O}_n, \mathbf{R}_T). \tag{4.131}$$

By (4.131) and (2.131) we have

$$\mathbf{q}_T \overset{p}{\Rightarrow} \mathbf{q}, \quad T \to \infty, \quad \text{where } \mathbf{q} \sim N(\mathbf{O}_n, \mathbf{R}). \tag{4.132}$$

According to (4.131), the distribution density $f(\mathbf{x}, \mathbf{R}_T)$ of $\mathbf{q}_T$, appearing in the integral (4.129), is given by

$$f(\mathbf{x}, \mathbf{R}_T) = (2\pi)^{-n/2}(\det \mathbf{R}_T)^{-1/2} \exp\left\{-\frac{1}{2}\mathbf{x}'\mathbf{R}_T^{-1}\mathbf{x}\right\}. \tag{4.133}$$

Let us show that the matrix $\mathbf{K}_T$, defined in (4.118), (4.126)–(4.131), is the consistent estimate of the matrix of m.s.e. of the regression parameter estimate.

Theorem 4.6. *If Assumptions 4.1, 2.9–2.12 hold true, then* $p\lim_{T\to\infty}\mathbf{K}_T = \mathbf{K}$, *where the matrix* $\mathbf{K}$ *is determined by (4.113), and* $\mathbf{U}$ *is the solution to problem (2.129).*

Proof. First we show that

$$p\lim_{T\to\infty}\mathbf{k}_T = \mathbf{k}, \tag{4.134}$$

where the matrix $\mathbf{k}_T$ is given by (4.127)–(4.131), $\mathbf{k} = \sigma^{-2}\mathbf{K} = \sigma^{-2}E\{\mathbf{U}\mathbf{U}'\}$.

For this purpose consider two quadratic programming problems, which are the generalizations of (4.86) and (4.87):

$$\frac{1}{2}\mathbf{y}'\mathbf{R}_T\mathbf{y} - \mathbf{q}'_T\mathbf{y} \to \min, \quad \tilde{\mathbf{G}}_{T1}\mathbf{y} \le \mathbf{b}_T^{(1)}, \tag{4.135}$$

$$\frac{1}{2}\mathbf{y}'\mathbf{R}\mathbf{y} - \mathbf{q}'\mathbf{y} \to \min, \quad \tilde{\mathbf{G}}_1\mathbf{y} \le \mathbf{b}^{(1)}. \tag{4.136}$$

Here $\tilde{\mathbf{G}}_{T1}$ and $\tilde{\mathbf{G}}_1$ are full-rank $(m_1 \times n)$ matrices (see Assumptions 2.10, 2.12). Denote the solutions to (4.135) and (4.136), respectively, by $\mathbf{s}(\mathbf{R}_T, \tilde{\mathbf{G}}_{T1}, \mathbf{b}_T^{(1)}, \mathbf{q}_T)$ and $\mathbf{s}(\mathbf{R}, \tilde{\mathbf{G}}_1, \mathbf{b}^{(1)}, \mathbf{q})$.

Define the $(n \times n)$ matrix of second moments of the solution to (4.135) as the matrix with elements

$$k_{ij}^T(\mathbf{R}_T, \tilde{\mathbf{G}}_{T1}, \mathbf{b}_T^{(1)}) = \int_{\Re^n} s_i(\mathbf{R}_T, \tilde{\mathbf{G}}_{T1}, \mathbf{b}_T^{(1)}, \mathbf{x}) s_j(\mathbf{R}_T, \tilde{\mathbf{G}}_{T1}, \mathbf{b}_T^{(1)}, \mathbf{x}) f(\mathbf{x}, \mathbf{R}_T) d\mathbf{x}, \tag{4.137}$$

where $s_i(\mathbf{R}_T, \tilde{\mathbf{G}}_{T1}, \mathbf{b}_T^{(1)}, \mathbf{x})$ is the ith component of $\mathbf{s}(\mathbf{R}_T, \tilde{\mathbf{G}}_{T1}, \mathbf{b}_T^{(1)}, \mathbf{x})$.

Comparing (4.87) and (4.136), we can easily see that the $(n \times n)$ matrix $\mathbf{k}$ is given by

$$\mathbf{k} = E\{\mathbf{s}(\mathbf{R}, \tilde{\mathbf{G}}_1, \mathbf{b}^{(1)}, \mathbf{q})\mathbf{s}(\mathbf{R}, \tilde{\mathbf{G}}_1, \mathbf{b}^{(1)}, \mathbf{q})'\}.$$

Consequently, by (4.132), we can write the (i, j)-element of the matrix $\mathbf{k}$ as

$$k_{ij}(\mathbf{R}, \tilde{\mathbf{G}}_1, \mathbf{b}^{(1)}) = \int_{\Re^n} s_i(\mathbf{R}, \tilde{\mathbf{G}}_1, \mathbf{b}^{(1)}, \mathbf{x}) s_j(\mathbf{R}, \tilde{\mathbf{G}}_1, \mathbf{b}^{(1)}, \mathbf{x}) f(\mathbf{x}, \mathbf{R}) d\mathbf{x}, \quad i, j = \overline{1, n}, \tag{4.138}$$

where $s_i(\mathbf{R}, \tilde{\mathbf{G}}_1, \mathbf{b}^{(1)}, \mathbf{x})$ is the ith component of $s(\mathbf{R}, \tilde{\mathbf{G}}_1, \mathbf{b}^{(1)}, \mathbf{x})$ and $f(\mathbf{x}, \mathbf{R})$ is the distribution density of $\mathbf{q}$,

$$f(\mathbf{x}, \mathbf{R}) = (2\pi)^{-n/2}(\det \mathbf{R})^{-1/2} \exp\left(-\frac{1}{2}\mathbf{x}'\mathbf{R}^{-1}\mathbf{x}\right). \tag{4.139}$$

By Assumption 2.9 and (2.141), (4.116), we have $\lim_{T\to\infty} \mathbf{R}_T = \mathbf{R}, \lim_{T\to\infty} \tilde{\mathbf{G}}_{T1} = \tilde{\mathbf{G}}_1$, $p\lim_{T\to\infty} \mathbf{b}_T^{(1)} = \mathbf{b}^{(1)} = \mathbf{O}_{m_1}$. Then, applying Lemma 4.4 to (4.137), we obtain from (4.137) and (4.138) that

$$p\lim_{T\to\infty} k_{ij}^T(\mathbf{R}_T, \tilde{\mathbf{G}}_{T1}, \mathbf{b}_T^{(1)}) = k_{ij}(\mathbf{R}, \tilde{\mathbf{G}}_1, \mathbf{b}^{(1)}), \quad i, j = \overline{1, n}. \tag{4.140}$$

Let us show that $k_{ij}^T(\mathbf{R}_T, \tilde{\mathbf{G}}_{T1}, \mathbf{b}_T^{(1)})$ and $K_{ij}^T(\mathbf{R}_T, \tilde{\mathbf{G}}_{T1}, \tilde{\mathbf{G}}_{T2}, \mathbf{b}_T^{(1)}, \mathbf{b}_T^{(2)})$ (see (4.129)) have the same limit as $T \to \infty$. Introduce the set

$$\omega(\tilde{\mathbf{G}}_{T1}, \mathbf{b}_T^{(1)}, \mathbf{b}_T^{(2)}) = \{\mathbf{x} : \mathbf{s}(\mathbf{R}_T, \tilde{\mathbf{G}}_{T1}, \mathbf{b}_T^{(1)}, \mathbf{x}) \leq \mathbf{b}_T^{(2)}, \quad \mathbf{x} \in \Re^n\}.$$

Observe that $\omega(\tilde{\mathbf{G}}_{T1}, \mathbf{b}_T^{(1)}, \mathbf{b}_{T1}^{(2)}) \subseteq \omega(\tilde{\mathbf{G}}_{T1}, \mathbf{b}_T^{(1)}, \mathbf{b}_{T2}^{(2)})$ provided that $\mathbf{b}_{T1}^{(2)} \leq \mathbf{b}_{T2}^{(2)}$.

Denote

$$\varphi_{ij}(\mathbf{R}_T, \tilde{\mathbf{G}}_{T1}, \mathbf{b}_T^{(1)}, \mathbf{x}) = s_i(\mathbf{R}_T, \tilde{\mathbf{G}}_{T1}, \mathbf{b}_T^{(1)}, \mathbf{x}) s_j(\mathbf{R}_T, \tilde{\mathbf{G}}_{T1}, \mathbf{b}_T^{(1)}, \mathbf{x}),$$

$$\Phi_{ij}(\mathbf{R}_T, \tilde{\mathbf{G}}_T, \mathbf{b}_T^{(1)}, \mathbf{b}_T^{(2)}, \mathbf{x}) = S_i(\mathbf{R}_T, \tilde{\mathbf{G}}_T, \mathbf{b}_T^{(1)}, \mathbf{b}_T^{(2)}, \mathbf{x}) S_j(\mathbf{R}_T, \tilde{\mathbf{G}}_T, \mathbf{b}_T^{(1)}, \mathbf{b}_T^{(2)}, \mathbf{x}),$$

$$K_{ij}^T(\mathbf{R}_T, \tilde{\mathbf{G}}_T, \mathbf{b}_T^{(1)}, \mathbf{b}_T^{(2)}) = K_{ij}^T(\mathbf{R}_T, \tilde{\mathbf{G}}_{T1}, \tilde{\mathbf{G}}_{T2}, \mathbf{b}_T^{(1)}, \mathbf{b}_T^{(2)}),$$

where $S_k(\mathbf{R}_T, \tilde{\mathbf{G}}_T, \mathbf{b}_T^{(1)}, \mathbf{b}_T^{(2)}, \mathbf{x}) = S_k(\mathbf{R}_T, \tilde{\mathbf{G}}_{T1}, \tilde{\mathbf{G}}_{T2}, \mathbf{b}_T^{(1)}, \mathbf{b}_T^{(2)}, \mathbf{x})$, $k = i, j$. Then by (4.129) and (4.137) we have

$$\begin{aligned}
&|K_{ij}^T(\mathbf{R}_T, \tilde{\mathbf{G}}_T, \mathbf{b}_T^{(1)}, \mathbf{b}_T^{(2)}) - k_{ij}^T(\mathbf{R}_T, \tilde{\mathbf{G}}_{T1}, \mathbf{b}_T^{(1)})| \\
&= \left| \int_{\omega(b_T^{(1)}, b_T^{(2)})} \varphi_{ij}(\mathbf{R}_T, \tilde{\mathbf{G}}_{T1}, \mathbf{b}_T^{(1)}, \mathbf{x}) f(\mathbf{x}, \mathbf{R}_T) d\mathbf{x} \right. \\
&\quad + \int_{\Re^n \backslash \omega(\mathbf{b}_T^{(1)}, \mathbf{b}_T^{(2)})} \varphi_{ij}(\mathbf{R}_T, \tilde{\mathbf{G}}_{T1}, \mathbf{b}_T^{(1)}, \mathbf{x}) f(\mathbf{x}, \mathbf{R}_T) d\mathbf{x} \\
&\quad - \int_{\omega(\mathbf{b}_T^{(1)}, \mathbf{b}_T^{(2)})} \Phi_{ij}(\mathbf{R}_T, \tilde{\mathbf{G}}_T, \mathbf{b}_T^{(1)}, \mathbf{b}_T^{(2)}, \mathbf{x}) f(\mathbf{x}, \mathbf{R}_T) d\mathbf{x} \\
&\quad \left. - \int_{\Re^n \backslash \omega(\mathbf{b}_T^{(1)}, \mathbf{b}_T^{(2)})} \Phi_{ij}(\mathbf{R}_T, \tilde{\mathbf{G}}_T, \mathbf{b}_T^{(1)}, \mathbf{b}_T^{(2)}, \mathbf{x}) f(\mathbf{x}, R_T) d\mathbf{x} \right| \\
&\quad - \Phi_{ij}(\mathbf{R}_T, \tilde{\mathbf{G}}_T, \mathbf{b}_T^{(1)}, \mathbf{b}_T^{(2)}, \mathbf{x})| f(\mathbf{x}, R_T) d\mathbf{x} = L_T(\mathbf{R}_T, \tilde{\mathbf{G}}_T, \mathbf{b}_T^{(1)}, \mathbf{b}_T^{(2)}),
\end{aligned} \tag{4.141}$$

since for $\mathbf{x} \in \omega(\tilde{\mathbf{G}}_{T1}, \mathbf{b}_T^{(1)}, \mathbf{b}_T^{(2)})$

$$s_i(\mathbf{R}_T, \tilde{\mathbf{G}}_{T1}, \mathbf{b}_T^{(1)}, \mathbf{x}) = S_i(\mathbf{R}_T, \tilde{\mathbf{G}}_T, \mathbf{b}_T^{(1)}, \mathbf{b}_T^{(2)}, \mathbf{x}), \quad i = \overline{1, n}.$$

Since $\mathbf{b}_T^{(2)} \to \infty$ we have for fixed $\mathbf{x}$

$$S_i(\mathbf{R}_T, \tilde{\mathbf{G}}_T, \mathbf{b}_T^{(1)}, \mathbf{b}_T^{(2)}, \mathbf{x}) \to s_i(\mathbf{R}_T, \tilde{\mathbf{G}}_{T1}, \mathbf{b}_T^{(1)}, \mathbf{x}) \quad \text{as } T \to \infty.$$

Thus, for any $\gamma > 0$ there exist $M > 0$ and $N > 0$ such that the inequality

$$\max_{\substack{||\mathbf{b}_T^{(1)}||\leq\varepsilon \\ ||\mathbf{b}_T^{(2)}||\geq M \\ ||\tilde{G}_T-\tilde{G}||\leq N}} \int_{\Re^n \setminus \omega(\mathbf{b}_T^{(1)},\mathbf{b}_T^{(2)})} |\varphi_{ij}(\mathbf{R}_T, \tilde{\mathbf{G}}_{T1}, \mathbf{b}_T^{(1)}, \mathbf{x}) - \Phi_{ij}(\mathbf{R}_T, \tilde{\mathbf{G}}_T, \mathbf{b}_T^{(1)}, \mathbf{b}_T^{(2)}, \mathbf{x})|$$
$$f(\mathbf{x}, \mathbf{R}_T)d\mathbf{x} \leq \gamma$$

holds true for given $\varepsilon > 0$.

In the inequality above, the vectors $\mathbf{b}_T^{(j)}$, $j = 1, 2$, whose components are defined by (4.128), have the asymptotic behavior (4.116). Therefore they satisfy inequalities (4.121), implying that

$$P\{L_T(\mathbf{R}_T, \tilde{\mathbf{G}}_T, \mathbf{b}_T^{(1)}, \mathbf{b}_T^{(2)}) \leq \gamma\}$$
$$\geq P\{||\mathbf{b}_T^{(1)}|| \leq \varepsilon, ||\mathbf{b}_T^{(2)}|| \geq M, ||\tilde{\mathbf{G}}_T - \tilde{\mathbf{G}}|| \leq N\}$$
$$= P\{||\mathbf{b}_T^{(1)}||\leq\varepsilon, ||\tilde{\mathbf{G}}_T-\tilde{\mathbf{G}}|| \leq N\} - P\{||\mathbf{b}_T^{(1)}|| \leq \varepsilon, ||\tilde{\mathbf{G}}_T-\tilde{\mathbf{G}}||\leq N, ||\mathbf{b}_T^{(2)}|| < M\}$$
$$\geq P\{||\mathbf{b}_T^{(1)}||\leq\varepsilon\} - P\{||\mathbf{b}_T^{(2)}|| < M\} - P\{||\tilde{\mathbf{G}}_T-\tilde{\mathbf{G}}|| > N\} \geq 1-\delta-\eta, \quad T > T_0,$$

where we used that by (2.141) $\lim_{T\to\infty} \tilde{\mathbf{G}}_T = \tilde{\mathbf{G}}$.

From the last inequality and (4.141) we have

$$P\{|K_{ij}^T(\mathbf{R}_T, \tilde{\mathbf{G}}_T, \mathbf{b}_T^{(1)}, \mathbf{b}_T^{(2)}) - k_{ij}^T(\mathbf{R}_T, \tilde{\mathbf{G}}_{T1}, \mathbf{b}_T^{(1)})| \leq \gamma\} \geq 1 - \delta - \eta, \quad T > T_0.$$

Thus,

$$p \lim_{T\to\infty} |K_{ij}^T(\mathbf{R}_T, \tilde{\mathbf{G}}_T, \mathbf{b}_T^{(1)}, \mathbf{b}_T^{(2)}) - k_{ij}^T(\mathbf{R}_T, \tilde{\mathbf{G}}_{T1}, \mathbf{b}_T^{(1)})| = 0. \tag{4.142}$$

By (4.142) and (4.140), $K_{ij}^T(\mathbf{R}_T, \tilde{\mathbf{G}}_T, \mathbf{b}_T^{(1)}, \mathbf{b}_T^{(2)})$ converges in probability to $k_{ij}(\mathbf{R}, \tilde{\mathbf{G}}_1, \mathbf{b}^{(1)})$, which proves (4.134). Then the statement of the theorem follows from Lemma 4.11. □

For the case considered in Sect. 4.5 the sample estimate of the matrix of m.s.e. of $\hat{\mathbf{K}}_T$ is given by

$$\hat{\mathbf{K}}_T = \hat{\sigma}_T^2 \hat{\mathbf{k}}_T, \tag{4.143}$$

where

$$\hat{\sigma}_T^2 = \left(T - n + \sum_{i\in I} \hat{\eta}_{iT}\right)^{-1} \sum_{t=1}^{T} (y_t - \mathbf{x}'_t \hat{\boldsymbol{\alpha}}_T)^2, \tag{4.144}$$

see (4.118), (4.115) and (4.128)–(4.131). Here $\hat{\boldsymbol{\alpha}}_T$ is the solution to problem (4.58) for a concrete sample, according to (4.58):

$$\hat{\eta}_{iT} = \begin{cases} 1 & \text{if } -\xi \le \mathbf{g}'_i\hat{\alpha}_T - b_i \le 0, \\ 0 & \text{if} \, \mathbf{g}'_i\hat{\alpha}_T - b_i < -\xi, \end{cases} \tag{4.145}$$

$$\hat{\mathbf{k}}_T = E\{|\hat{\mathbf{u}}_T\hat{\mathbf{u}}'_T|\hat{\boldsymbol{\alpha}}_T, \hat{\sigma}_T\}, \tag{4.146}$$

where $\hat{\mathbf{u}}_T$ is the solution to

$$\frac{1}{2}\mathbf{y}'\mathbf{R}_T - \mathbf{q}'_T\mathbf{y} \to \min, \quad \tilde{\mathbf{G}}_T\mathbf{y} \le \hat{\mathbf{b}}_T, \quad \mathbf{q}_T \sim N(\mathbf{O}_n, \mathbf{R}_T). \tag{4.147}$$

In (4.145) ξ is the accuracy of computer calculations, and $\hat{\mathbf{b}}_T \in \Re^m$ in (4.147) is the vector with components (see (4.128))

$$\hat{b}_{iT} = \frac{\bar{e}_{iT}(b_i - \mathbf{g}'_i\hat{\boldsymbol{\alpha}}_T)}{\hat{\sigma}_T}(1 - \hat{\eta}_{iT}), \quad i \in I. \tag{4.148}$$

To calculate $\hat{\mathbf{k}}_T$ one can use the Monte Carlo method. For this we need to solve iteratively the problem (4.147).

4.6 Calculation of Sample Estimate of the Matrix of m.s.e. Regression Parameters Estimates for Three Inequality Constraints

In Sect. 4.5, we suggested the general method for calculating the sample estimate of the matrix of the m.s.e. of the estimate $\hat{\mathbf{K}}_T$, based on Monte Carlo method. In this section, we describe the method which allows to calculate $\hat{\mathbf{K}}_T$ precisely under Assumptions 4.1 on the normality of the noise in the regression, and the assumptions that the dimension n of the regression parameter is larger than m, where $m \le 3$ is the number of inequality constraints. This case appears quite often in practice. In our notation, we need to take in (4.148) $I = \{1, 2, 3\}$. For calculations we use the sequence of linear transformations of the space of regression parameters, which allows to reduce the dimension of problem (4.147), i.e., to replace the problem with arbitrary number of variables $n \ge 3$ with three constraints to the problem with three variables and constraints.

4.6.1 *Transformation of the Original Problem*

In order to simplify the notation we will drop T and the sign "^" (unless otherwise specified). Then the problem (4.147) can be written as

$$\frac{1}{2}\mathbf{y}'\mathbf{R}\mathbf{y} - \mathbf{q}'\mathbf{y} \to \min, \quad \tilde{\mathbf{G}}\mathbf{y} \le \mathbf{b}, \quad \mathbf{q} \sim N(\mathbf{O}_n, \mathbf{R}). \tag{4.149}$$

Let us transform (4.149). Let $\mathbf{M} = \mathrm{diag}(\mu_i)$, $i = \overline{1, n}$, where μ_i is the ith eigenvalue of $\mathbf{R}$, $\mathbf{C}$ is the orthogonal $(n \times n)$ matrix such that $\mathbf{C}'\mathbf{R}\mathbf{C} = \mathbf{M}$, (note that $\mathbf{C}'\mathbf{C} = \mathbf{J}_n$). Putting

$$\mathbf{X} = \mathbf{M}^{1/2}\mathbf{C}'\mathbf{y}, \tag{4.150}$$

we obtain the transformed problem (4.149):

$$\frac{1}{2}\mathbf{X}'\mathbf{X} - (\mathbf{P}^*)'\mathbf{X} \to \min, \quad \mathbf{A}\mathbf{X} \leq \mathbf{b}, \tag{4.151}$$

where $\mathbf{P}^* \sim N(\mathbf{O}_n, \mathbf{J}_n), \mathbf{A} = \tilde{\mathbf{G}}\mathbf{C}\mathbf{M}^{-1/2}$.

According to Assumption 2.10, matrix $\mathbf{A}$ has a full rank. Then its orthogonal decomposition according to Lawson and Hanson (1974, Chap. 2), has the form

$$\mathbf{A} = \mathbf{S}\mathbf{H}, \quad \mathbf{S} = [\mathbf{s} \vdots \mathbf{O}_{m,n-m}], \; m \leq n, \tag{4.152}$$

where $\mathbf{s}$ is the triangular $(m \times m)$ matrix, with zero elements located above the main diagonal, and $\mathbf{H}$ is orthogonal $(n \times n)$ matrix $(\mathbf{H}'\mathbf{H} = \mathbf{J}_n)$. For $m = 3$

$$\mathbf{s} = \begin{bmatrix} s_{11} & 0 & 0 \\ s_{21} & s_{22} & 0 \\ s_{31} & s_{32} & s_{33} \end{bmatrix}. \tag{4.153}$$

Let $\mathbf{Z} = \mathbf{H}\mathbf{X}$. Taking into account the transformations above, we obtain from (4.151)

$$\frac{1}{2}\mathbf{Z}'\mathbf{Z} - \mathbf{P}'\mathbf{Z} \to \min, \quad \mathbf{S}\mathbf{Z} \leq \mathbf{b}, \tag{4.154}$$

where $\mathbf{P} = \mathbf{H}\mathbf{P}^* \sim N(\mathbf{O}_n, \mathbf{J}_n)$.

Denote by $\mathbf{V}$ the solution to (4.154), by Z_i, V_i, P_i ith components of vectors $\mathbf{Z}, \mathbf{V}, \mathbf{P}$, respectively. It is easy to show that $V_i = P_i, i = \overline{m+1, n}$ is the solution to the problem (4.154), and the remaining components of $\mathbf{Z}$, namely $V_i, i = \overline{1, m}$, are the solutions to

$$\frac{1}{2}\bar{\mathbf{Z}}'\bar{\mathbf{Z}} - \bar{\mathbf{P}}'\bar{\mathbf{Z}} \to \min, \quad \mathbf{s}\bar{\mathbf{Z}} \leq \mathbf{b}, \tag{4.155}$$

where $\bar{\mathbf{Z}} = [Z_1 \ldots Z_m]', \bar{\mathbf{P}} = [P_1 \ldots P_m]', \bar{\mathbf{P}} \sim N(\mathbf{O}_m, \mathbf{J}_m)$.

Let $\overline{\mathbf{Z}} = \tilde{\mathbf{\Pi}}\mathbf{z}$, where $\tilde{\mathbf{\Pi}} = \mathrm{diag}(\mathrm{sign}\, s_{ii})$, $i = \overline{1, m}$. By Assumption 2.10 the matrix $\mathbf{B} = \mathbf{s}\tilde{\mathbf{\Pi}}$ is non-degenerate, and its elements $b_{ij} \neq 0$, $i, j = \overline{1, m}$, $i \geq j$, differ from the corresponding elements s_{ij} of the matrix $\mathbf{s}$ only by the sign. Moreover, $b_{ii} = s_{ii}\mathrm{sign}(s_{ii}) > 0, i = \overline{1, m}$. If $m = 3$, matrix $\mathbf{B}$ takes the form (cf. (4.153) for $\mathbf{s}$):

$$\mathbf{B} = \begin{bmatrix} b_{11} & 0 & 0 \\ b_{21} & b_{22} & 0 \\ b_{31} & b_{32} & b_{33} \end{bmatrix}. \tag{4.156}$$

Taking into account the introduced transformation, we obtain from (4.155)

$$\frac{1}{2}\mathbf{z}'\mathbf{z} - \mathbf{p}'\mathbf{z} \to \min, \quad \mathbf{p} \sim N(\mathbf{O}_m, \mathbf{J}_m), \tag{4.157}$$

$$\mathbf{Bz} \leq \mathbf{b}, \tag{4.158}$$

where $\mathbf{p} = \tilde{\mathbf{\Pi}}\bar{\mathbf{P}}$.

By (4.148) we have in (4.158):

$$\mathbf{b} \geq \mathbf{O}_m. \tag{4.159}$$

The admissible region given by inequalities (4.158), is bounded by the planes ω_i, $i = 1, 2, 3$, defined, respectively, by the equations

$$\mathbf{B}'_1\mathbf{z} = b_{11}z_1 = b_1, \tag{4.160}$$

$$\mathbf{B}'_2\mathbf{z} = b_{21}z_1 + b_{22}z_2 = b_2, \tag{4.161}$$

$$\mathbf{B}'_3\mathbf{z} = b_{31}z_1 + b_{32}z_2 + b_{33}z_3 = b_3, \tag{4.162}$$

where $\mathbf{B}'_i$ is the ith row of $\mathbf{B}$, $i = \overline{1,3}$; b_i are the components of $\mathbf{b}$; z_i are the components of $\mathbf{z}$, $i = \overline{1,3}$.

The cut of the admissible region by the plane parallel to the coordinate plane z_1Oz_2, is represented on Fig. 4.1. Here Φ is the angle between the plane ω_2 and the abscissa $\mathrm{tg}\Phi = -b_{21}/b_{22}$, $|\Phi| \leq \pi/2$. When $\Phi = -\pi/2$, two inequalities coincide, when $\Phi = \pi/2$ the admissible region is situated between the planes ω_1 and ω_2, which are parallel to each other and perpendicular to X-axis.

Denote the solution to problem (4.157), (4.158) by $\mathbf{v}$. It is defined for all realizations of $\mathbf{p}$ for fixed $\mathbf{b}$. According to the calculations above the solution to (4.149) is $\mathbf{u} = \mathbf{CM}^{-1/2}\mathbf{H}'\mathbf{\Pi}\mathbf{V}$, where $\mathbf{\Pi} = \begin{bmatrix} \tilde{\mathbf{\Pi}} & \mathbf{O}_{m,n-m} \\ \mathbf{O}_{n-m,m} & \mathbf{J}_{n-m} \end{bmatrix}$. Hence, the solution to (4.147) is $\hat{\mathbf{u}}_T = \mathbf{C}_T\mathbf{M}_T^{-1/2}\mathbf{H}'_T\mathbf{\Pi}_T\mathbf{V}$. Then, according to (4.146),

$$\hat{\mathbf{k}}_T = \mathbf{C}_T\mathbf{M}_T^{-1/2}\mathbf{H}'_T\mathbf{\Pi}_T\hat{\mathbf{K}}_\mathbf{V}\mathbf{\Pi}_T\mathbf{H}_T\mathbf{M}_T^{-1/2}\mathbf{C}'_T, \tag{4.163}$$

where

$$\hat{\mathbf{K}}_\mathbf{V} = E\{\mathbf{VV}'\} = \begin{bmatrix} \hat{\mathbf{M}}_\mathbf{v}[m] & \mathbf{O}_{m,n-m} \\ \mathbf{O}_{n-m,m} & \mathbf{J}_{n-m} \end{bmatrix} \tag{4.164}$$

and $\hat{\mathbf{M}}_\mathbf{v}[m] = E\{\mathbf{vv}'\}$.

From (4.143) and (4.163) we get the expression for the sample estimate of the matrix of m.s.e. of the estimate

$$\hat{\mathbf{K}}_T = \hat{\sigma}_T^2\mathbf{C}_T\mathbf{M}_T^{-1/2}\mathbf{H}'_T\mathbf{\Pi}_T\hat{\mathbf{K}}_\mathbf{V}\mathbf{\Pi}_T\mathbf{H}_T\mathbf{M}_T^{-1/2}\mathbf{C}'_T. \tag{4.165}$$

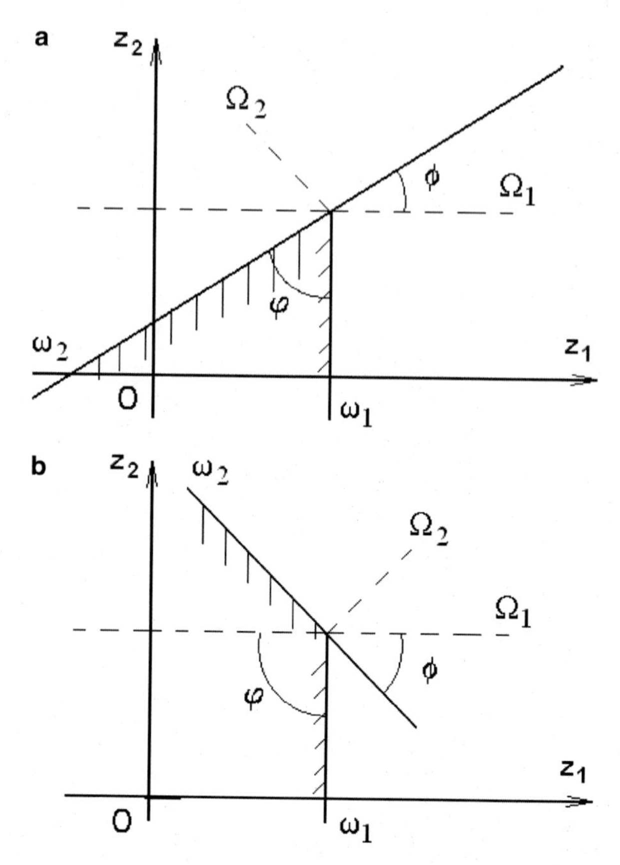

Fig. 4.1 Section of the admissible domain in coordinates $(z_1; z_2; z_3)$ by the plane $z_3 = \text{const}$

As it follows by (4.165) and (4.164), to find $\hat{\mathbf{K}}_T$ one needs to calculate $\hat{\sigma}_T^2$ (by (4.144)) and matrix $\hat{\mathbf{M}}_v[m]$, the remaining values in (4.165) can be calculated using known algorithms of linear algebra, see for example, Lawson and Hanson (1974). Below we find $\mathbf{M}_v[m]$ for $m = 3$.

4.6.2 *Finding Matrix* $\mathbf{M}_v[3]$

To calculate the elements of $\mathbf{M}_v[3]$, it is necessary to determine the density of the unit mass distribution at the boundary of the admissible region, given by the constraints (4.158), i.e., (see Fig. 4.1) on the planes ω_i, $i = \overline{1,3}$, in their intersection point $\omega_{1,2,3}$, on the faces $\omega_{12}, \omega_{13}, \omega_{23}$, which are the intersections of planes ω_1 and ω_2, ω_1 and ω_3, ω_2 and ω_3, respectively, and inside the admissible region ω_4, allocated on Fig. 4.1 by dotted lines.

As it is shown on Fig. 4.1, the portion of the unit mass on the plane ω_1 is defined by those realizations of $\mathbf{p}$, for which the constraint $\mathbf{B}'_1\mathbf{z} \leq b_1$ is active. For planes ω_2 and ω_3 the situation is similar. The mass, determined by realizations of $\mathbf{p}$ enclosed between planes Ω_1 and Ω_2, orthogonal, respectively, to planes ω_1 and ω_2, is concentrated on the face ω_{12}. The mass distribution on the planes ω_{13} and ω_{23} is determined similarly. At the point $\omega_{1,2,3}$ the mass concentration is determined by the simultaneous activity of all constraints in (4.158). The mass concentration in the domain ω_4 is determined by the realizations of $\mathbf{p}$, for which all constraints are inactive.

From above, the distribution function $F(\mathbf{z})$ of $\mathbf{v}$ (the solution to problem (4.157), (4.158)), is not continuous. Therefore elements $m_{ij}[3]$, $i, j = 1, 2$ of the matrix $\mathbf{M}_{\mathbf{v}}[3]$ are determined by the Lebesgue-Stieltjes integral

$$m_{ij}[3] = \int_{\mathbf{z}\in\Re^3} z_i z_j \, \mathrm{d}F(\mathbf{z}) = \sum_{k=1}^{4} \int_{z\in\omega_k^{(\mathbf{z})}} z_i z_j \, \mathrm{d}F_k(\mathbf{z}) + \sum_{\substack{k=1 \\ k\neq l}}^{3} \sum_{\substack{l=1 \\ l>k}}^{3} \int_{z\in\omega_{kl}^{(\mathbf{z})}} z_i z_j \, \mathrm{d}F_{kl}(\mathbf{z})$$
$$+ \int_{\omega_{1,2,3}^{(z)}} z_i z_j \, \mathrm{d}F_{1,2,3}(\mathbf{z}) = \sum_{k=1}^{4} m_{ij}^{(k)}[3] + \sum_{\substack{k=1 \\ k\neq l}}^{3} \sum_{\substack{l=1 \\ l>k}}^{3} m_{ij}^{(k,l)}[3]$$
$$+ m_{ij}^{(1,2,3)}[3], \ i, j = \overline{1, 3}, \tag{4.166}$$

where $F_k(\mathbf{z})$, $F_{kl}(\mathbf{z})$, $F_{1,2,3}(\mathbf{z})$ are the distribution functions of portions of the unit mass, respectively, on the face ω_k, on the edge ω_{kl}, at the point $\omega_{1,2,3}$ of intersection of three faces; $\omega_k^{(\mathbf{z})}$, $\omega_{ki}^{(\mathbf{z})}$ are the sets of points on the face ω_k and on the edge ω_{ki}, respectively, which are located in the admissible region (4.158) and determined in coordinates of $\mathbf{z}$; $\omega_{1,2,3}^{(z)}$ is the point of $\omega_{1,2,3}$ with the coordinates $\mathbf{z}^* = (z_1^*, z_2^*, z_3^*)'$ ($\mathbf{z}^*$ is the solution to the system of (4.160)–(4.162)). Such a solution always exists, since the matrix $\mathbf{B}$ is non-degenerate.

According to (4.166), we have

$$m_{ij}^{(k)}[3] = \int_{z\in\omega_k^{(\mathbf{z})}} z_i z_j \varphi_k(z) \mathrm{d}\omega_k^{(\mathbf{z})}, \quad m_{ij}^{(k,l)}[3] = \int_{z\in\omega_{kl}^{(z)}} z_i z_j \varphi_{kl}(z) \mathrm{d}\omega_{kl}^{(\mathbf{z})},$$
$$m_{ij}^{(1,2,3)}[3] = z_i z_j p_{1,2,3}, \quad i, j = \overline{1, 3},$$

where $\varphi_k(\mathbf{z})$ and $\varphi_{kl}(\mathbf{z})$ are distribution densities of the fractions of the unit mass on the face ω_k and on the edge ω_{kl}, and $p_{1,2,3} \neq 0$ is the fraction of the unit mass located in the point of intersection of faces ω_k, $k = \overline{1, 2, 3}$.

Introduce the matrices

$$\mathbf{M}^{(k)}[3] = \left[m_{ij}^{(k)}[3]\right], \quad k = \overline{1, 4},$$
$$\mathbf{M}^{(k,l)}(3) = \left[m_{ij}^{(k,l)}(3)\right], \quad k, l = \overline{1, 3},$$
$$\mathbf{M}^{(1,2,3)}[3] = \left[m_{ij}^{(1,2,3)}[3]\right], \quad i, j = \overline{1, 3}.$$

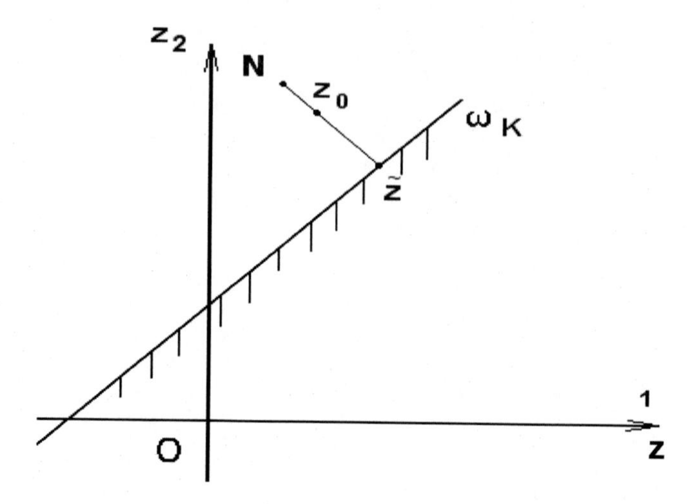

Fig. 4.2 Projection of the unit mass on the boundary of the admissible domain in coordinates $(z_1; z_2; z_3)$ (a section by the plane $z_3 = \text{const}$)

By (4.166)

$$\mathbf{M}_\nu[3] = \sum_{k=1}^{4} \mathbf{M}^{(k)}[3] + \sum_{\substack{k=1 \\ k \neq l}}^{3} \sum_{\substack{l=1 \\ l>k}}^{3} \mathbf{M}^{(k,l)}[3] + \mathbf{M}^{(1,2,3)}[3]. \tag{4.167}$$

To calculate matrices $\mathbf{M}^{(k)}[3]$, $k = \overline{1,4}$, we examine the case when there is one linear constraint $\mathbf{B}'_k\mathbf{z} \leq b_k$, $\mathbf{B}_k = [b_{k1}\ b_{k2}\ b_{k3}]'$, $k = \overline{1,3}$. In order to illustrate further transformations let us examine Fig. 4.2, where the cut of plane ω_k by the plane, perpendicular to the axis Oz_3 is drawn.

Let the solution z_0 of (4.157) be located outside the admissible region, formed by the plane ω_k determined by the equation $\mathbf{B}'_k\mathbf{z} \leq b_k$, i.e., outside the shaded area in Fig. 4.2. Then, the solution $\tilde{\mathbf{z}}$ to (4.157) is the projection of $\mathbf{z}_0$ onto ω_k. Obviously, $\tilde{\mathbf{z}}$ is the solution to this problem for all cases when $\mathbf{p}$ is located on the normal N to the plane ω_k at the point $\tilde{\mathbf{z}}$. Thus, in the neighborhood of $\tilde{\mathbf{z}}$, belonging to the plane ω_k, we have the mass

$$\mathrm{d}F_k(\tilde{\mathbf{z}}) = \varphi_k(\tilde{\mathbf{z}})\mathrm{d}\omega_k, \quad \tilde{\mathbf{z}} \in \omega_k, \tag{4.168}$$

where

$$\varphi_k(\tilde{\mathbf{z}}) = \int_0^\infty f(\mathbf{z}(t))||\mathbf{B}_k||\mathrm{d}t, \tag{4.169}$$

$f(\mathbf{z})$ is the density of normal distribution, $\mathbf{z}(t)$ is the normal to the plane ω_k, $\mathbf{z}(t) = \tilde{\mathbf{z}} + \mathbf{B}_k t, 0 \leq t \leq \infty$. According to (4.150), we have $f(\mathbf{z}) = (1/2\pi)\exp\{-(\mathbf{z}'\mathbf{z}/2)\}$.

Therefore,

$$f(\mathbf{z}(t)) = \frac{1}{2\pi}\exp\left\{-\frac{2b_k t + ||\mathbf{B}_k||^2\, t^2}{2}\right\}\exp\left\{-\frac{\tilde{\mathbf{z}}'\tilde{\mathbf{z}}}{2}\right\}.$$

Substituting this expression into (4.169) and making some transformations we obtain the density of the mass distribution at the point $\mathbf{z} = \tilde{\mathbf{z}}$ (note that $\mathbf{B}'_k\mathbf{z} = b_k$):

$$\varphi_k(\tilde{\mathbf{z}}) = \frac{a_k}{2\pi} \exp\left\{-\frac{b_k^2}{2||\mathbf{B}_k||^2}\right\} \exp\left\{-\frac{\tilde{\mathbf{z}}'\tilde{\mathbf{z}}}{2}\right\}, \tag{4.170}$$

where

$$a_k = \frac{1}{\sqrt{2\pi}} \int_{b_k/||\mathbf{B}_k||}^{\infty} e^{-x^2/2}\,\mathrm{d}x, \quad k = \overline{1,3}. \tag{4.171}$$

In further calculations we use the linear transformation of variables. For convenience we give below one useful property.

Statement 4.1. Let $\mathbf{x} \in \Re^n$, $\mathbf{y} \in \Re^n$, be related as $\mathbf{y} = \mathbf{A}\mathbf{x}$, where $\mathbf{A}$ is the orthogonal matrix. Introduce two $(n \times n)$ matrices: $\mathbf{M} = [M_{ij}]$ and $\mathbf{N} = [N_{ij}]$, $i, j = \overline{1,n}$, with elements

$$M_{ij} = \int_{\Gamma_\mathbf{y}} y_i y_j \phi(\mathbf{y})\mathrm{d}\Gamma_\mathbf{y}, \quad N_{ij} = \int_{\Gamma_\mathbf{x}} x_i x_j f(\mathbf{x})\mathrm{d}\Gamma_\mathbf{x}.$$

Here $\varphi(\mathbf{y}) = f(\mathbf{A}'\mathbf{y})$, $f(\mathbf{x}) = g(h(\mathbf{x}))$, where $g(\mathbf{u})$ is a monotone decreasing differentiable function, $h(\mathbf{x}) = \mathbf{x}'\mathbf{x}$, $\mathbf{G}$ is the $(m \times n)$ matrix, $\mathbf{b} \in \Re^m$, $\Gamma_\mathbf{x} = \{\mathbf{x} : \mathbf{G}\mathbf{x} \le \mathbf{b}, \mathbf{x} \in \Re^n\}$, $\Gamma_\mathbf{y} = \{\mathbf{y} : \mathbf{G}\mathbf{A}^{-1}\mathbf{y} \le \mathbf{b}, \mathbf{y} \in \Re^n\}$. Then $\mathbf{M} = \mathbf{A}\mathbf{N}\mathbf{A}'$.

Proof. The component i of the vector y is given by $y_i = \sum_{k=1}^n a_{ik}x_k$, where x_k is the kth component of $\mathbf{x}$. Then

$$y_i y_j = \sum_{k=1}^{n}\sum_{s=1}^{n} a_{ik}x_k x_s a_{js}. \tag{4.172}$$

Let us multiply both parts of (4.172) by $f(\mathbf{x})\mathrm{d}\Gamma_\mathbf{x}$ and integrate the left and the right-hand sides over the domain $\Gamma_\mathbf{x}$. We obtain

$$\int_{\Gamma_x} y_i y_j f(\mathbf{x})\mathrm{d}\Gamma_\mathbf{x} = \sum_{k=1}^{n}\sum_{s=1}^{n} a_{ik}a_{js} \int_{\Gamma_x} x_k x_s f(\mathbf{x})\mathrm{d}\Gamma_\mathbf{x}. \tag{4.173}$$

Taking into account the fact that $f(\mathbf{x}) = \phi(\mathbf{y})$, $(\det \mathbf{A}^{-1}) = (\det \mathbf{A})^{-1} = 1$ and $\mathrm{d}\Gamma_\mathbf{x} = (\det \mathbf{A})^{-1}\mathrm{d}\Gamma_\mathbf{y}$ we obtain from (4.173)

$$M_{ij} = \int_{\Gamma_y} y_i y_j f(\mathbf{y})\mathrm{d}\Gamma_\mathbf{y} = \sum_{k=1}^{n}\sum_{s=1}^{n} a_{ik}N_{ks}a_{js},$$

proving Statement 4.1. □

For example, one can take in Statement 4.1 as $f(\mathbf{x})$ the density of multivariate normal distribution, or Student's distribution.

Let

$$\psi(y) = \frac{1}{\sqrt{2\pi}} \exp\left(-\frac{y^2}{2}\right),$$

$$\Psi(y) = \frac{1}{\sqrt{2\pi}} \int_{-\infty}^{y} \exp\left(-\frac{x^2}{2}\right) \mathrm{d}x. \tag{4.174}$$

Let us calculate the matrices $\mathbf{M}^{(k)}$, $\mathbf{M}^{(kl)}$, $\mathbf{M}^{(1,2,3)}$ from (4.167).

Calculation of $\mathbf{M}^{(1)}[3]$. According to (4.170) the distribution density of the fraction of the unit mass on the face ω_1 (given by (4.160) with $\tilde{\mathbf{z}} = (z_1^*, z_2, z_3)'$, where $z_1^* = b_1/b_{11} = b_1/||\mathbf{B}_1|| > 0$), is given by

$$\tilde{\varphi}_1(\tilde{\mathbf{z}}) = \frac{a_1}{2\pi} \exp\left\{-\frac{z_2^2 + z_3^2}{2}\right\} = \varphi_1(\mathbf{z}). \tag{4.175}$$

Here a_1 is determined by (4.171) with $k = 1$.

Points on the plane ω_1, satisfying inequalities (4.158), belong to the set

$$\omega_1^{(\mathbf{z})} = \{\mathbf{z} \in \Re^3 = z_1^*, -\infty < z_2 \leq z_2^*, -\infty < z_3 \leq b_{33}^{-1}(b_3 - b_{31}z_1^* - b_{32}z_2) = L_{31}(z_2)\}. \tag{4.176}$$

In what follows the indices of the linear function $L_{ij}(\cdot)$ indicate the following: i is the number of the variable, determined by this function, $i \leq 3$; j is the function number.

From (4.175) and (4.176) we obtain the expression for elements of the matrix $\mathbf{M}^{(1)}[3] = [m_{ij}^{(1)}[3]]$, $i, j = \overline{1,3}$:

$$m_{ij}^{(1)}[3] = \frac{a_1}{2\pi} \int_{-\infty}^{z_2^*} z_i z_j e^{-z_2^2/2} \int_{-\infty}^{L_{31}(z_2)} e^{-z_3^2/2} dz_3 dz_2, \quad i, j = \overline{1,3}, \ z_1 = z_1^* = \text{const},$$

where $L_{31}(z_2) = b_{33}^{-1}(b_3 - b_{31}z_1^* - b_{32}z_2)$.

Calculation of $\mathbf{M}^{(2)}[3]$. Consider the plane ω_2, given by (4.161). We make the change of variable

$$\mathbf{z} = \mathbf{D}(\mathbf{z}, \boldsymbol{\beta})\boldsymbol{\beta}, \quad \boldsymbol{\beta} \in \Re^3, \tag{4.177}$$

where $\mathbf{D}(\mathbf{z}, \boldsymbol{\beta})$ is the orthogonal matrix which transforms the vector $\boldsymbol{\beta}$ into the vector $\mathbf{z}$. Write $\mathbf{D}(\mathbf{z}, \boldsymbol{\beta})$ in the form

$$\mathbf{D}(\mathbf{z}, \boldsymbol{\beta}) = \begin{bmatrix} \bar{\mathbf{D}}(\mathbf{z}, \boldsymbol{\beta}) & O_{21} \\ O_{12} & 1 \end{bmatrix},$$

where $\bar{\mathbf{D}}(\mathbf{z}, \boldsymbol{\beta}) = [d_{ij}]$, $i, j = 1, 2$ is the orthogonal matrix, such that $\mathbf{B}'_2\mathbf{D}(z, \boldsymbol{\beta}) = \left[0 \vdots ||\mathbf{B}_2|| \vdots 0\right]$. Then elements of $\bar{\mathbf{D}}(\mathbf{z}, \boldsymbol{\beta})$ are determined by

$$d_{11} = \cos\Phi, \quad d_{12} = -\sin\Phi, \quad d_{21} = -d_{12}, \quad d_{22} = d_{11}, \tag{4.178}$$

where Φ is the angle of the rotation of axes.

In the new coordinate system the point of the intersection of constraints is

$$\boldsymbol{\beta}^* = \mathbf{D}'(\mathbf{z},\boldsymbol{\beta})\mathbf{z}^* = [\beta_1^*\ \beta_2^*\ \beta_3^{*\prime}].$$

According to (4.170) and (4.177), in the new coordinate system the mass distribution on the plane ω_2 has the form

$$\tilde{\varphi}_2(\tilde{\mathbf{z}}) = \varphi_2(\mathbf{z}) = \Theta_2(\boldsymbol{\beta}) = \frac{1}{2\pi}a_2 \exp\left(\frac{(\beta_2^*)^2}{2}\right)\exp\left(-\frac{\boldsymbol{\beta}'\boldsymbol{\beta}}{2}\right),$$

where $\tilde{\mathbf{z}} = \mathbf{z}$ satisfies (4.160), and a_2 is given by (4.171).

Taking into account that $\beta_2 = \beta_2^* = \text{const}$, we get

$$\Theta_2(\boldsymbol{\beta}) = \frac{1}{2\pi}a_2 \exp\left\{-\frac{\beta_1^2 + \beta_3^2}{2}\right\}. \tag{4.179}$$

Define in new coordinates the part of ω_2 which belongs to the admissible domain. For this we find the equation of the edge ω_{23}. According to (4.158) and (4.177), ω_{23} is described by the system of equations:

$$\begin{cases}\mathbf{B}'_2\mathbf{z} = \beta_2 = \beta_2^*,\\ \mathbf{B}'_3\mathbf{z} = \mathbf{r}'_3(\boldsymbol{\beta})\boldsymbol{\beta} = r_{31}(\boldsymbol{\beta})\beta_1 + r_{32}(\boldsymbol{\beta})\beta_2 + r_{33}(\boldsymbol{\beta})\beta_3 = b_3,\end{cases} \tag{4.180}$$

where $\mathbf{r}_3(\boldsymbol{\beta}) = \mathbf{D}'(\mathbf{z},\boldsymbol{\beta})\mathbf{B}_3$ is the vector with components $r_{3i}(\boldsymbol{\beta})$, $i = \overline{1,3}$.

From (4.180) we obtain the equation for ω_{23}

$$r_{31}(\boldsymbol{\beta})\beta_1 + r_{33}(\boldsymbol{\beta})\beta_3 = R_3(\boldsymbol{\beta}) = b_3 - r_{32}(\boldsymbol{\beta})\beta_2^*.$$

Thus, in new coordinates the points of ω_2 which belong to the admissible domain are given by

$$\begin{aligned}\omega_2^{(\boldsymbol{\beta})} = \{&\boldsymbol{\beta} \in \Re^3 : -\infty < \beta_1 = \beta_1^*, \beta_2 = \beta_2^*, -\infty < \beta_3 \le r_{33}^{-1}(\boldsymbol{\beta})(R_3(\boldsymbol{\beta})\\ &-r_{31}(\boldsymbol{\beta})\beta_1) = L_{32}(\beta_1)\}.\end{aligned} \tag{4.181}$$

Denote $\mathbf{M}(\boldsymbol{\beta}) = E\{\boldsymbol{\beta}\boldsymbol{\beta}'\}$. Since the distribution density β is determined by (4.179), we obtain by (4.181) the expression for the matrix elements $\mathbf{M}(\boldsymbol{\beta}) = [m_{ij}(\boldsymbol{\beta})]$, $i,j = \overline{1,3}$:

$$\begin{aligned}m_{ij}(\boldsymbol{\beta}) &= \frac{a_2}{2\pi}\int_{-\infty}^{\beta_1^*}\beta_i\beta_j e^{-\beta_1^2/2}\,\mathrm{d}\beta_1 \int_{-\infty}^{L_{32}(\beta_1)} e^{-\beta_3^2/2}\,\mathrm{d}\beta_3, \quad i,j = \overline{1,3},\\ \beta_2 &= \beta_2^* = \text{const},\end{aligned}$$

where $L_{32}(\beta)_1 = r_{33}^{-1}(\boldsymbol{\beta})(R_3(\boldsymbol{\beta}) - r_{31}(\boldsymbol{\beta})\beta_1)$.

According to Statement 4.1 and (4.177), $\mathbf{M}^{(2)}[3] = \mathbf{D}(\mathbf{z}, \boldsymbol{\beta})\mathbf{M}(\boldsymbol{\beta})\mathbf{D}'(\mathbf{z}, \boldsymbol{\beta})$.

Calculation of $\mathbf{M}^{(3)}[3]$. Consider the plane ω_3 given by (4.162). Make the change of variables

$$\mathbf{z} = \mathbf{D}(\mathbf{z}, \boldsymbol{\delta})\boldsymbol{\delta}, \boldsymbol{\delta} \in \Re^3, \tag{4.182}$$

where $\mathbf{D}(\mathbf{z}, \boldsymbol{\delta})$ is the orthogonal matrix, satisfying

$$\mathbf{B}'_3 z = \mathbf{B}'_3\mathbf{D}(\mathbf{z}, \boldsymbol{\delta})\boldsymbol{\delta} = [0 \vdots 0 \vdots ||\mathbf{B}_3||]\boldsymbol{\delta} = b_3 \tag{4.183}$$

and $\boldsymbol{\delta} = [\delta_1 \ \delta_2 \ \delta_3]'$.

Denote

$$\mathbf{r}(\boldsymbol{\delta}) = \mathbf{B}\mathbf{D}(\mathbf{z}, \boldsymbol{\delta}) = [r_{ij}(\boldsymbol{\delta})], \quad i, j = \overline{1, 3}, \tag{4.184}$$

where the matrix $\mathbf{B}$ is determined in (4.156). We can write $\mathbf{r}(\boldsymbol{\delta})$ as follows

$$\mathbf{r}(\boldsymbol{\delta}) = \begin{bmatrix} \bar{\mathbf{r}}(\boldsymbol{\delta}) & \vdots & r_{13}(\boldsymbol{\delta}) \\ & \vdots & r_{23}(\boldsymbol{\delta}) \\ \cdots & \cdots & \cdots \\ \mathbf{O}_{12} & \vdots & r_{33}(\boldsymbol{\delta}) \end{bmatrix}, \tag{4.185}$$

where $\bar{\mathbf{r}}(\boldsymbol{\delta})$ is the (2×2) matrix, $r_{33}(\boldsymbol{\delta}) = ||\mathbf{B}_3||$ (see (4.183)). Since the matrix $\mathbf{r}(\boldsymbol{\delta})$ is non-degenerate, $\bar{\mathbf{r}}(\boldsymbol{\delta})$ is also non-degenerate, implying that its orthogonal decomposition is of the form:

$$\bar{\mathbf{r}}(\boldsymbol{\delta}) = \mathbf{s}(\boldsymbol{\delta})\mathbf{h}(\boldsymbol{\delta}), \tag{4.186}$$

where $\mathbf{s}(\boldsymbol{\delta}) = \begin{bmatrix} s_{11}(\boldsymbol{\delta}) & 0 \\ s_{21}(\boldsymbol{\delta}) & s_{22}(\boldsymbol{\delta}) \end{bmatrix}$, $\mathbf{h}(\boldsymbol{\delta})$ is the orthogonal matrix. Introduce the matrices

$$\boldsymbol{\Pi}(\boldsymbol{\delta}) = -\begin{bmatrix} \text{sign}\, s_{11}(\boldsymbol{\delta}) & 0 \\ 0 & \text{sign}\, s_{22}(\boldsymbol{\delta}) \end{bmatrix}, \quad \mathbf{D}(\boldsymbol{\delta}, \boldsymbol{\gamma}) = \begin{bmatrix} \mathbf{h}'(\boldsymbol{\delta})\boldsymbol{\Pi}(\boldsymbol{\delta}) & \mathbf{O}_{21} \\ \mathbf{O}_{12} & 1 \end{bmatrix}, \tag{4.187}$$

and put $\boldsymbol{\delta} = \mathbf{D}(\boldsymbol{\delta}, \boldsymbol{\gamma})\boldsymbol{\gamma}$. Then by (4.184)–(4.187) we obtain from (4.158)

$$\mathbf{B}\mathbf{z} = \mathbf{r}(\boldsymbol{\delta})\mathbf{D}(\boldsymbol{\delta}, \boldsymbol{\gamma})\boldsymbol{\gamma} = \mathbf{B}\mathbf{D}(\mathbf{z}, \boldsymbol{\gamma})\boldsymbol{\gamma} \leq \mathbf{b}, \tag{4.188}$$

where

$$\mathbf{r}(\boldsymbol{\delta})\mathbf{D}(\boldsymbol{\delta}, \boldsymbol{\gamma}) = \begin{bmatrix} s(\boldsymbol{\delta})\Pi(\boldsymbol{\delta}) & \vdots & r_{13}(\boldsymbol{\delta}) \\ & \vdots & r_{23}(\boldsymbol{\delta}) \\ \cdots & \cdots & \cdots \\ \mathbf{O}_{12} & \vdots & r_{33}(\boldsymbol{\delta}) \end{bmatrix}, \quad \mathbf{D}(\mathbf{z}, \boldsymbol{\gamma}) = \mathbf{D}(\mathbf{z}, \boldsymbol{\delta})\mathbf{D}(\boldsymbol{\delta}, \boldsymbol{\gamma}). \tag{4.189}$$

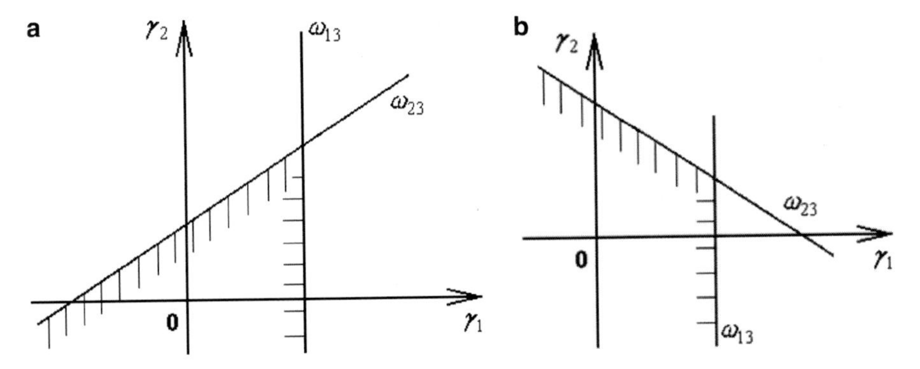

Fig. 4.3 Section of the admissible domain in coordinates $(\gamma_1; \gamma_2; \gamma_3)$ by the plane $\gamma_3 = \text{const}$

Let $\mathbf{r}(\boldsymbol{\gamma}) = \mathbf{r}(\boldsymbol{\delta})\mathbf{D}(\boldsymbol{\delta}, \boldsymbol{\gamma}) = [r_{ij}(\boldsymbol{\gamma})]$, $i, j = \overline{1,3}$. Moreover, according to (4.185), we have

$$r_{i3}(\boldsymbol{\gamma}) = r_{i3}(\boldsymbol{\delta}), \quad i = \overline{1,3}. \tag{4.190}$$

Then by (4.188) the system of inequalities

$$\begin{cases} r_{11}(\boldsymbol{\gamma})\gamma_1 + r_{13}(\boldsymbol{\gamma})\gamma_3 \leq b_1, \\ r_{21}(\boldsymbol{\gamma})\gamma_1 + r_{22}(\boldsymbol{\gamma})\gamma_2 + r_{23}(\boldsymbol{\gamma})\gamma_3 \leq b_2, \\ r_{33}(\boldsymbol{\gamma})\gamma_3 \leq b_3 \end{cases} \tag{4.191}$$

determines the admissible area (4.158) in new coordinates.

Taking in the inequalities above $\gamma_3 = \gamma_3^* = b_3/r_{33}(\boldsymbol{\gamma}) = b_3/||\mathbf{B}_3||$ we get the inequalities which describe the admissible values of γ on ω_3 (see Fig. 4.3):

$$\begin{cases} r_{11}(\boldsymbol{\gamma})\gamma_1 \leq R_1(\boldsymbol{\gamma}) = b_1 - r_{13}(\boldsymbol{\gamma})\gamma_3^*, \\ r_{21}(\boldsymbol{\gamma})\gamma_1 + r_{22}(\boldsymbol{\gamma})\gamma_2 \leq R_2(\boldsymbol{\gamma}) = b_2 - r_{23}(\gamma)\gamma_3^*. \end{cases} \tag{4.192}$$

Here the edges ω_{13} and ω_{23} correspond to the cases when we have equalities in the first and in the second constraints, respectively.

Let us determine the distribution density of the fraction of the unit mass on ω_3. According to (4.170), we have

$$\tilde{\varphi}_3(\tilde{\mathbf{z}}) = \frac{a_3}{2\pi} \exp\left\{-\frac{b_3^2}{2||\mathbf{B}_3||^2}\right\} \exp\left\{-\frac{\tilde{\mathbf{z}}'\tilde{\mathbf{z}}}{2}\right\}, \tag{4.193}$$

where $\tilde{\mathbf{z}}$ is the projection of $\mathbf{z}$ on ω_3, a_3 is determined by (4.171). We have

$$\mathbf{z} = \mathbf{D}(\mathbf{z}, \boldsymbol{\gamma})\boldsymbol{\gamma}, \tag{4.194}$$

where matrix $\mathbf{D}(\mathbf{z},\boldsymbol{\gamma})$ is defined in (4.189). It is easy to see that $\mathbf{D}'(\mathbf{z},\boldsymbol{\gamma})\mathbf{D}(\mathbf{z},\boldsymbol{\gamma}) = \mathbf{J}_3$. Hence, according to (4.194), $\mathbf{z}'\mathbf{z} = \tilde{\mathbf{z}}'\tilde{\mathbf{z}} = \boldsymbol{\gamma}'\boldsymbol{\gamma}$. Then from (4.193) we obtain the distribution density of the fraction of the unit mass on ω_3.

$$\tilde{\varphi}_3(\tilde{\mathbf{z}}) = \Theta_3(\boldsymbol{\gamma}) = \frac{a_3}{2\pi}\exp\left\{-\frac{b_3^2}{2||\mathbf{B}_3||^2}\right\}\exp\left\{-\frac{\gamma_1^2+\gamma_2^2}{2}\right\}, \tag{4.195}$$

where $\mathbf{z} = \tilde{\mathbf{z}}$ satisfies (4.160).

Denote the point of the intersection of all constraints (or, equivalently, of edges ω_{13} and ω_{23}) in new coordinates by $\boldsymbol{\gamma}^* = [\gamma_1^*, \gamma_2^*, \gamma_3^*]' = \mathbf{D}'(\mathbf{z},\boldsymbol{\gamma})\mathbf{z}^*$. Then by (4.192) the range of the values of γ_1 and γ_2 on the plane ω_3 is

$$\omega_3^{(\boldsymbol{\gamma})} = \{\boldsymbol{\gamma} \in \Re^3 : -\infty < \gamma_1 \le \gamma_1^*, -\infty < \gamma_2 \le r_{22}^{-1}(\boldsymbol{\gamma})(R_2(\boldsymbol{\gamma}) - r_{21}(\boldsymbol{\gamma})\gamma_1) = L_{23}(\gamma_1)\}. \tag{4.196}$$

Let $\mathbf{M}(\boldsymbol{\gamma}) = E\{\boldsymbol{\gamma}\boldsymbol{\gamma}'\} = [m_{ij}(\boldsymbol{\gamma})], i, j = \overline{1,3}$. By (4.196) and (4.193) the elements of $\mathbf{M}(\boldsymbol{\gamma})$ are given by

$$m_{ij}(\gamma) = \frac{a_3}{2\pi}\int_{-\infty}^{\gamma_1^*} \gamma_i\gamma_j e^{-\gamma_1^2/2}\,d\gamma_1 \int_{-\infty}^{L_{23}(\gamma_1)} e^{-\gamma_2^2/2} d\gamma_2, \quad i,j = \overline{1,3}, \gamma_3 = \gamma_3^* = \text{const},$$

where $L_{23}(\gamma_1) = r_{22}^{-1}(\boldsymbol{\gamma})(R_2(\boldsymbol{\gamma}) - r_{21}(\boldsymbol{\gamma})\gamma_1)$.

According to Statement 4.1, we have $\mathbf{M}^{(3)}[3] = \mathbf{D}(\mathbf{z},\boldsymbol{\gamma})\mathbf{M}(\boldsymbol{\gamma})\mathbf{D}'(\mathbf{z},\boldsymbol{\gamma})$.

Calculation of $\mathbf{M}^{(4)}[3]$. The distribution density of the unit mass within the admissible area is

$$\varphi_4(\mathbf{z}) = \frac{1}{\left(\sqrt{2\pi}\right)^3}\exp\left\{-\frac{\mathbf{z}'\mathbf{z}}{2}\right\}, \tag{4.197}$$

and the range of the variables defined by inequalities (4.158) is given by

$$\omega_4^{(\mathbf{z})} = \{\mathbf{z} \in \Re^3 : -\infty < z_1 \le z_1^*, -\infty < z_2 \le b_{22}^{-1}(b_2 - b_{21}z_1) = L_{24}(z_1),$$
$$-\infty < z_3 \le b_{33}^{-1}(b_1 - b_{31}z_1 - b_{32}z_2) = L_{35}(z_1, z_2)\}. \tag{4.198}$$

By (4.197) and (4.198), elements of the matrix $\mathbf{M}^{(4)}[3] = [m_{ij}^{(4)}[3]], i, j = \overline{1,3}$ are of the form

$$m_{ij}^{(4)}[3] = \frac{1}{\left(\sqrt{2\pi}\right)^3}\int_{-\infty}^{z_1^*} z_i z_j e^{-z_1^2/2}\,dz_1 \int_{-\infty}^{L_{24}(z_1)} e^{-z_2^2/2}$$
$$\times\, dz_2 \int_{-\infty}^{L_{35}(z_1,z_2)} e^{-z_3^2/2}\,dz_3, \quad i,j = \overline{1,3}.$$

Calculation of $\mathbf{M}^{(12)}[3]$. For the edge ω_{12} the range of variables is given by

$$\omega_{12}^{(\mathbf{z})} = \{\mathbf{z} \in \Re^3 : z_1 = z_1^*, z_2 = z_2^*, -\infty < z_3 \le z_3^*\}. \tag{4.199}$$

Consider the cut of the area determined in (4.158) by the plane Ω_0 orthogonal to the axis Oz_3 when $z_3 \in \omega_{12}^{(z)}$ (see Fig. 4.1). Denote by Ω_1 and Ω_2 the planes orthogonal to planes ω_1 and ω_2, respectively. On Fig. 4.1, $\Omega_1, \Omega_2, \omega_1, \omega_2$ mean the lines, obtained by intersecting, respectively, the planes $\Omega_1, \Omega_2, \omega_1, \omega_2$ with Ω_0. According to Fig. 4.1, the mass, enclosed between Ω_1 and Ω_2 is concentrated at the point (z_1^*, z_2^*, z_3), located on Ω_0. We obtain from (4.158) that the equation of the line ω_2 is $z_2 = kz_1 + b_2^*$, where $k = -b_{21}/b_{22}$, $b_2^* = b_2/b_{22}$. Then the equation of the line Ω_2, orthogonal to the line ω_2 and passing through the point (z_1^*, z_2^*) is

$$z_1 = L_{16}(z_2) = k(l - z_2), \quad l = z_2^* + z_1^*/k.$$

Hence, the entire mass, located in the domain $O = \{\mathbf{z} = [z_1 z_2] : z_1 \geq k(l - z_2), z_2 \geq z_2^*; \mathbf{z} \in \Re^2\}$ is projected into the point (z_1^*, z_2^*) on the plane Ω_0. The distribution density of the projected mass is standard three-dimensional normal density, see (4.197). Then for $(z_1^*, z_2^*, z_3) \in \omega_{12}^{(z)}$ the fraction of the unit mass has the density

$$\varphi_{12}(\mathbf{z}) = \varphi_{12}(z_3) = \psi(z_3)\left[\frac{1}{\sqrt{2\pi}}\int_{z_2^*}^{\infty} \exp\left(-\frac{z_2^2}{2}\right)(1 - \Psi[L_{16}(z_2)])\mathrm{d}z_2\right].$$

Hence we obtain the expression for elements of the matrix $\mathbf{M}^{(12)}[3] = [m_{ij}^{(12)}[3]]$, $i, j = \overline{1,3}$:

$$m_{ij}^{(12)}[3] = \int_{-\infty}^{z_3^*} z_i z_j \varphi_{12}(z_3)\mathrm{d}z_3, \quad i, j = \overline{1,3},\ z_i = z_i^* = \text{const},\ i, j = 1, 2.$$

Calculation of $\mathbf{M}^{(13)}[3]$. The range of $\boldsymbol{\gamma}$ on the edge ω_{13} (which is the intersection of ω_1 and ω_3) is

$$\omega_{13}^{(\gamma)} = \{\boldsymbol{\gamma} \in \Re^3 : \gamma_1 = \gamma_1^*, -\infty < \gamma_2 \leq \gamma_2^*, \gamma_3 = \gamma_3^*\}. \tag{4.200}$$

The mass, located between the lines, formed by the cut of planes Ω_1 and Ω_3 by the plane, parallel to the coordinate plane γ_2Oγ_1, is projected into an arbitrary point $\boldsymbol{\gamma} \in \omega_{13}^{(\gamma)}$. Let us find equations for Ω_1 and Ω_3.

When $r_{13}(\gamma) \neq 0$ we get the equation for ω_1 from (4.191):

$$\gamma_3 = \frac{r_{11}(\boldsymbol{\gamma})}{r_{13}(\boldsymbol{\gamma})}\gamma_1 + \frac{b_3}{r_{13}(\boldsymbol{\gamma})}.$$

Then the equation of the plane Ω_1, orthogonal to the plane ω_1, is

$$\gamma_3 = -\frac{r_{13}(\boldsymbol{\gamma})}{r_{11}(\boldsymbol{\gamma})}\gamma_1 + \frac{r_{13}(\boldsymbol{\gamma})}{r_{11}(\boldsymbol{\gamma})}\gamma_1^* + \gamma_3^* = L_{37}(\gamma_1).$$

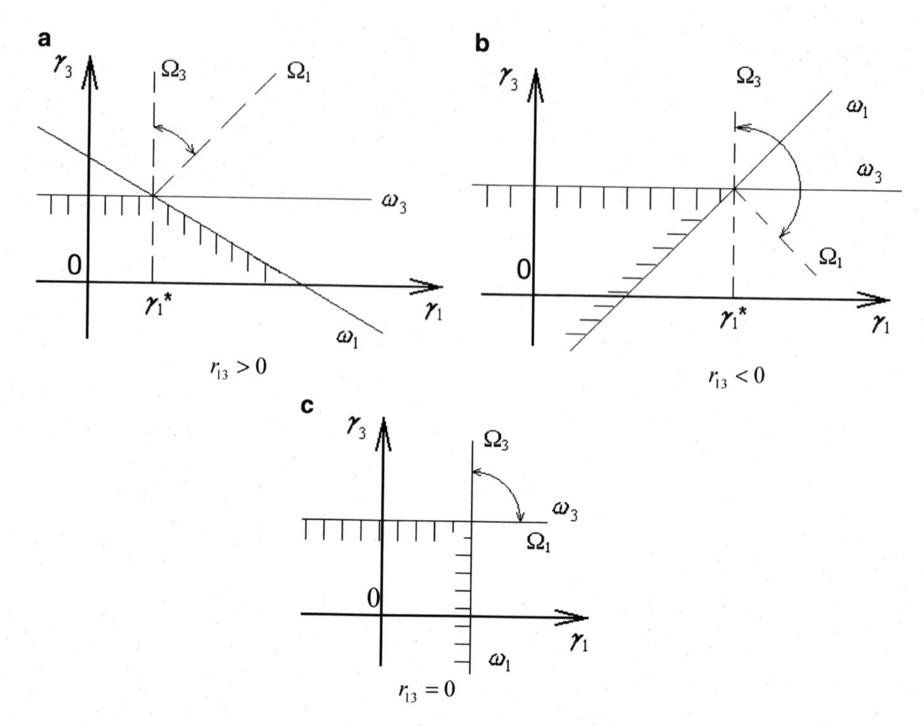

Fig. 4.4 Section of the admissible domain in coordinates $(r_1; r_2; r_3)$ by the plane $r_2 = \text{const}$

The equation of the plane Ω_3 orthogonal to the plane ω_3 and parallel to the coordinate axis $O\gamma_2$ is $\gamma_1 = \gamma_1^*$. Thus, the mass projected into the point $\boldsymbol{\gamma} \in \omega_{13}^{(\gamma)}$, is located in the following area:

$$\{\boldsymbol{\gamma} \in \Re^3 : \gamma_1^* \leq \gamma_1 < \infty, -\infty < \gamma_2 \leq \gamma_2^*, L_{37}(\gamma_1) \leq \gamma_3 < \infty\}. \tag{4.201}$$

The considerations above are illustrated in Fig. 4.4. It is easy to see that (4.201) is valid for all values of $r_{13}(\boldsymbol{\gamma})$, including $r_{13}(\boldsymbol{\gamma}) = 0$.

By (4.201) the distribution density of the fraction of the unit mass on the edge $\omega_{13}^{(\gamma)}$ is

$$\varphi_{13}(\gamma_2) = \psi(\gamma_2)\left[\frac{1}{\sqrt{2\pi}}\int_{\gamma_1^*}^{\infty} \exp\left(-\frac{\gamma_1^2}{2}\right)(1 - \Psi[L_{37}(\gamma_1)])\mathrm{d}\gamma_1\right].$$

Denote $\mathbf{M}^{(13)}(\boldsymbol{\gamma}) = E\{\boldsymbol{\gamma}\boldsymbol{\gamma}'\} = [m_{ij}^{(13)}(\boldsymbol{\gamma})]$, $\boldsymbol{\gamma} \in \omega_{13}^{(\gamma)}$, $i, j = \overline{1,3}$. The elements of $\mathbf{M}^{(13)}(\boldsymbol{\gamma})$ are given by

$$m_{ij}^{(13)}(\boldsymbol{\gamma}) = \int_{-\infty}^{\gamma_2^*} \gamma_i \gamma_j \varphi_{13}(\gamma_2)\mathrm{d}\gamma_2, \quad i, j = \overline{1,3},\ \gamma_i = \gamma_i^* = \text{const},\ i, j = 1, 3.$$

Using the relation $\mathbf{z} = \mathbf{D}(z, \boldsymbol{\gamma})\boldsymbol{\gamma}$ and Statement 4.1, we have

$$\mathbf{M}^{(13)}[3] = \mathbf{D}(z, \boldsymbol{\gamma})\mathbf{M}^{(13)}(\boldsymbol{\gamma})\mathbf{D}'(z, \boldsymbol{\gamma}).$$

Calculation of $\mathbf{M}^{(23)}[3]$. Introduce the transformation

$$\boldsymbol{\gamma} = \mathbf{D}(\boldsymbol{\gamma}, \boldsymbol{\eta})\boldsymbol{\eta}, \tag{4.202}$$

where $\mathbf{D}(\boldsymbol{\gamma}, \boldsymbol{\eta}) = \begin{bmatrix} \bar{\mathbf{D}}(\boldsymbol{\gamma}, \boldsymbol{\eta}) & \mathbf{O}_{21} \\ \mathbf{O}_{12} & 1 \end{bmatrix} = [r_{ij}(\eta)],\ i, j = \overline{1, 3}$. Here $\bar{\mathbf{D}}(\boldsymbol{\gamma}, \boldsymbol{\eta})$ is the orthogonal 2×2 matrix. We choose $\bar{\mathbf{D}}(\boldsymbol{\gamma}, \boldsymbol{\eta})$ from the condition

$$[r_{21}(\boldsymbol{\eta})\vdots r_{22}(\boldsymbol{\eta})] = [r_{21}(\boldsymbol{\gamma})\vdots r_{22}(\boldsymbol{\gamma})]\bar{\mathbf{D}}(\boldsymbol{\gamma}, \boldsymbol{\eta}) = \left[\sqrt{(r_{21}(\boldsymbol{\gamma}))^2 + (r_{22}(\boldsymbol{\gamma}))^2}\vdots 0\right], \tag{4.203}$$

where $r_{21}(\boldsymbol{\gamma})$ and $r_{22}(\boldsymbol{\gamma})$ are coefficients of the equation of the plane ω_2, see (4.191), $r_{21}(\boldsymbol{\eta})$ and $r_{22}(\boldsymbol{\eta})$ are the first two coefficients of the same plane equation in new coordinates (η_1, η_2, η_3).

By (4.202) and (4.191) the equation of the edge ω_{23} is

$$\begin{cases} r_{21}(\boldsymbol{\eta})\eta_1 + r_{23}(\boldsymbol{\eta})\eta_3 = b_2, \\ r_{33}(\boldsymbol{\eta})\eta_3 = b_3, \end{cases} \tag{4.204}$$

where $r_{21}(\boldsymbol{\eta}) = \sqrt{(r_{21}(\boldsymbol{\gamma}))^2 + (r_{22}(\boldsymbol{\gamma}))^2}$ (see (4.203)), and by (4.190) $r_{i3}(\boldsymbol{\eta}) = r_{i3}(\boldsymbol{\gamma}) = r_{i3}(\boldsymbol{\delta}),\ i = 2, 3$. Here coefficients $r_{i3}(\boldsymbol{\delta}),\ i = 2, 3$, are determined by (4.189).

Further considerations are analogous to those used for calculating $\mathbf{M}^{(13)}$.

By (4.204) the equation of the plane ω_2 at $r_{23}(\boldsymbol{\eta}) \neq 0$ is

$$\eta_3 = \frac{r_{21}(\boldsymbol{\eta})}{r_{23}(\boldsymbol{\eta})}\eta_1 + \frac{b_1}{r_{23}(\boldsymbol{\gamma})}.$$

Then the equation of the plane Ω_2 orthogonal to ω_2 is

$$\eta_3 = -\frac{r_{23}(\boldsymbol{\eta})}{r_{21}(\boldsymbol{\eta})}\eta_1 + \frac{r_{23}(\boldsymbol{\gamma})}{r_{21}(\boldsymbol{\gamma})}\eta_1^* + \eta_3^* = L_{38}(\eta_1),$$

where η_1^* η_3^* are the components of $\boldsymbol{\eta}^* = \mathbf{D}'(\boldsymbol{\gamma}, \boldsymbol{\eta})\boldsymbol{\gamma}^*$.

The equation of the plane Ω_3 orthogonal to ω_3 and parallel to coordinate axis $O\eta_2$ is $\eta_1 = \eta_1^*$.

Points of the edge ω_{23}, located in the admissible area, belong to the set

$$\omega_{13}^{(\boldsymbol{\eta})} = \{\boldsymbol{\eta} \in \Re^3 : \eta_1 = \eta_1^*, -\infty < \eta_2 \leq \eta_2^*, \eta_3 = \eta_3^*\}. \tag{4.205}$$

The mass, which is located in the domain

$$\{\boldsymbol{\eta} \in \Re^3 : \eta_1^* \leq, \eta_1 < \infty, -\infty < \eta_2 \leq \eta_2^*, L_{38}(\eta_1) \leq \eta_3 < \infty\}, \tag{4.206}$$

is projected into the point $\boldsymbol{\eta} \in \omega_{13}^{(\boldsymbol{\eta})}$.

It is easy to check both for this case and for the edge ω_{13}, that the expression (4.203) holds for all values of $r_{21}(\boldsymbol{\eta})$.

By (4.206), the distribution density of the portion of the unit mass on the edge ω_{23} is

$$\varphi_{23}(\eta_2) = \psi(\eta_2)\left[\frac{1}{\sqrt{2\pi}}\int_{\eta_1^*}^{\infty} e^{-\eta_1^2/2}(1 - \Psi[L_{38}(\eta_1)])\mathrm{d}\eta_1\right].$$

The elements of $\mathbf{M}^{(23)}(\boldsymbol{\eta}) = E\{\boldsymbol{\eta}\boldsymbol{\eta}'\} = [m_{ij}^{(23)}(\boldsymbol{\eta})]$, $\boldsymbol{\eta} \in \omega_{23}, i, j = \overline{1,3}$ are

$$m_{ij}^{(23)}(\boldsymbol{\eta}) = \int_{-\infty}^{\eta_2^*} \eta_i \eta_j \varphi_{23}(\eta_2)\mathrm{d}\eta_2, \quad i, j = \overline{1,3},\ \eta_i = \eta_i^* = \text{const},\ i, j = 1, 3.$$

Using $\mathbf{z} = \mathbf{D}(\mathbf{z}, \boldsymbol{\gamma})\boldsymbol{\gamma}$ and (4.202) we obtain

$$\mathbf{z} = \mathbf{D}(\mathbf{z}, \boldsymbol{\eta})\boldsymbol{\eta}, \quad \mathbf{D}(\mathbf{z}, \boldsymbol{\eta}) = \mathbf{D}(\mathbf{z}, \boldsymbol{\gamma})\mathbf{D}(\boldsymbol{\gamma}, \boldsymbol{\eta}). \tag{4.207}$$

It is easy to see that the matrix $\mathbf{D}(\mathbf{z}, \boldsymbol{\eta})$ is orthogonal. According to Statement 4.1 we have by (4.207)

$$\mathbf{M}^{(23)}[3] = \mathbf{D}(\mathbf{z}, \boldsymbol{\eta})\mathbf{M}^{(23)}(\boldsymbol{\eta})\mathbf{D}'(\mathbf{z}, \boldsymbol{\eta}).$$

Calculation of $\mathbf{M}^{(1,2,3)}[3]$. Let us calculate $p_{1,2,3}$, i.e., the mass, which is projected to the point z^*. It is obvious that

$$p_{1,2,3} = 1 - \bar{p}_{1,2,3}, \tag{4.208}$$

where $\bar{p}_{1,2,3}$ is the mass located in the admissible domain, given by inequalities (4.158) on faces ω_i, $i = \overline{1,3}$, in area ω_4 and on all edges:

$$\bar{p}_{1,2,3} = \sum_{i=1}^{4} p_i + p_{12} + p_{13} + p_{23},$$

where p_i is the mass concentrated on ω_i, p_{12}, p_{13}, p_{23} are the masses concentrated on the edges ω_{12}, ω_{13}, ω_{23}, respectively.

We have from above:

$$p_1 = \int_{z \in \omega_1^{(z)}} \varphi_1(\mathbf{z})\mathrm{d}\omega_1^{(\mathbf{z})} = a_1 \int_{-\infty}^{z_2^*} \psi(x)\Psi[L_{31}(x)]\mathrm{d}x,$$

$$p_2 = \int_{\mathbf{z} \in \omega_2^{(\mathbf{z})}} \varphi_2(\mathbf{z})\mathrm{d}\omega_2^{(\mathbf{z})} = \int_{\boldsymbol{\beta} \in \omega_2^{(\beta)}} \Theta_2(\boldsymbol{\beta})\mathrm{d}\omega_2^{(\boldsymbol{\beta})} = a_2 \int_{-\infty}^{\beta_1^*} \psi(x)\Psi[L_{32}(x)]\mathrm{d}x,$$

$$p_3 = \int_{\mathbf{z}\in\omega_2^{(\mathbf{z})}} \varphi_3(\mathbf{z})\mathrm{d}\omega_3^{(\mathbf{z})} = \int_{\boldsymbol{\gamma}\in\omega_3^{(\boldsymbol{\gamma})}} \Theta_3(\boldsymbol{\gamma})\mathrm{d}\omega_3^{(\boldsymbol{\gamma})} = a_3 \int_{-\infty}^{\gamma_1^*} \psi(x)\Psi[L_{23}(x)]\mathrm{d}x,$$

$$p_4 = \int_{\mathbf{z}\in\omega_4^{(\mathbf{z})}} \varphi_4(\mathbf{z})\mathrm{d}\omega_4^{(\mathbf{z})} = \frac{1}{\left(\sqrt{2\pi}\right)^3} \int_{-\infty}^{z_1^*} e^{-z_1^2/2}\,\mathrm{d}z_1 \int_{-\infty}^{L_{24}(z_1)} e^{-z_2^2/2}$$

$$\mathrm{d}z_2 \int_{-\infty}^{L_{35}(z_1,z_2)} e^{-z_3^2/2}\,\mathrm{d}z_3,$$

$$p_{12} = \int_{\mathbf{z}\in\omega_{12}^{(\mathbf{z})}} \varphi_{12}(\mathbf{z})\mathrm{d}\omega_{12}^{(\mathbf{z})} = \int_{-\infty}^{z_3^*} \varphi_{12}(z_3)\mathrm{d}z_3,$$

$$p_{13} = \int_{\mathbf{z}\in\omega_{13}^{(\mathbf{z})}} \varphi_{13}(\mathbf{z})\mathrm{d}\omega_{13}^{(\mathbf{z})} = \int_{-\infty}^{\gamma_2^*} \varphi_{13}(\gamma_2)\mathrm{d}\gamma_2,$$

$$p_{23} = \int_{\mathbf{z}\in\omega_{23}^{(\mathbf{z})}} \varphi_{23}(\mathbf{z})\mathrm{d}\omega_{23}^{(\mathbf{z})} = \int_{-\infty}^{\eta_2^*} \varphi_{23}(\eta_2)\mathrm{d}\eta_2.$$

These expressions allow to calculate $\bar{p}_{1,2,3}$ in (4.208). The elements of the matrix $\mathbf{M}^{(1,2,3)}[3] = [m_{ij}^{(1,2,3)}[3]]$, $i,j = \overline{1,3}$ are determined by

$$m_{ij}^{(1,2,3)}[3] = z_i^* z_j^* p_{1,2,3}, \quad i,j = \overline{1,3}.$$

4.7 Sample Estimates of the Matrix of m.s.e. of Parameter Estimates When the Number of Inequality Constraints Is less than Three

From the relations obtained in Sect. 4.6 one can get the formulas when the number of constraints is $m = 1, 2$. In this case, the admissible area is defined by lines ω_1 and ω_2 (case $m = 2$), see Fig. 4.1, and by the line ω_1 (case $m = 1$). The matrices s and B are $(m \times m)$ matrices, located in the upper left corners, of, respectively, matrices (4.153) and (4.156).

4.7.1 *Case m = 2*

In this case, the unit mass is distributed in the admissible area in the sets ω_k (in the line ω_1 when $k = 1$, in the line ω_2 when $k = 2$, and in the set within the admissible area when $k = 3$) and in the point (1,2) which is the intersection of ω_1 and ω_2. Then the symmetric matrix $\mathbf{M}_\upsilon[3]$ in (4.167) can be written as

$$\mathbf{M}_{\nu}[2] = \sum_{k=1}^{3} \mathbf{M}^{(k)}[2] + \mathbf{M}^{(1,2)}[2]. \tag{4.209}$$

Elements of the matrices in the right-hand side of (4.209) are calculated as follows.

Calculation of $\mathbf{M}^{(1)}[2] = [m_{ij}^{(1)}[2]]$, $i, j = 1, 2$. Since for $m = 2$ we have in (4.176) $L_{31}(z_2) = \infty$, we obtain from the expression for $m_{ij}^{(1)}$ in Sect. 4.6

$$m_{11}^{(1)}[2] = \frac{a_1}{\sqrt{2\pi}} \int_{-\infty}^{z_2^*} (z_1^*)^2 \exp\left\{-\frac{z_2^2}{2}\right\} \mathrm{d}z_2,$$

$$m_{22}^{(1)}[2] = \frac{a_1}{\sqrt{2\pi}} \int_{-\infty}^{z_2^*} z_2^2 \exp\left\{-\frac{z_2^2}{2}\right\} \mathrm{d}z_2,$$

$$m_{12}^{(1)}[2] = \frac{a_1}{\sqrt{2\pi}} \int_{-\infty}^{z_2^*} z_1^* z_2 \exp\left\{-\frac{z_2^2}{2}\right\} \mathrm{d}z_2, \tag{4.210}$$

where z_1^*, z_2^* are the components of $\mathbf{z}^* \in \Re^2$, which is the solution to $\mathbf{Bz} = \mathbf{b}$, and a_1 is determined by (4.171) for $k = 1$, $||\mathbf{B}_1|| = b_{11}$.

Calculation of $\mathbf{M}^{(2)}[2]$. Make the change of variables $\mathbf{z} = \overline{\mathbf{D}}(\mathbf{z}, \boldsymbol{\beta})\boldsymbol{\beta}$, where $\boldsymbol{\beta} \in \Re^2$, and the matrix $\overline{\mathbf{D}}(\mathbf{z}, \boldsymbol{\beta}) = [d_{ij}]$, $i, j = 1, 2$ is defined by (4.178).

Then

$$\mathbf{M}^{(2)}[2] = \overline{\mathbf{D}}(\mathbf{z}, \boldsymbol{\beta})\mathbf{M}(\boldsymbol{\beta})\overline{\mathbf{D}}'(\mathbf{z}, \boldsymbol{\beta}), \quad \mathbf{M}(\boldsymbol{\beta}) = [m_{ij}(\boldsymbol{\beta})], \quad i, j = 1, 2. \tag{4.211}$$

Taking into account that for $m = 2$ in (4.181) $L_{32}(z_2) = \infty$, we obtain from the expression for $m_{ij}(\boldsymbol{\beta})$ in the Sect. 4.6 that

$$m_{11}(\boldsymbol{\beta}) = \frac{a_2}{\sqrt{2\pi}} \left(-\beta_1^* e^{-(\beta_1^*)^2/2} + \int_{-\infty}^{\beta_1^*} e^{-\beta_1^2/2} \,\mathrm{d}\beta_1 \right),$$

$$m_{22}(\boldsymbol{\beta}) = \frac{a_2(\beta_2^*)^2}{\sqrt{2\pi}} \int_{-\infty}^{\beta_1^*} e^{-\beta_1^2/2} \,\mathrm{d}\beta_1,$$

$$m_{12}(\boldsymbol{\beta}) = \frac{a_2\beta_2^*}{\sqrt{2\pi}} \int_{-\infty}^{\beta_1^*} x e^{-\beta_1^2/2} \,\mathrm{d}\beta_1, \tag{4.212}$$

where $\boldsymbol{\beta}^* = \overline{\mathbf{D}}'(\mathbf{z}, \boldsymbol{\beta})\mathbf{z}^*$, β_1^*, β_2^* are the components of $\boldsymbol{\beta}^*$, a_2 is determined by (4.171) for $k = 2$, $\mathbf{B}_2 = [b_{21}\ b_{22}]'$.

Calculation of $\mathbf{M}^{(3)}[2] = [m_{ij}^{(3)}[2]]$, $i, j = 1, 2$. Calculations below are particular cases of the calculations made for matrix $\mathbf{M}^{(4)}[3]$, see Sect. 4.6. In this case, in (4.198) we have $L_{35}(z_1, z_2) = \infty$. Let us rewrite the linear function in (4.198) $L_{24}(z_1) = b_{22}^{-1}(b_2 - b_{21}z_1)$ as $L_{24}(z_1) = kz_1 + b_2^*$, where $k = -b_{21}/b_{22}$, $b_2^* = b_2/b_{22}$. Then (see (4.174) for the notation)

$$m_{11}^{(3)}[2] = \frac{1}{\sqrt{2\pi}} \int_{-\infty}^{z_1^*} z_1^2 e^{-z_1^2/2} \Psi_0(kz_1 + b_2^*) \mathrm{d}z_1,$$

$$m_{22}^{(3)}[2] = \frac{1}{\sqrt{2\pi}} \int_{-\infty}^{z_1^*} e^{-z_1^2/2} \Psi_0(kz_1 + b_2^*) \mathrm{d}z_1,$$

$$m_{12}^{(3)}[2] = \frac{1}{\sqrt{2\pi}} \int_{-\infty}^{z_1^*} z_1 e^{-z_1^2/2} \Psi_0(kz_1 + b_2^*) \mathrm{d}z_1. \quad (4.213)$$

Calculation of $\mathbf{M}^{(1,2)}[2] = [m_{ij}^{(1,2)}[2]]$, $i, j = 1, 2$. In this case,

$$m_{ij}^{(1,2)}[2] = z_i^* z_j^* p_3, \quad i, j = 1, 2, \quad (4.214)$$

where p_3 is the mass, concentrated at the point of intersection of lines ω_1 and ω_2.

According to Fig. 4.1, p_3 is the mass enclosed between the normals Ω_1 and Ω_2 to lines ω_1 and ω_2, respectively. From (4.161) we obtain the equation for ω_2: $z_2 = kz_1 + b_2^*$. Then the equation of the line Ω_2, orthogonal to ω_2 and passing through the point $z = z^*$, is $z_1 = k(l - z_2)$, where $l = z_2^* + z_1^*/k$. Hence, the entire mass, located in the area $O = \{z : z_1 \geq k(l - z_2), z_2 \geq z_2^*; z \in \Re^2\}$, is projected to the point ω_3. Thus

$$p_3 = \frac{1}{\sqrt{2\pi}} \int_{z_2^*}^{\infty} \exp\left(-\frac{z_2^2}{2}\right) (1 - \Psi_0(k(l - z_2))) \mathrm{d}z_2.$$

4.7.2 Case $m = 1$

Since in this case we do not have ω_2, we put in the formulas of Sect. 4.7.1 $z_2^* = k = b_2^* = \infty$. Then in (4.209) we have $\mathbf{M}_\nu[2] = \mathbf{M}^{(1)}[2] + \mathbf{M}^{(3)}[2]$. Using (4.210) and (4.214) one can show that $\mathbf{M}_\nu[2] = \mathrm{diag}(M_\nu[1], 1)$, with

$$M_\nu[1] = m_{11}^{(1)}[2] + m_{11}^{(3)}[2] = a_1(z_1^*)^2 + \frac{1}{\sqrt{2\pi}} \int_{-\infty}^{z_1^*} z_1^2 \exp\left\{-\frac{z_1^2}{2}\right\} \mathrm{d}z_1. \quad (4.215)$$

4.7.3 Comparison of the Estimate of the Matrix of m.s.e. of the Regression Parameter Estimate Obtained with and Without Inequality Constraints for $m = 1, 2$

The estimate for the matrix of m.s.e. of the regression parameter estimate is given by (see (4.165))

$$\mathbf{K}_T = \sigma_T^2 \mathbf{C}_T \mathbf{M}_T^{-1/2} \mathbf{H}'_T \mathbf{\Pi}_T \mathbf{K}_\mathbf{V} \mathbf{\Pi}_T \mathbf{H}_T \mathbf{M}_T^{-1/2} \mathbf{C}'_T, \quad (4.216)$$

where (cf. (4.164))

$$\mathbf{K_V} = \begin{bmatrix} \mathbf{M}_\nu[m] & \mathbf{O}_{m,n-m} \\ \mathbf{O}_{n-m,m} & \mathbf{J}_{n-m} \end{bmatrix}, \tag{4.217}$$

and $\mathbf{M}_\nu[m]$ is a matrix whose elements are functions of the random vector $\mathbf{b}_T$ with components given in (4.128). Denote estimate for the matrix of m.s.e. of the regression parameter estimate by the LS method without constraints by $\tilde{\mathbf{K}}_T$. In order to determine the effect of the influence of constraints on the accuracy of estimation of $\tilde{\mathbf{K}}_T$ and $\mathbf{K}_T$ we use in both cases the same estimate for σ^2. We can take any estimate for σ^2 obtained as the solution to (4.58) with or without constraints. To be more precise, we use the estimate σ_T^2, defined in (4.115). Then for $\mathbf{K_V} = \mathbf{J}_n$ we get from (4.216) $\tilde{\mathbf{K}}_T = \sigma_T^2 \mathbf{R}_T^{-1}$.

Let $m = 1$. For this case the following theorem takes place.

Theorem 4.7. *If $m = 1$, and Assumptions 2.9, 4.1, 4.2 hold true, then: (a) the matrix $\tilde{\mathbf{K}}_T - \mathbf{K}_T$ is non-negative definite for $n > 1$, (b) $\tilde{\mathbf{K}}_T - \mathbf{K}_T$ is positive definite for $n = 1$, (c) if the first component of the n-dimensional vector $\mathbf{W}_T = \mathbf{\Pi}_T \mathbf{H}_T \mathbf{M}_T^{-1/2} \mathbf{C}'_T \mathbf{x}$ is not equal to zero, then for an arbitrary non-zero vector $\mathbf{x} \in \Re^n$ we have $\mathbf{x}'(\tilde{\mathbf{K}}_T - \mathbf{K}_T)\mathbf{x} > 0$ for $n > 1$.*

Proof. By (4.217) we have

$$\mathbf{K_V} = \text{diag}(M_\nu[1]\ 1\ \dots\ 1), \tag{4.218}$$

where the scalar $M_\nu[1]$ is given in (4.215). Here, according to (4.171), $a_1 = \left(\sqrt{2\pi}\right)^{-1} \int_{z_1^*}^{\infty} \exp\left\{-\frac{z_1^2}{2}\right\} \mathrm{d}z_1$, where $z_1^* = b_{1T}/||\mathbf{B}_1||$, and $b_{1T} = b_T \in \Re^1$ is a random variable. According to (4.159) we have $b_T \geq 0$. (Recall that we dropped in the index T in (4.171) and (4.159)).

Let us find the upper bound for $M_\nu[1]$. Since b_T is non-negative, we have $z_1^* \geq 0$. Then

$$a_1(z_1^*)^2 = (z_1^*)^2 \left(\sqrt{2\pi}\right)^{-1} \int_{z_1^*}^{\infty} \exp\left\{-\frac{z_1^2}{2}\right\} \mathrm{d}z_1 < \left(\sqrt{2\pi}\right)^{-1} \int_{z_1^*}^{\infty} z_1^2 \exp\left\{-\frac{z_1^2}{2}\right\} \mathrm{d}z_1.$$

From here and (4.215) we get

$$M_\nu[1] < 1. \tag{4.219}$$

From (4.215) we obtain the lower bound for $M_\nu[1]$:

$$M_\nu[1] = a_1(z_1^*)^2 + \frac{1}{\sqrt{2\pi}} \int_{-\infty}^{z_1^*} z_1^2 \exp\left\{-\frac{z_1^2}{2}\right\} \mathrm{d}z_1 \geq \frac{1}{2},$$

as $z_1^* \geq 0$. Moreover, $M_\nu[1] = 1/2$ if $z_1^* = 0$.

By (4.216) $\tilde{\mathbf{K}}_T - \mathbf{K}_T = \sigma_T^2 \mathbf{L}_T(\mathbf{J}_n - \mathbf{K_V})\mathbf{L}'_\mathbf{T}$, where $\mathbf{L}_T = \mathbf{C}_T \mathbf{M}_T^{-1/2} \mathbf{H}'_T \mathbf{\Pi}_T$, implying together with (4.218), (4.219) the statements (a) and (b).

To prove (c) consider the quadratic form $s_T = \sigma_T^2 \mathbf{x}'\mathbf{L}_T(\mathbf{J}_n - \mathbf{K_V})\mathbf{L}'_T\mathbf{x} = \sigma_T^2 \mathbf{W}'_T(\mathbf{J}_n - \mathbf{K_V})\mathbf{W}_T$, where $\mathbf{x}$ is an arbitrary nonzero vector, $\mathbf{W}_T = \mathbf{L}'_T\mathbf{x}$. By

(4.218) and (4.219) $s_T = \sigma_T^2(1 - M_\nu[1])w_{1T}^2$, where w_{1T} is the first component of the vector $\mathbf{W}_T$. Moreover, $s_T > 0$ if $w_{nT} \neq 0$, which proves (c). □

Let us substitute in (4.128) the values b_{iT}, $i \in I$, as well as $\boldsymbol{\alpha}_T$ and σ_T (4.216) with their true values $\boldsymbol{\alpha}^0$ and σ. Moreover, put in (4.128) $\eta_{iT} = 0$, $i \in I$. Then (4.216) and (4.128) determine the true matrices of the m.s.e. of parameter estimates $\mathbf{K}_T^0$ (with constraints) and $\tilde{\mathbf{K}}_T^0$ (without constraints). It is obvious that Theorem 4.7 remains valid also for $\mathbf{K}_T^0$ and $\tilde{\mathbf{K}}_T^0$.

Let us compare the estimate $\mathbf{K}_T$, which takes into account all constraints for each sample, with the truncated estimate $\mathbf{K}_T^{\text{tr}}$ of the matrix $\mathbf{K}_T^0$. According to (4.14) we have for the linear regression

$$\mathbf{K}_T^{\text{tr}} = \sum_{l=1}^{L} \mathbf{K}_{lT}\gamma_{lT}, \quad \sum_{l=1}^{L} \gamma_{lT} = 1. \tag{4.220}$$

Here $L = 2^m$ is the number of all combinations of active and inactive constraints up to ξ, including all inactive constraints case,

$$\mathbf{K}_{lT} = \sigma_T^2 \mathbf{k}_{lT}, \quad k_{lT} = E\{\mathbf{u}_{lT}\mathbf{u}'_{lT}\}, \tag{4.221}$$

where $\mathbf{u}_{lT}$ is a solution to the problem

$$\frac{1}{2}\mathbf{y}'\mathbf{R}_T\mathbf{y} - \mathbf{q}'_T\mathbf{y} \to \min, \quad \tilde{\mathbf{g}}'_{iT}\mathbf{y} \leq 0, \quad i \in I_l \subseteq I, \; \mathbf{q}_T \sim N(\mathbf{O}_n, \mathbf{R}_T). \tag{4.222}$$

In (4.222) $\tilde{\mathbf{g}}'_{iT}$ is the ith row of the matrix $\tilde{\mathbf{G}}_T$, I_l is the set of numbers of the lth combination of active and inactive constraints up to ξ (which is completely determined by the combination of active constraints up to ξ).

For the case $m = 1$ we have for $L = 2$ (see (4.220))

$$\mathbf{K}_T^{\text{tr}} = \tilde{\boldsymbol{\kappa}}_T\gamma_{1T} + \mathbf{K}_{1T}\gamma_{2T}, \quad \gamma_{1T} + \gamma_{2T} = 1,$$

where the matrices $\tilde{\boldsymbol{\kappa}}_T$ and $\mathbf{K}_{1T}$ are defined by (4.220)–(4.222), when $I(\boldsymbol{\alpha}_T) = \varnothing$ and $I(\boldsymbol{\alpha}_T) = \{1\}$, respectively. Hence, $\boldsymbol{\kappa}_{1T} = \hat{\boldsymbol{\kappa}}_T$, where $\hat{\boldsymbol{\kappa}}_T$ is a sample estimate of the matrix $\boldsymbol{\kappa}_T$, calculated by (4.216), (4.217) for $m = 1$ and the right-hand side of the constraint $\hat{b}_{1T} = 0$.

We have $\boldsymbol{\kappa}_T^{\text{tr}} - \boldsymbol{\kappa}_T = (\tilde{\boldsymbol{\kappa}}_T - \boldsymbol{\kappa}_T)\gamma_{1T} + (\boldsymbol{\kappa}_{1T} - \boldsymbol{\kappa}_T)\gamma_{2T}$. In this expression the first term is a non-negative definite matrix according to Theorem 4.7 and the inequality $\gamma_{1T} \geq 0$. Let $\gamma_{2T} = 1$ (i.e., the constraint is active up to ξ), which holds true when $\hat{b}_{1T} = 0$. Then $\hat{\mathbf{K}}_T^{\text{tr}} - \hat{\mathbf{K}}_T = \mathbf{K}_{1T} - \hat{\mathbf{K}}_T = \mathbf{O}_{nn}$, since in this case $\boldsymbol{\kappa}_{1T} = \hat{\boldsymbol{\kappa}}_T$. Thus, $\hat{\mathbf{K}}_T^{\text{tr}} - \hat{\mathbf{K}}_T$ is the non-negative definite matrix for all possible samples.

Let us formulate the obtained result.

Theorem 4.8. *If $m = 1$, and Assumptions 2.9, 4.1, 4.2 hold true, then the difference $\mathbf{K}_T^{\text{tr}} - \mathbf{K}_T$ is the non-negative definite matrix.*

According to Theorem 4.8 the estimate $\mathbf{K}_T$ of the matrix of m.s.e. of the parameter estimate obtained by taking into account the constraints on parameters is more precise than the truncated estimate $\mathbf{K}_T^{\text{tr}}$.

Let us compare the accuracy of parameter estimation with the accuracy of the LS method when the number of constraints is $m = 2$. To avoid tedious calculations we determine the accuracy of estimation with the constraints my means of the truncated matrix $\mathbf{K}_T^{\text{tr}}$. According to (4.220) for $L = 4$ we have

$$\mathbf{K}_T^{tr} = \tilde{\boldsymbol{\kappa}}_T \gamma_{1T} + \mathbf{K}_{1T}\gamma_{2T} + \mathbf{K}_{2T}\gamma_{3T} + \mathbf{K}_{3T}\gamma_{4T}, \quad \gamma_{1T} + \gamma_{2T} + \gamma_{3T} + \gamma_{4T} = 1 \quad (4.223)$$

where the matrices $\tilde{\boldsymbol{\kappa}}_T$ and $\mathbf{K}_{iT}$, $i = \overline{1,2}$ are defined by (4.220)–(4.222) for $I_1 = \varnothing$ and $I_l = \{i\}, l = i + 1$, respectively, and the matrix $\mathbf{K}_{3T}$ is determined by (4.220)–(4.222) with $I_4 = \{1, 2\}$. Hence

$$\mathbf{K}_{iT} = \hat{\boldsymbol{\kappa}}_T, \quad i = 1, 2, 3, \quad (4.224)$$

where $\hat{\boldsymbol{\kappa}}_T$is the sample estimate of $\boldsymbol{\kappa}_T$, calculated by (4.164), (4.165) for $m = 1$ and the right-hand side of the constraint $\hat{b}_{iT} = 0$, $i = 1, 2$. For $i = 3$ in (4.224) $\hat{\boldsymbol{\kappa}}_T$ is also calculated by (4.164), (4.165), but with $m = 2$ and $\hat{\mathbf{b}}_T = \mathbf{0}_2$.

Consider the matrix $\tilde{\mathbf{K}}_T - \mathbf{K}_{3T}$. According to (4.164) and (4.165), the difference between $\tilde{\mathbf{K}}_T$ and $\mathbf{K}_{3T} = \tilde{\mathbf{K}}$ is determined by $\mathbf{J}_2 - \hat{\mathbf{M}}_V[2]$. By (4.209), elements of the matrix $\hat{\mathbf{M}}_V[2] = [m_{ij}]$, $i, j = 1, 2$ are given by

$$m_{ij}[2] = \sum_{k=1}^{3} m_{ij}^{(k)}[2] + m_{ij}^{(1,2)}[2], \quad i, j = 1, 2, \quad (4.225)$$

where the values in the right-hand side of (4.225) are determined by (4.210)–(4.214), and (4.178), where $Z_1^* = Z_2^* = 0$. From these formulas and (4.225) we obtain

$$m_{11} = \frac{1}{4}\sin^2\varphi + \frac{1}{2\pi}\left(\varphi - \frac{1}{2}\sin 2\varphi\right),$$

$$m_{22} = \frac{1}{4}(1 + \cos^2\varphi) + \frac{1}{2\pi}\left(\varphi + \frac{1}{2}\sin 2\varphi\right),$$

$$m_{12} = -\frac{\cos\varphi\sin\varphi}{4} + \frac{\sin^2\varphi}{2\pi}, \quad (4.226)$$

where φ is the angle between the lines ω_1 and ω_2, see Fig. 4.1. φ and the angle Φ, appearing in (4.178), are related by $\Phi = \varphi - (\pi/2)$. When $\varphi = \pi$ two inequality constraints are transforming into one, see Fig. 4.1.

Lemma 4.11. *Let the elements of the matrix* $\hat{\mathbf{M}}_{\mathrm{v}}[2]$ *be determined by (4.225). Then the matrix* $\mathbf{A} = \mathbf{J}_2 - \hat{\mathbf{M}}_{\mathrm{v}}[2]$ *is positive definite for* $0 \leq \phi < \pi$ *and is equal to the matrix* $\operatorname{diag}(1, 0)$ *for* $\varphi = \pi$.

Proof. Let us show that $\mathrm{tr}\mathbf{A} > 0$ and $\det \mathbf{A} = \Delta_{\mathrm{A}} > 0$. By (4.226) we obtain $\mathrm{tr}\mathbf{A} = 2 - ((1/2) + (\varphi/\pi)) \geq 1/2$. Further, after transformation, we get from (4.226)

$$\Delta_{\mathrm{A}}(\varphi) = 1/2 + s_1(\varphi) + s_2(\varphi) + s_3(\varphi),$$

where

$$s_1(\varphi) = -\frac{(\pi+1)\sin 2\varphi}{16\pi}, \quad s_2(\varphi) = \frac{(\pi^2 \sin^2 \varphi)}{16\pi^2}, \quad s_3(\varphi) = -\frac{3\varphi}{4\pi} + \frac{\varphi^2}{4\pi^2}.$$

We divide the interval φ into two intervals: $i_1 = [0; \pi/2)$ and $i_2 = [\pi/2; \pi]$.

In the interval i_1 we have $dS_3(\varphi)/d\varphi < 0$, and $s_3(\varphi)$ decreases from $s_3(0) = 0$ to $s_3(\pi/2) = -5/6$. Then

$$\Delta_{\mathrm{A}}(\varphi) \geq -\frac{\pi+1}{16\pi} + s_3(\varphi) + \frac{1}{2} = 0,1051, \quad 0 \leq \varphi \leq \frac{\pi}{2}. \tag{4.227}$$

In the interval i_2 we have $\Delta_{\mathrm{A}}(\varphi) \geq \bar{\Delta}_{\mathrm{A}}(\varphi) = (1/2) + d\sin^2\varphi - (3\varphi + 4\pi) + (\varphi^2/4\pi^2)$, where $d = ((\pi^2 - 4)/16\pi^2) = 0.03714$. For this interval $(d\bar{\Delta}_{\mathrm{A}}(\varphi)/d\varphi) = d\sin 2\varphi - (1/4\pi) < 0$, as $\sin 2\varphi \leq 0$, $\varphi \in [\pi/2, \pi]$. Consequently, $\Delta_{\mathrm{A}}(\varphi) \geq \bar{\Delta}_{\mathrm{A}}(\varphi) > 0$ for $\pi/2 \leq \varphi < \pi$, as $\bar{\Delta}_{\mathrm{A}}(\varphi)$ decreases monotone in i_2 from $\bar{\Delta}_{\mathrm{A}}(\pi/2) = 0.2246$ to $\bar{\Delta}_{\mathrm{A}}(\pi) = 0$. From above and (4.227), taking into account that $\Delta_{\mathrm{A}}(\pi) = 0$, we obtain the statement of the lemma. □

Theorem 4.9. *If $m = 2$ and Assumptions 2.9, 4.1, 4.2 hold true, then the matrix $\tilde{\mathbf{K}}_T - \mathbf{K}_T^{\mathrm{tr}}$ is non-negative definite.*

Proof. From (4.223) we have

$$\tilde{\mathbf{K}}_T - \mathbf{K}_T^{\mathrm{tr}} = \sum_{i=1}^{3} (\tilde{\boldsymbol{\kappa}}_T - \boldsymbol{\kappa}_{iT})\gamma_{i+1,T} \tag{4.228}$$

By Theorem 4.7, (4.224), and $\gamma_{iT} \geq 0$, $i = 1, 2,,$ it follows that $(\tilde{\boldsymbol{\kappa}}_T - \boldsymbol{\kappa}_{iT})\gamma_{i+1,T}$ is a non-negative definite matrix.

It follows from Lemma 4.8 that the matrix $\mathbf{J}_n - \hat{\mathbf{K}}_{\mathbf{v}}$, where $\hat{K}_v = \begin{vmatrix} \hat{\mathbf{M}}_v[2] & \mathbf{O}_{2,n-2} \\ \mathbf{O}_{n-2,2} & \mathbf{J}_{n-2} \end{vmatrix}$ (see 4.164), is positive definite for $n = 2$, and is non-negative definite for $n > 2$. Then, according to (4.224) and (4.165), we have $\tilde{\boldsymbol{\kappa}}_T - \boldsymbol{\kappa}_{3T} = \hat{\sigma}_T^2 \mathbf{L}_T(\mathbf{J}_n - \hat{\mathbf{K}}_V)\mathbf{L}_T'$, where the matrix $\mathbf{L}_T = \mathbf{C}_T \mathbf{M}_T^{-1/2} \mathbf{H}_T' \prod_T$ is non-degenerate, as the product of non-degenerate matrices. Hence, since $\gamma_{4T} \geq 0$, we see that the matrix $(\tilde{\boldsymbol{\kappa}}_T - \boldsymbol{\kappa}_{3T})_{\gamma 4T}$ is the non-negative definite, and the theorem follows by (4.228). □

Chapter 5
Asymptotic Properties of Recurrent Estimates of Parameters of Nonlinear Regression with Constraints

In Sect. 1.2 we consider iterative procedures of estimation of multidimensional nonlinear regression parameters under inequality constraints. Here we investigate asymptotic properties of iterative approximations of estimates, obtained on each iteration. Moreover, we allow the situation when the multidimensional regressor has a trend, for example, it may increase to infinity. We start with a special case in which the constraints are absent.

5.1 Estimation in the Absence of Constraints

In Sect. 1.2.1.2, we described algorithms of estimation without constraints, and minimized the function (1.4) with respect to $\boldsymbol{\alpha}$. Such algorithms are the particular cases in the general scheme of estimation under constraints. We give the representations for the cost function of the problem, and the algorithms for calculation of estimates, which take into account the sample volume T.

According to (1.4) the cost function is

$$S_T^*(\boldsymbol{\alpha}) = \frac{1}{2}\sum_{t=1}^{T}(y_t - f_t(\boldsymbol{\alpha}))^2.$$

(Note that in Chap. 1 we used the notation $S_T^*(\boldsymbol{\alpha})$ for $S(\boldsymbol{\alpha})$.)

According to Sect. 1.2.1.2, the iterative process of estimation can be written as

$$\boldsymbol{\alpha}_{k+1,T} = \boldsymbol{\alpha}_{kT} + \gamma_{kT}\mathbf{X}_{kT}, \quad k = 0, 1, 2, \ldots, \tag{5.1}$$

where $\boldsymbol{\alpha}_{kT}$ is the regression parameter on the kth iteration of estimation,

$$\mathbf{X}_{kT} = -(\mathbf{D}'(\boldsymbol{\alpha}_{kT})\mathbf{D}(\boldsymbol{\alpha}_{kT}) + v_{kT}\mathbf{A}_T(\boldsymbol{\alpha}_{kT}))^{-1}\nabla S_T^*(\boldsymbol{\alpha}_{kT}).$$

Here $\mathbf{D}_T(\boldsymbol{\alpha}) = \partial f_t(\boldsymbol{\alpha})/\partial\alpha_j,\, t = \overline{1,T},\, j = \overline{1,n}$ is a $T \times n$ matrix.

P.S. Knopov and A.S. Korkhin, *Regression Analysis Under A Priori Parameter Restrictions*, Springer Optimization and Its Applications 54,
DOI 10.1007/978-1-4614-0574-0_5, © Springer Science+Business Media, LLC 2012

Since the regressors might have a trend, the assumptions on the regression functions used in Chap. 2 need to be replaced by the assumption given below.

Assumption 5.1. A. Functions $f_t(\alpha), t = \overline{1,T}$ are twice differentiable on $\Re^n$.
B. There is a diagonal matrix E_T of dimension $n \times n$ with positive elements on the main diagonal E_{Ti}, which possesses the following properties:

1. $E_{Ti} \to \infty$ as $T \to \infty, k = \overline{1,n}$
2. $(\partial f_t(\alpha)/\partial \alpha_i)^2 / E_{Ti}^2 \to 0$ as $T \to \infty, i = \overline{1,n}$, uniformly in $\alpha \in \Theta$, where Θ is any compact set
3. $\bar{\mathbf{R}}_T(\alpha) = \mathbf{E}_T^{-1}\mathbf{D}'_T(\alpha)\mathbf{D}_T(\alpha)\mathbf{E}_T \to \bar{\mathbf{R}}(\alpha)$ as $T \to \infty$, uniformly in $\alpha \in \Theta$, and $\bar{\mathbf{R}}(\alpha^0)$ is a non-degenerate matrix
4. $\lim_{T\to\infty} \sum_{t=1}^{T} |\partial^2 f_t(\alpha)/\partial \alpha_k \partial \alpha_l| / E_{Tk} E_{Tl} < \infty$, uniformly in $\alpha \in \Theta$, $k, l = \overline{1,n}$
5. $E_{Tk}^{-2} E_{Tl}^{-2} \sum_{t=1}^{T} (\partial^2 f_t(\alpha)/\partial \alpha_k \partial \alpha_l)^2 \to 0$ as $T \to \infty$, uniformly in $\alpha \in \Theta$, $k, l = \overline{1,n}$
6. $\lim_{T\to\infty} \sqrt{T}/E_{Tk} < \infty, k = \overline{1,n}$

As an example of the matrix $\mathbf{E}_T$ which satisfies Assumption 5.1.B one can take a matrix with

$$E_{Ti}^2 = \sum_{t=1}^{T} \left(\frac{\partial f_1(\alpha)}{\partial \alpha_i}\bigg|_{\alpha=\alpha^0} \right)^2 i = \overline{1,n},\ or\ \mathbf{E}_T = \sqrt{T}\mathbf{J}_n.$$

For a noise ε_t in the regression model $y_t = f_t(\alpha^0) + \varepsilon_t,\ t = 1, 2, \ldots$ we use Assumption 2.8, in which we do not assume that $\varepsilon_t,\ t = 1, 2, \ldots,$ are identically distributed, but only that their dispersions σ^2 are equal.

Using the notation introduced above, the formulas for calculation of X_{kT} can be rewritten as follows:

$$\mathbf{X}_{kT} = \sqrt{T}\mathbf{E}_T^{-1}\bar{\bar{\mathbf{R}}}_T^{-1}(\alpha_{kT})\bar{\mathbf{q}}_T(\alpha_{kT}), \tag{5.2}$$

where

$$\bar{\bar{\mathbf{R}}}_T(\alpha_{kT}) = \bar{\mathbf{R}}_T(\alpha_{kT}) + v_{kT}\mathbf{E}_T^{-1}\mathbf{A}_T\mathbf{E}_T^{-1},$$

$$\bar{\mathbf{R}}_T(\alpha_{kT}) = \mathbf{E}_T^{-1}\mathbf{D}'_T(\alpha_{kT})\mathbf{D}_T(\alpha_{kT})\mathbf{E}_T^{-1}, \tag{5.3}$$

$$\bar{\mathbf{q}}_T(\alpha_{kT}) = \sqrt{T}\mathbf{E}_T^{-1}\mathbf{q}_T(\alpha_{kT}), \tag{5.4}$$

where

$$\mathbf{q}_T(\alpha_{kT}) = -T^{-1}\nabla S_T^*(\alpha_{kT}), \quad S_T^*(\alpha) = TS_T(\alpha) = \frac{1}{2}\sum_{t=1}^{T}(y_t - f_t(\alpha))^2, \tag{5.5}$$

and $\mathbf{A}_T$ is a positive definite matrix, properties of which are specified in Condition 5.1 below.

Algorithms of estimation (see Sect. 1.2.1.1) differ in the choice of the step multiplier γ_{kT} and the regularization parameter v_{kT}. In the present chapter, we study

Algorithms B5.1and B5.2 in which γ_{kT} is determined by minimization of the cost function. Below, we give a short description of the algorithms taking into account the volume of the sample.

Algorithm B5.1

The parameter γ_{kT} is found by Armijo condition: consecutive division of the unit into two equal parts until the inequality

$$S_T^*(\boldsymbol{\alpha}_{kT} + \gamma \mathbf{X}_{kT}) \leq S_T^*(\boldsymbol{\alpha}_{kT}) - \gamma^2 \mathbf{X}'_{kT}(\mathbf{D}'(\boldsymbol{\alpha}_{kT})\mathbf{D}(\boldsymbol{\alpha}_{kT}) + v_{kT}\mathbf{A}_T(\boldsymbol{\alpha}_{kT}))\mathbf{X}_{kT} \tag{5.6}$$

holds true. Further the parameter v_{kT} is specified:

$$v_{k+1,T} = \begin{cases} v_{kT} & \text{if } \gamma_{kT} = 1, \\ v_{kT} & \text{if } \gamma_{kT} < 1, cv_{kT} > V, \\ cv_{kT} & \text{if } \gamma_{kT} < 1, cv_{kT} \leq V. \end{cases} \tag{5.7}$$

where $c > 1$, $V = \text{const}$, $v_{0T} = v_0 = \text{const}$.

Algorithm B5.2

The parameter $v_{kT} = 0$, while γ_{kT} is found as in Algorithm B5.2.
Since we study the asymptotic properties of iterative calculation of regression parameter estimates, we need to impose some restrictions on the matrix $\mathbf{A}_T(\boldsymbol{\alpha}_{kT})$, introduced in (5.3).

Condition 5.1 The matrix $\mathbf{A}_T(\boldsymbol{\alpha}_{kT})$ is positive definite. Its elements can be constants, or they might depend on $\boldsymbol{\alpha}_{kT}$. In the latter case, if $\boldsymbol{\alpha}_{kT}$ is a consistent estimate of $\boldsymbol{\alpha}^0$, then

$$p \lim_{T\to\infty} \mathbf{E}_T^{-1}\mathbf{A}_T(\boldsymbol{\alpha}_{kT})\mathbf{E}_T^{-1} = \mathbf{O}_{nn}. \tag{5.8}$$

In particular, Condition 5.1 is fulfilled for the matrix $\mathbf{A}_T(\boldsymbol{\alpha}_{kT}) = \text{diag}(r_{ii}^T(\boldsymbol{\alpha}_{kT}))$, $i = \overline{1,n}$, where $r_{ii}^T(\boldsymbol{\alpha}_{kT})$ is a diagonal element of the matrix $\mathbf{R}_T(\boldsymbol{\alpha}_{kT}) = T^{-1}\mathbf{D}'_T(\boldsymbol{\alpha}_{kT})\mathbf{D}_T(\boldsymbol{\alpha}_{kT})$.

According to the above expressions, parameters γ_{kT} and v_{kT} depend on a random vector Y, i.e. on the values of the dependent variable, since independent variables are deterministic.

Suppose that in (5.1) $\boldsymbol{\alpha}_{0T}$ is random. Then the sequences $\boldsymbol{\alpha}_{kT}, k = 1, 2, \ldots; v_{kT}$, $k = 1, 2, \ldots; \gamma_{kT}, k = 0, 1, 2, \ldots$ formed by Algorithms B5.1 and B5.2, are sequences of random variables.

Lemma 5.1. *Suppose that for any finite T we can fix arbitrary random vectors $\mathbf{a}_T$, $\mathbf{b}_{1T}, \mathbf{b}_{2T}$, satisfying with probability 1*

$$||\mathbf{a}_T|| \leq ||\mathbf{b}_{1T}|| + ||\mathbf{b}_{2T}||, \tag{5.9}$$

and, moreover, for any numbers $\eta_i, 0 < \eta_i < 1, i = 1,2$ *there exist* $\varepsilon_i > 0, \tau_i > 0$, *such that*

$$P\{||\mathbf{b}_{iT}|| \geq \varepsilon_i\} < \eta_i, \quad T > \tau_i, i = 1,2. \tag{5.10}$$

Then for any η_i, $0 < \eta_i < 1$, $i = 1,2$, *there exist* $\varepsilon > 0$ *and* $\tau > 0$, *such that*

$$P\{||\mathbf{a}_T|| \geq \varepsilon\} < \eta, \quad T > \tau. \tag{5.11}$$

Proof. We have for $\varepsilon = \varepsilon_1 + \varepsilon_2$

$$P\{||\mathbf{a}_T|| < \varepsilon\} \geq P\{||\mathbf{b}_{1T}|| + ||\mathbf{b}_{2T}|| < \varepsilon\} \geq P\{||\mathbf{b}_{1T}|| < \varepsilon_1, ||\mathbf{b}_{2T}|| < \varepsilon_2\}$$
$$\geq P\{||\mathbf{b}_{1T}|| < \varepsilon_1\} - P\{||\mathbf{b}_{2T}|| \geq \varepsilon_2\} > 1 - \eta_1 - \eta_2 = 1 - \eta, \quad T > \tau = \max(\tau_1, \tau_2),$$

implying (5.11). □

Lemma 5.2. *Suppose that Assumptions 2.8, 5.1, and Condition 5.1 for the matrix* $\mathbf{A}_T(\boldsymbol{\alpha}_{kT})$ *are fulfilled. Let* $\boldsymbol{\alpha}_{0T}$ *be a random variable,* $\boldsymbol{\alpha}_{kT}$ *is a consistent estimate of the regression parameter* $\boldsymbol{\alpha}^0 \in \Theta$, *where* Θ *is some compact set. Assume that for a given value* $\Delta_k > 0$ *there exist* $\omega_k > 0$ *and* $T_k > 0$ *such that* $P\{||\mathbf{U}_{kT}|| \geq \omega_k\} < \Delta_k$ *for* $T > T_k$, *where* $\mathbf{U}_{kT} = \mathbf{E}_T(\boldsymbol{\alpha}_{kT} - \boldsymbol{\alpha}^0)$. *Then for given* $\Delta_{k+1} > 0$ *there exist* $\omega_{k+1} > 0$ *and* $T_{k+1} > 0$ *such that*

$$P\{||\mathbf{U}_{k+1,T}|| \geq \omega_{k+1}\} < \Delta_{k+1}, \quad T > T_{k+1}, \tag{5.12}$$

where $\mathbf{U}_{k+1,T} = \mathbf{E}_T(\boldsymbol{\alpha}_{k+1,T} - \boldsymbol{\alpha}^0)$, $\boldsymbol{\alpha}_{k+1,T}$ *is defined by (5.1) and Algorithm B5.1.*

Proof. From (5.1), (5.2) and (5.4) we have

$$\boldsymbol{\alpha}_{k+1,T} = \boldsymbol{\alpha}_{kT} - \gamma_{kT}\mathbf{E}_T^{-1}\bar{\bar{\mathbf{R}}}^{-1}(\boldsymbol{\alpha}_{kT})\mathbf{E}_T^{-1}\nabla S_T^*(\boldsymbol{\alpha}_{kT}). \tag{5.13}$$

Similarly to (2.26),

$$\mathbf{E}_T^{-1}\nabla S_T^*(\boldsymbol{\alpha}_{kT}) = \bar{\Phi}_T\mathbf{E}_T(\boldsymbol{\alpha}_{kT} - \boldsymbol{\alpha}^0) - \mathbf{Q}_T, \tag{5.14}$$

where $\mathbf{Q}_T = \mathbf{E}_T^{-1}\mathbf{D}'_T(\boldsymbol{\alpha}^0)\boldsymbol{\varepsilon}_T$, $\boldsymbol{\varepsilon}_T = [\varepsilon_1 \ldots \varepsilon_T]'$, the matrix $\mathbf{E}_T$ is defined in Assumption 5.1.B, $\bar{\boldsymbol{\Phi}}_T = T\mathbf{E}_T^{-1}\boldsymbol{\Phi}_T\mathbf{E}_T^{-1}$, where $\boldsymbol{\Phi}_T$ is defined by expressions (2.26), (2.27). If Assumptions 5.1.B.1–5.1.B.3 are satisfied, then by Theorem 2.1 (Anderson 1971)

$$\mathbf{Q}_T \overset{p}{\Rightarrow} \mathbf{Q} \sim N(\mathbf{O}_n, \sigma^2\bar{\mathbf{R}}(\boldsymbol{\alpha}^0)), \quad T \to \infty. \tag{5.15}$$

Using (2.27), we can write the elements of $\bar{\boldsymbol{\Phi}}_T$:

$$\bar{\Phi}_{kl}^T(\boldsymbol{\xi}_{1T}) = \bar{r}_{kl}^T(\boldsymbol{\xi}_{1T}) + (E_{Tk}E_{Tl})^{-1}\sum_{t=1}^{T}(y_t - f_t(\boldsymbol{\xi}_{1T}))\left.\frac{\partial^2 f_t(\boldsymbol{\alpha})}{\partial\alpha_k\partial\alpha_l}\right|_{\boldsymbol{\alpha}=\boldsymbol{\xi}_{1T}}, \tag{5.16}$$

where $||\boldsymbol{\xi}_{1T} - \boldsymbol{\alpha}^0|| \leq ||\boldsymbol{\alpha}_{kT} - \boldsymbol{\alpha}^0||$, $\bar{r}_{kl}^T(\boldsymbol{\alpha})$ is an element of matrix $\bar{\mathbf{R}}_T(\boldsymbol{\alpha})$, see (5.3).

We can find the limit in probability of $\bar{\boldsymbol{\Phi}}_T$ in the same way as it was done for $\boldsymbol{\Phi}_T$ in Sect. 2.2, see (2.30). Using the mean value theorem, we get the expression for the second summand in the formula for $\bar{\Phi}^T_{kl}(\boldsymbol{\xi}_{1T})$ (which we denote by B_T):

$$B_T = B_{1T} - B_{2T} = (E_{Tk}E_{Tl})^{-1}\sum_{t=1}^{T}\varepsilon_t \left.\frac{\partial^2 f_t(\boldsymbol{\alpha})}{\partial\alpha_k\partial\alpha_l}\right|_{\boldsymbol{\alpha}=\boldsymbol{\xi}_{1T}}$$

$$-(E_{Tk}E_{Tl})^{-1}\sum_{t=1}^{T}(\nabla f_t(\boldsymbol{\xi}_{2t}))'(\boldsymbol{\xi}_{2t}-\boldsymbol{\alpha}^0)\left.\frac{\partial^2 f_t(\boldsymbol{\alpha})}{\partial\alpha_k\partial\alpha_l}\right|_{\boldsymbol{\alpha}=\boldsymbol{\xi}_{1t}},$$

where $||\boldsymbol{\xi}_{2T}-\boldsymbol{\alpha}^0||\leq||\boldsymbol{\xi}_{1T}-\boldsymbol{\alpha}^0||\leq||\boldsymbol{\alpha}_{kT}-\boldsymbol{\alpha}^0||$. From Assumption 5.1.B.5 and Chebyshev inequality, it follows that $p\lim_{T\to\infty}(E_{Tk}E_{Tl})^{-1}\sum_{t=1}^{T}\varepsilon_t(\partial^2 f_t(\boldsymbol{\alpha})/\partial\alpha_k\partial\alpha_l)=0$ uniformly in $\boldsymbol{\alpha}\in\Theta$. Then according to Wilks (1962, Chap. 4), $p\lim_{T\to\infty}B_{1T}=0$, and we have by consistency of $\boldsymbol{\alpha}_{kT}$ that $p\lim_{T\to\infty}\boldsymbol{\xi}_{1T}=\boldsymbol{\alpha}^0$.

After necessary transformations we get

$$B_{2T} = (E_{Tk}E_{Tl})^{-1}\sum_{i=1}^{n}\Delta\xi_{1t,i}\sum_{t=1}^{T}\frac{\partial^2 f_t(\boldsymbol{\alpha})}{\partial\alpha_k\partial\alpha_l}\frac{\partial f_t(\boldsymbol{\alpha})}{\partial\alpha_i} = \sum_{i=1}^{n}B_{2T,i},$$

where $\Delta\xi_{1t,i}=\xi_{1t,i}-\boldsymbol{\alpha}^0$, $\xi_{1t,i}$ is the ith component of $\boldsymbol{\xi}_{1t}$. We have

$$B_{2T,i}\leq(E_{Ti}|\Delta\xi_{1t,i}|)\frac{\max_{t=\overline{1,T}}|(\partial f_t(\boldsymbol{\alpha})/\partial\alpha_i)|_{\boldsymbol{\alpha}=\boldsymbol{\xi}_{2t}}|}{E_{Ti}}\left|\sum_{t=1}^{T}\left|(E_{Tk}E_{Tl})^{-1}\left.\frac{\partial^2 f_t(\boldsymbol{\alpha})}{\partial\alpha_k\partial\alpha_l}\right|_{\boldsymbol{\alpha}=\boldsymbol{\xi}_{1t}}\right|\right|.$$

According to Assumptions 5.1.B.1–5.1.B.2 and Lemma 2.1 (Anderson 1971),

$$\frac{\max_{t=\overline{1,T}}(\partial f_t(\boldsymbol{\alpha})/\partial\alpha_i)^2}{E_{Ti}^2}\to 0 \text{ uniformly in } \boldsymbol{\alpha}\in\Theta \text{ as } T\to\infty.$$

Then, since by consistency of $\boldsymbol{\alpha}_{kT}$ we get $p\lim_{T\to\infty}\boldsymbol{\xi}_{2T}=\boldsymbol{\alpha}^0$, the second multiplier in the right-hand side of the inequality for $B_{2T,i}$ converges in probability to 0. Similarly, by Assumption 5.1.B.4 the third multiplier is bounded.

We have $E_{Ti}|\Delta\xi_{1t,i}|\leq||\mathbf{U}_{kT}||$. By the condition of the lemma for $||\mathbf{U}_{kT}||$ we obtain $P\{E_{Ti}|\Delta\xi_{1t,i}|\geq\omega_k\}<\Delta_k$ for any $T>T_k$. Then it is easy to show that the product of the first and the second multipliers in the right-hand side of the inequality for $B_{2T,i}$ converges in probability to 0 and, hence, the right-hand side part of the inequality also converges in probability to 0, implying $p\lim_{T\to\infty}B_{2T,i}=0$. Therefore, since $p\lim_{T\to\infty}B_{1T}=0$, we arrive at $p\lim_{T\to\infty}B_T=0$.

By Assumption 5.1.B.3 and the consistency of $\boldsymbol{\xi}_{1T}$ we have $p\lim_{T\to\infty}\bar{r}^T_{kl}(\boldsymbol{\xi}_{1T})=\bar{r}_{kl}(\boldsymbol{\alpha}^0)$ (that is, the (k,l)th element of the matrix $\bar{\mathbf{R}}(\boldsymbol{\alpha}^0)$). Hence, $\bar{\Phi}^T_{kl}(\boldsymbol{\xi}_{1T})$ converges to 0 in probability, which holds for arbitrary k and l.

Therefore,

$$p \lim_{T\to\infty} \bar{\Phi}_T = \bar{R}(\alpha^0). \tag{5.17}$$

From (5.13) and (5.14) we have

$$\mathbf{U}_{k+1,T} = \mathbf{U}_{kT} + \gamma_{kT}\bar{\bar{\mathbf{R}}}_T^{-1}(\alpha_{kT})(\mathbf{Q}_T - \bar{\Phi}_T\mathbf{U}_{kT}). \tag{5.18}$$

We get from (5.2), (5.3)

$$p \lim_{T\to\infty} \bar{\bar{\mathbf{R}}}_T(\alpha_{kT}) = \bar{\mathbf{R}}(\alpha^0), \tag{5.19}$$

where we used Assumption 5.1.3, Condition 5.1, and boundedness of v_{kT} (see (5.7)) and γ_{kT}. Then we derive from (5.18)

$$\mathbf{U}_{k+1,T} = \mathbf{B}_{1T} + \mathbf{B}_{2T},$$

where

$$\begin{aligned}\mathbf{B}_{1T} &= \gamma_{kT}\bar{\bar{\mathbf{R}}}_T^{-1}(\alpha_{kT})\mathbf{Q}_T,\\ \mathbf{B}_{2T} &= (1-\gamma_{kT})\mathbf{U}_{kT} + \gamma_{kT}(\mathbf{J}_n - \mathbf{M}_T(\alpha_{kT}))\mathbf{U}_{kT}.\end{aligned} \tag{5.20}$$

Here

$$\mathbf{M}_T(\alpha_{kT}) = \bar{\bar{\mathbf{R}}}_T^{-1}(\alpha_{kT})\bar{\Phi}_T. \tag{5.21}$$

Therefore,

$$||\mathbf{U}_{k+1,T}|| = ||\mathbf{B}_{1T}|| + ||\mathbf{B}_{2T}||. \tag{5.22}$$

By (5.15) and (5.19), the random variable $\bar{\bar{\mathbf{R}}}_T^{-1}(\alpha_{kT})\mathbf{Q}_T$ has a limit distribution. Then for any $\delta_1 > 0$ there exist positive values e_1 and τ_1, such that

$$P\{||\mathbf{B}_{1T}|| < e_1\} \geq P\{||\bar{\bar{\mathbf{R}}}_T^{-1}(\alpha_{kT})\mathbf{Q}_T(\alpha_{kT})|| < e_1\} \geq 1-\delta_1, \quad T > \tau_1. \tag{5.23}$$

From (5.20) we get

$$||\mathbf{B}_{2T}|| \leq (1-\gamma_{kT})||\mathbf{U}_{kT}|| + \gamma_{kT}||(\mathbf{J}_n - \mathbf{M}(\alpha_{kT}))\mathbf{U}_{kT}||.$$

By the conditions of the lemma for $||\mathbf{U}_{kT}||$,

$$P\{(1-\gamma_{kT})||\mathbf{U}_{kT}|| < e_{21}\} \geq P\{||\mathbf{U}_{kT}|| < e_{21}\} \geq 1-\delta_{21}, \quad T > \tau_{21}, \tag{5.24}$$

where $e_{21} = \omega_k$, $\delta_{21} = \Delta_k$, $\tau_{21} = T_k$.

According to (5.17) and (5.19), the matrix $\mathbf{J}_n - \mathbf{M}(\alpha_{kT})$ converges in probability to a zero matrix as $T \to \infty$, i.e. for any $\bar{\delta} > 0$, $\bar{e}_{22} > 0$,

$$P\{||\mathbf{J}_n - \mathbf{M}(\alpha_{kT})|| < \bar{e}_{22}\} \geq 1-\bar{\delta}, \quad T > \bar{\tau}.$$

Then, since $\gamma_{kT} \leq 1$ we have

$$\begin{aligned} &P\{\gamma_{kt}||(\mathbf{J}_n - \mathbf{M}_T(\boldsymbol{\alpha}_{kT}))\mathbf{U}_{kt}|| < e_{22}\} \\ &\geq P\{||\mathbf{J}_n - \mathbf{M}_T(\boldsymbol{\alpha}_{kT})|| \cdot ||\mathbf{U}_{kt}|| < e_{22}\} \\ &\geq P\left\{||\mathbf{J}_n - \mathbf{M}_T(\boldsymbol{\alpha}_{kT})|| < \frac{e_{22}}{e_{21}}\right\} - P\{||\mathbf{U}_{kt}|| \geq e_{21}\} \\ &\geq 1 - \delta_{22}, \quad T > \tau_{22} = \max(\tau_{21}, \bar{\tau}), \end{aligned}$$

where $\delta_{22} = \delta_{21} + \bar{\delta}$.

Further, put in (5.9) $\mathbf{a}_T = \mathbf{B}_{2T}, \mathbf{b}_{1T} = (1 - \gamma_{kT})||\mathbf{U}_{kT}||$, $\mathbf{b}_{2T} = \gamma_{kT}||(\mathbf{J}_n - \mathbf{M}_T(\boldsymbol{\alpha}_{kT}))\mathbf{U}_{kt}||$ and in (5.10) $\varepsilon_i = e_{2i}, \eta_i = \delta_{2i}, \tau_i = \tau_{2i}, i = 1, 2$. Then from (5.11) we have

$$P\{||\mathbf{B}_{2T}|| \geq e_2\} < \delta_2, \quad T > \tau_2 = \max(\tau_{21}, \tau_{22}), \tag{5.25}$$

where $e_2 = e_{21} + e_{22}, \delta_2 = \delta_{21} + \delta_{22}$.

Similarly, using (5.9)–(5.11) for (5.22), (5.23), (5.25) and putting $\varepsilon_i = e_i, \eta_i = \delta_i, i = 1, 2, \omega_{k+1} = e_1 + e_2, \Delta_{k+1} = \delta_1 + \delta_2$ we obtain the statement of the lemma. □

Let us proceed to the main results.

Theorem 5.1. *Suppose that Assumptions 2.8, 5.1 and the Condition 5.1 are fulfilled. Assume that the initial value of iterative process is a consistent estimate* $\boldsymbol{\alpha}_{0T}$ *of* $\boldsymbol{\alpha}^0 \in \boldsymbol{\Theta}$, *where* $\boldsymbol{\Theta}$ *is a compact set. Then the values* $\boldsymbol{\alpha}_{kT}$, $k = 1, 2, \ldots$, *obtained by Algorithm B5.1, are the consistent estimates of* $\boldsymbol{\alpha}^0$.

Proof. Consider the expression (5.1). Let k be an arbitrary value, and let $\boldsymbol{\alpha}_{kT}$ be a consistent estimate of $\boldsymbol{\alpha}^0$. From (5.2), (5.4), (5.5), we have $\mathbf{X}_{kT} = -\sqrt{T}\mathbf{E}_T^{-1}\bar{\bar{\mathbf{R}}}_T^{-1}(\boldsymbol{\alpha}_{kT})\mathbf{E}_T^{-1}\nabla S_T^*(\boldsymbol{\alpha}_{kT})$. According to (5.14), (5.15), (5.17) and (5.19) we obtain

$$p \lim_{T\to\infty} \mathbf{X}_{kT} = p \lim_{T\to\infty} \mathbf{E}_T^{-1}\bar{\bar{\mathbf{R}}}_T^{-1}(\boldsymbol{\alpha}_{kT})(\mathbf{Q}_T - \bar{\Phi}_T\mathbf{U}_{kT}) = \mathbf{O}_n.$$

From the last expression and (5.1), we obtain the consistency of $\boldsymbol{\alpha}_{k+1,T}$. Since k is arbitrary, $\boldsymbol{\alpha}_{0T}$ is consistent, and v_{0T} is constant, we get (5.2). Theorem is proved. □

Theorem 5.2. *Let conditions of Theorem 5.1 be satisfied, and assume that* $\mathbf{U}_{0T} = \mathbf{E}_T(\boldsymbol{\alpha}_{0T} - \boldsymbol{\alpha}^0)$ *has a limit distribution. Then* $\tilde{\mathbf{U}}_{kT} = \mathbf{E}_T(\tilde{\boldsymbol{\alpha}}_{kT} - \boldsymbol{\alpha}^0)$, *k=1,2..., where* $\tilde{\boldsymbol{\alpha}}_{kT} = \boldsymbol{\alpha}_{k-1,T} + \mathbf{X}_{k-1,T}$, *are asymptotically normal, i.e. they converge as* $T \to \infty$ *in distribution to a random variable* $\mathbf{U} \sim N(\mathbf{O}_n, \sigma^2\mathbf{R}^{-1}(\boldsymbol{\alpha}^0))$. *Here* $\boldsymbol{\alpha}_{kT}$ *is the sequence of values formed by Algorithm B5.1.*

Proof. Take an arbitrary $k \geq 1$. Then $\boldsymbol{\alpha}_{kT}$ is the consistent estimate of $\boldsymbol{\alpha}^0$ (see Theorem 5.1). According to Lemma 5.2 and the fact that $\mathbf{U}_{0T}$ has a limit distribution, (5.12) holds for any $k \geq 0$. The arguments above allow us to investigate the limit properties of $\tilde{\boldsymbol{\alpha}}_{k+1,T}$.

We have

$$\tilde{\mathbf{U}}_{k+1,T} = (\mathbf{J}_n - \mathbf{M}_T(\boldsymbol{\alpha}_{kT}))\mathbf{U}_{kT} + \bar{\bar{\mathbf{R}}}_T^{-1}(\boldsymbol{\alpha}_{kT})\mathbf{Q}_T, \tag{5.26}$$

where $\mathbf{M}_T(\boldsymbol{\alpha}_{kT})$ is defined by (5.21).

The matrix $\mathbf{J}_n - \mathbf{M}_T(\boldsymbol{\alpha}_{kT})$ converges in probability to a zero matrix, which implies the convergence in probability of the first term in (5.26) to a zero vector (cf. (5.12)). According to (5.15) and (5.19), the second term in (5.26) converges in distribution to a random variable $\mathbf{U} \sim N(\mathbf{O}_n, \sigma^2\mathbf{R}^{-1}(\boldsymbol{\alpha}^0))$. Thus, $\tilde{\mathbf{U}}_{k+1,T}$ also converges in distribution to $\mathbf{U}$. Theorem is proved. □

It is easy to see that Theorems 5.1 and 5.2 hold true for Algorithm B5.2 if the conditions of Theorem 5.2 are supplemented by the next one: the matrix $\bar{\mathbf{R}}_T(\boldsymbol{\alpha}_{kT})$ is positive definite for any k and T (which is true under Assumption 1.5, Sect. 1.2.1.1, if we replace $\mathbf{D}(\boldsymbol{\alpha})$ by $\mathbf{D}_T(\boldsymbol{\alpha})$).

It is necessary to note that for finite k the sequences $\tilde{\boldsymbol{\alpha}}_{kT}$ and $\boldsymbol{\alpha}_{kT}$, $k = 1, 2, \ldots$ are close enough. In the beginning of the iterative process the step multiplier equals 1, implying $\tilde{\boldsymbol{\alpha}}_{kT} = \boldsymbol{\alpha}_{kT}$. Under large division of the step, the size $\gamma_{kT}\mathbf{X}_{kT}$ is small, implying $\tilde{\boldsymbol{\alpha}}_{kT} \approx \boldsymbol{\alpha}_{kT}$. The same correspondence takes place for large k independently from the size of the step multiplier, because $\mathbf{X}_{kT} \to \mathbf{O}_n$ as $k \to \infty$. Such a property of $\mathbf{X}_{kT}$ takes place according to Theorem 1.3 (estimation under constraints using Algorithm A5.1).

A little modification of Algorithms B5.1 and B5.2 allows to obtain the sequence $\boldsymbol{\alpha}_{kT}$, $k = 1, 2, \ldots$ with the same properties as $\tilde{\boldsymbol{\alpha}}_{kT}$, $k = 1, 2, \ldots$. From this point of view the following result is useful.

Corollary 5.1. *Let conditions of Theorem 5.2 be fulfilled,* k_0 *be the number of the iteration for which* $\gamma_{k_0T} = 1$, $\forall T (k_0 \geq 0)$ *and starting from* $\boldsymbol{\alpha}_{k_0T}$ *Algorithm B5.1 converges. Then for the kth iteration with* $k > k_0$ *the limit distribution of* $\mathbf{U}_{kT} = \mathbf{E}_T(\boldsymbol{\alpha}_{kT} - \boldsymbol{\alpha}^0)$ *coincides with the distribution* $\mathbf{U} \sim N(\mathbf{O}_n, \sigma^2\mathbf{R}^{-1}(\boldsymbol{\alpha}^0))$.

Proof. From (5.18) we derive

$$\mathbf{U}_{k_0+1,T} = (\mathbf{J}_n\mathbf{M}_T(\boldsymbol{\alpha}_{kT}))\mathbf{U}_{k_0T} + \bar{\bar{\mathbf{R}}}(\boldsymbol{\alpha}_{KT})\mathbf{Q}_T. \tag{5.27}$$

Since the matrix $\mathbf{J}_n - \mathbf{M}_T(\boldsymbol{\alpha}_{kT})$ converges in probability, the first term in (5.27) also converges to a zero vector. Taking into account that $\bar{\bar{\mathbf{R}}}^{-1}(\boldsymbol{\alpha}_{kT})\mathbf{Q}_T \stackrel{p}{\Rightarrow} \mathbf{U}$, we obtain $\mathbf{U}_{k_0+1,T} \stackrel{p}{\Rightarrow} \mathbf{U}$. Then from (5.15)–(5.18) we derive $\mathbf{U}_{k+1,T} \stackrel{p}{\Rightarrow} \mathbf{U}$ for $k > k_0 + 1$. □

According to the algorithms investigated above, the iterative process of calculating the estimate for the multi-dimensional regression parameter allows to achieve (with prescribed accuracy) the neighbourhood of the point in which the sufficient condition for the extremum of the cost function $S_T(\boldsymbol{\alpha})$ in the estimation problem holds true. Moreover, there is no guarantee that the point where the calculation process breaks (we call it for simplicity the *machine minimum*) is close or equal to the true global minimum. It is known that the solution to the estimation problem, corresponding to the global minimum of $S_T(\boldsymbol{\alpha})$, is the consistent estimate of the regression parameter, see, for example Jennrich (1969) and Malinvaud (1969), which is asymptotically normal (Ivanov 2007).

According to Corollary 5.1, machine local minima of $S_T(\boldsymbol{\alpha})$ are asymptotically equivalent, if calculation process begins with a random variable which has a limit distribution. In many cases (Demidenko 1989) $S_T(\boldsymbol{\alpha})$ has 2–3 local minima and for any initial approximation it is necessary to find the global minimum. If the initial estimate of the parameter has a limit distribution, then, as follows from the results proved in this section, the estimate corresponding to any local minimum is consistent. In this case, it is desirable to sort local minima and choose one that has the best (in some sense) matrix consisting of sum of squares of parameter estimates. As a possible criterion for the choice of such matrix one can take its determinant, i.e. the generalized dispersion. Other possible criterion follows from the interpretation of the obtained result from the point of view of the stability of regression parameter estimations. In Demidenko (1989, Chap. 6, Sect. 4), it is suggested to use the norm of the gradient $\partial\alpha_{iT}/\partial\mathbf{Y}$, $i = \overline{1,n}$ (where α_{iT} is the ith component of $\boldsymbol{\alpha}_T$, $\mathbf{Y}$ is T-dimensional vector of observed values of the dependent variable). It is shown that if $\mathbf{E}_T^2 = T\mathbf{J}_n$, then $||\partial\alpha_{iT}/\partial\mathbf{Y}|| \approx (\mathbf{R}_T^{-1}(\boldsymbol{\alpha}_T))_{ii}$. Thus, choosing as a solution to the argument of the machine minimum of the sum of squares, which corresponds to the minimal trace of the matrix $\mathbf{R}_T^{-1}(\boldsymbol{\alpha}_T)$, we get the solution which is steady enough.

By consistency of the estimate at the point of the machine local minimum $\boldsymbol{\alpha}_{k^*,T}$ (here k^* is the iteration number on which the desired accuracy of estimation is achieved), instead of a matrix $\sigma_T^2\mathbf{R}^{-1}(\boldsymbol{\alpha}_{k^*,T})$ of m.s.e. of estimates of parameters we can use more exact estimates (here σ_T^2 is the estimate of the dispersion σ^2 corresponding to the remainders calculated for $\boldsymbol{\alpha} = \boldsymbol{\alpha}_{k^*,T}$). For instance, one can use the estimation that takes into account second derivatives of the regression function (Demidenko 1981, Sect. 8.3), or an interesting method which requires more detailed analysis of the regression function (Ivanov 1997, Chap. 4).

The described properties of the iterative process are based on the assumption about consistency of the initial approximation and the existence of its limit distribution. The approximation, which satisfies this condition, can be derived by other methods of parameter estimation. It is possible to specify the area of applied statistics where such initial approximation exists – modeling of distributed lags. It can be found by the method of instrumental variables which is applied, for example, for parameter estimation in dynamic systems (in econometric and in time series analysis they are also called distributed lags), see Ljung (1987, Sect. 7.6) and Hannan (1976, Chap. 7).

5.2 Estimation with Inequality Constraints

Consider the statistical properties of iterative calculations performed by Algorithms A1.1 and A1.2. Recall that these algorithms describe the estimation of parameters of a nonlinear regression with inequality constraints. Here we give a short description of Algorithms A1.1, A1.2 modifications. Denote by T the sample volume.

Algorithm A5.1

We specify regression parameters according to (5.1); the step multiplier used in (5.1) can be found by (1.36), which can be written as

$$u(\boldsymbol{\alpha}_{kT} + \gamma_{kT}\mathbf{X}_{kT}, \Psi_T) - u(\boldsymbol{\alpha}_{kT}, \Psi_T) \le -\gamma_{kT}^2 \mathbf{X}'_{kT}(\mathbf{D}'(\boldsymbol{\alpha}_{kT})\mathbf{D}(\boldsymbol{\alpha}_{kT}) + v_{kT}\mathbf{A}_T(\boldsymbol{\alpha}_{kT}))\mathbf{X}_{kT}, \quad (5.28)$$

where $u(\boldsymbol{\alpha}_{kT}, \Psi_T) = S_T^*(\boldsymbol{\alpha}) + \Psi_T \Phi(\boldsymbol{\alpha}_{kT})$, and the matrix $\bar{\tilde{\mathbf{R}}}_T(\boldsymbol{\alpha}_{kT})$ is defined by (5.3). Then the regularization parameter v_{kT} is specified by (5.7). Vector $\mathbf{X}_{kT}$ is the solution to the quadratic programming problem which arises from (1.32) by multiplying the cost function by T^{-1} and replacing $S(\boldsymbol{\alpha})$ by $S_T^*(\boldsymbol{\alpha})$, and $\boldsymbol{\alpha}$ by $\boldsymbol{\alpha}_{kT}$:

$$\begin{cases} \frac{1}{2}\mathbf{X}'\tilde{\mathbf{R}}_T(\boldsymbol{\alpha}_{kT})\mathbf{X} - \mathbf{q}'_T(\boldsymbol{\alpha}_{kT})\mathbf{X} \to \min, \\ \mathbf{G}(\boldsymbol{\alpha}_{kT})\mathbf{X} \le -\mathbf{g}(\boldsymbol{\alpha}_{kT}). \end{cases} \quad (5.29)$$

Here $\tilde{\mathbf{R}}_T(\boldsymbol{\alpha}_{kT}) = T^{-1}(\mathbf{D}'(\boldsymbol{\alpha}_{kT})\mathbf{D}(\boldsymbol{\alpha}_{kT}) + v_{kT}\mathbf{A}_T(\boldsymbol{\alpha}_{kT}))$, $\nabla \mathbf{g}'_i(\boldsymbol{\alpha}_{kT})$; $i \in I$, $\mathbf{g}(\boldsymbol{\alpha}_{kT}) = [\mathbf{g}'_i(\boldsymbol{\alpha}_{kT})]$, $i \in I$, are the lines of the matrix $\mathbf{G}(\boldsymbol{\alpha}_{kT})$. According to (5.5), we have $\mathbf{q}_T(\boldsymbol{\alpha}_{kT}) = -\nabla S_T(\boldsymbol{\alpha}_{kT})$, where $S_T(\boldsymbol{\alpha}) = (1/2T)\sum_{t=1}^{T}(y_t - f_t(\boldsymbol{\alpha}))^2$.

In contrast to (1.32), the supplementary problem (5.29) includes all linearized constraints (i.e. the case $\delta = \infty$), which simplifies further investigations. In practice, in the auxiliary problem all constraints are taken into account.

Algorithm A5.2

We perform the iterative specification of the estimation of regression parameters by (5.1). The regularization parameter v_{kT} identically equals to 0. The step multiplier γ_{kT} is defined by (5.6), (5.28). In this algorithm we take $\tilde{\mathbf{R}}_T(\boldsymbol{\alpha}_{kT}) = T^{-1}\mathbf{D}'(\boldsymbol{\alpha}_{kT})\mathbf{D}(\boldsymbol{\alpha}_{kT})$ in case of problem (5.29).

We add to Assumptions 2.2.A and 2.4 about properties of constraints which were used in Sect. 2.2, the assumptions on properties of the noise ε_t and regression functions (see Sect. 5.1). Also, we add one more assumption:

Assumption 5.2. There exists an $(m \times m)$ diagonal matrix $\bar{\mathbf{E}}_T$ with positive elements on the main diagonal, such that for any $\boldsymbol{\alpha} \in \Re^n$ we have $p\lim_{T\to\infty} \bar{\mathbf{E}}_T \mathbf{G}(\boldsymbol{\alpha})\mathbf{E}_T^{-1} = \bar{\mathbf{G}}(\boldsymbol{\alpha})$, where $(m \times n)$ matrix $\mathbf{G}(\boldsymbol{\alpha})$ has an element $\nabla g_i'(\boldsymbol{\alpha})$ in its ith line. Further, $\lim_{T\to\infty} \sqrt{T}/\bar{E}_{Tk} < \infty$, $k = \overline{1,n}$, where $\bar{E}_{Tk}$ is the kth element on the main diagonal of the matrix $\bar{\mathbf{E}}_T$.

The assumption above generalizes Assumption 2.12, see Sect. 2.5. We assume everywhere below that Assumption 1.2 (on the initial approximation and the solvability of auxiliary problem (5.29)) holds true.

As before, we assume that the true value of the vector of regression parameters satisfies the inequality constraints and the equalities (2.14):

$$g_i(\boldsymbol{\alpha}^0) = 0,\ i \in I_1^0, \quad g_i(\boldsymbol{\alpha}^0) < 0, \quad i \in I_2^0, \quad I_1^0 \cup I_2^0 = I = \{1, ..., m\}.$$

Assume also that in I_1^0 that $m_1 < n$, where n is the number of regression parameters.

We need some auxiliary statements which characterize statistical properties of iterative procedures of calculations based on Algorithms A5.1 and A5.2.

Lemma 5.3. *Suppose that the approximation $\boldsymbol{\alpha}_{kT}$ obtained by Algorithms A5.1 and A5.2 on the kth iteration is the consistent estimate of $\boldsymbol{\alpha}^0 \in \Theta$, where Θ is a compact set. Suppose also that Condition 5.1 and Assumptions 2.2A, 2.8, 5.1, 5.2 (in case of using Algorithm A5.2, are fulfilled. Assume that the matrix $\mathbf{D}_T(\boldsymbol{\alpha})$ is of full rank in any compact set and for any given $\Delta_k > 0$ there exist $\omega_k > 0$ and $T_k > 0$ such that*

$$P\{||\mathbf{U}_{kT}|| \geq \omega_k\} < \Delta_k, \quad T > T_k, \mathbf{U}_{kT} = \mathbf{E}_T(\boldsymbol{\alpha}_{kT} - \boldsymbol{\alpha}^0). \tag{5.30}$$

Then for the solution to auxiliary problem (5.29) on $k + 1$th iteration $\mathbf{X}_{kT}$ we have

$$p \lim_{T \to \infty} \mathbf{X}_{kT} = p \lim_{T \to \infty} \bar{\mathbf{X}}_{kT} = \mathbf{O}_n, \tag{5.31}$$

where $\bar{\mathbf{X}}_{kT} = \left(\sqrt{T}\right)^{-1} \mathbf{E}_T \mathbf{X}_{kT}$.

Proof. Change the variables:

$$\mathbf{X} = \sqrt{T}\mathbf{E}_T^{-1}\bar{\mathbf{X}}. \tag{5.32}$$

Then by (5.29)

$$\frac{1}{2}\bar{\mathbf{X}}'\bar{\bar{\mathbf{R}}}_T(\boldsymbol{\alpha}_{kT})\bar{\mathbf{X}} - \bar{\mathbf{q}}_T'(\boldsymbol{\alpha}_{kT})\bar{\mathbf{X}} \to \min, \quad \bar{\mathbf{G}}_T(\boldsymbol{\alpha}_{kT})\bar{\mathbf{X}} \leq -\bar{\mathbf{g}}_T(\boldsymbol{\alpha}_{kT}), \tag{5.33}$$

where $\bar{\bar{\mathbf{R}}}_T(\boldsymbol{\alpha}_{kT})$ is defined in (5.3), $\bar{\mathbf{q}}_T'(\boldsymbol{\alpha}_{kT})$ is defined in (5.4) and

$$\bar{\mathbf{G}}_T(\boldsymbol{\alpha}) = \bar{\mathbf{E}}_T\mathbf{G}(\boldsymbol{\alpha})\mathbf{E}_T^{-1}, \quad \bar{\mathbf{g}}_T(\boldsymbol{\alpha}) = \left(\sqrt{T}\right)^{-1} \bar{\mathbf{E}}_T\mathbf{g}(\boldsymbol{\alpha}), \tag{5.34}$$

where matrices $\bar{\mathbf{E}}_T$, $\mathbf{G}(\boldsymbol{\alpha})$, $\mathbf{E}_T^{-1}$ are defined in Assumption 5.2.

Since the matrix $\bar{\bar{\mathbf{R}}}_T(\boldsymbol{\alpha}_{kT})$ is positive definite, we have

$$\bar{\bar{\mathbf{R}}}_T(\boldsymbol{\alpha}_{kT}) = \tilde{\mathbf{M}}_T'(\boldsymbol{\alpha}_{kT})\tilde{\mathbf{M}}_T(\boldsymbol{\alpha}_{kT}). \tag{5.35}$$

Here $\tilde{\mathbf{M}}_T(\boldsymbol{\alpha}_{kT}) = \tilde{\Lambda}_T^{1/2}(\boldsymbol{\alpha}_{kT})\tilde{\mathbf{C}}_T(\boldsymbol{\alpha}_{kT})$, where $\tilde{\Lambda}_T^{1/2}(\boldsymbol{\alpha}_{kT})$ is a diagonal matrix with values $\bar{\bar{\mathbf{R}}}_T(\boldsymbol{\alpha}_{kT})$ on the main diagonal, $\tilde{\mathbf{C}}_T(\boldsymbol{\alpha}_{kT})$ is an orthogonal matrix.

Put

$$\tilde{\mathbf{X}} = \tilde{\mathbf{M}}_T(\boldsymbol{\alpha}_{kT})\bar{\mathbf{X}}. \tag{5.36}$$

Then we obtain from (5.33)

$$\frac{1}{2}\tilde{\mathbf{X}}'\tilde{\mathbf{X}} - \tilde{\bar{\mathbf{q}}}_T(\boldsymbol{\alpha}_{kT})\tilde{X} \to \min, \quad \bar{\tilde{\mathbf{G}}}_T(\boldsymbol{\alpha}_{kT})\tilde{X} \leq -\bar{\mathbf{g}}_T(\boldsymbol{\alpha}_{kT}), \tag{5.37}$$

where

$$\bar{\bar{\mathbf{q}}}_T(\alpha_{kT}) = (\tilde{\mathbf{M}}_T^{-1}(\alpha_{kT}))'\bar{\mathbf{q}}_T(\alpha_{kT}), \quad \bar{\bar{\mathbf{G}}}_T(\alpha_{kT}) = \bar{\mathbf{G}}_T(\alpha_{kT})\tilde{\mathbf{M}}_T^{-1}(\alpha_{kT}). \tag{5.38}$$

According to Lemmas 2.6 and 4.8, the solution to (5.37) is given by

$$\tilde{\mathbf{X}}_{kT} = F(\bar{\bar{\mathbf{q}}}_T(\alpha_{kT}), \quad \bar{\bar{\mathbf{G}}}_T(\alpha_{kT}), \quad \bar{\mathbf{g}}_T(\alpha_{kT})). \tag{5.39}$$

It is continuous in $\bar{\bar{\mathbf{q}}}_T(\alpha_{kT})$ $\bar{\bar{\mathbf{G}}}_T(\alpha_{kT})$, $\bar{\mathbf{g}}_T(\alpha_{kT})$, which in turn are continuous functions of $\boldsymbol{\alpha}_{kT}$.

From (5.4), (5.5), and taking into account (5.14), we obtain after some transformations that

$$\bar{\mathbf{q}}_T(\alpha_{kT}) = \left(\sqrt{T}\right)^{-1}(\mathbf{Q}_T - \bar{\Phi}_T\mathbf{U}_{kT}). \tag{5.40}$$

Further, by (5.40), taking into account (5.15), (5.17), (5.30) and the consistency of α_{kT} (cf. conditions of the theorem) we arrive at

$$p\lim_{T\to\infty} \bar{\mathbf{q}}_T(\alpha_{kT}) = \mathbf{O}_n. \tag{5.41}$$

Under Assumptions 2.2A and 5.2,

$$p\lim_{T\to\infty} \bar{\mathbf{G}}_T(\alpha_{kT}) = \bar{\mathbf{G}}(\alpha^0). \tag{5.42}$$

By Lemma 2.4 and (5.19),

$$p\lim_{T\to\infty} \tilde{\mathbf{M}}_T(\alpha_{kT}) = \mathbf{M}(\alpha^0), \tag{5.43}$$

where $\mathbf{M}(\alpha^0) = \Lambda^{1/2}(\alpha^0)\mathbf{C}(\alpha^0)$. Here $\Lambda(\alpha^0)$ is a diagonal matrix with elements $\mathbf{R}(\alpha^0)$ on its main diagonal, and $\mathbf{C}(\alpha^0)$ is the modal matrix of $\mathbf{R}(\alpha^0)$. Then from (5.38) and (5.41)–(5.43) it follows that

$$p\lim_{T\to\infty} \bar{\bar{\mathbf{q}}}_T(\alpha_{kT}) = \mathbf{O}_n, \quad p\lim_{T\to\infty} \bar{\bar{\mathbf{G}}}_T(\alpha_{kT}) = \bar{\bar{\mathbf{G}}}(\alpha^0). \tag{5.44}$$

Now we estimate the limit (in probability) of $\bar{\mathbf{g}}_T(\alpha_{kT})$. Using the mean value theorem, we have for the ith components of $\mathbf{g}(\alpha_{kT})$

$$g_i(\alpha_{kT}) = g_i(\alpha^0) + \nabla g_i'(\boldsymbol{\xi}_{ikT})(\alpha_{kT} - \alpha^0),$$

where the random variable $\xi_{ikT} \in \Re^n$ is such that $||\xi_{ikT} - \alpha^0|| \leq ||\alpha_{kT} - \alpha^0||$. Under the conditions of the lemma and the consistency of $\boldsymbol{\alpha}_{kT}$, we get $p\lim_{T\to\infty} \xi_{ikT} = \alpha^0$.

Multiplying both parts of the expression for $g_i(\boldsymbol{\alpha}_{kT})$ on $\left(\sqrt{T}\right)^{-1} \bar{E}_{Ti}$, where $\bar{E}_{Ti}$ is the element on the main diagonal of matrix $\bar{\mathbf{E}}_T$ defined in Assumption 5.2, we obtain for the ith component $\bar{g}_{Ti}(\boldsymbol{\alpha})$ of the vector $\bar{\mathbf{g}}_T(\boldsymbol{\alpha})$, $i \in I$,

$$\bar{g}_{Ti}(\boldsymbol{\alpha}_{kT}) = \bar{g}_{Ti}(\boldsymbol{\alpha}^0) + \left(\sqrt{T}\right)^{-1} \bar{E}_{Ti} \nabla g_i'(\boldsymbol{\xi}_{ikT}) \mathbf{E}_T^{-1} \mathbf{U}_{kT}. \tag{5.45}$$

According to Assumption 5.2 on the continuity of functions $g_i(\boldsymbol{\alpha}), i \in I$ (Assumption 2.2.A), and using the consistency of $\boldsymbol{\xi}_{iT}$, we get

$$p \lim_{T\to\infty} \bar{E}_{Ti} \nabla g_i'(\boldsymbol{\xi}_{ikT}) \mathbf{E}_T^{-1} \mathbf{U}_{kT} = \bar{\mathbf{G}}_i'(\boldsymbol{\alpha}^0), \tag{5.46}$$

where $\bar{\mathbf{G}}_i'(\boldsymbol{\alpha}^0)$ is the ith line of matrix $\bar{\mathbf{G}}(\boldsymbol{\alpha}^0)$.

Then taking into account (5.30) we get from (5.45)

$$p \lim_{T\to\infty} \bar{g}_{Ti}(\boldsymbol{\alpha}_{kT}) = 0, \; i \in I_1^0,$$

$$p \lim_{T\to\infty} \bar{g}_{Ti}(\boldsymbol{\alpha}_{kT}) = \lim_{T\to\infty} \bar{g}_{Ti}(\boldsymbol{\alpha}^0) = g_i(\boldsymbol{\alpha}^0) \lim_{T\to\infty} \frac{\mathbf{E}_{Ti}}{\sqrt{T}} = g_i^0(\boldsymbol{\alpha}^0), i \in I_2^0. \tag{5.47}$$

By Assumption 5.2, $g_i(\boldsymbol{\alpha}^0) \in (-\infty, 0], i \in I_1^0$.

Taking into account (5.44) and (5.47), we derive from (5.39)

$$p \lim_{T\to\infty} \tilde{\mathbf{X}}_{kT} = F(\mathbf{O}_n, \bar{\bar{\mathbf{G}}}(\boldsymbol{\alpha}^0), \mathbf{g}^0(\boldsymbol{\alpha}^0)) = \tilde{\mathbf{X}}_k, \tag{5.48}$$

where $\mathbf{g}^0(\boldsymbol{\alpha}^0) = [g_i^0(\boldsymbol{\alpha}^0)], i \in I$. According to (5.47), the components $g^0(\boldsymbol{\alpha}^0)$ with indexes $i \in I_2^0$ belong to non-positive semi-axis, while the components with indexes $i \in I_1^0$ are equal to 0.

One can show that the vector $\tilde{\mathbf{X}}_k$ is the solution to the problem

$$\frac{1}{2}\tilde{\mathbf{X}}'\tilde{\mathbf{X}} \to \min, \quad \bar{\bar{\mathbf{G}}}(\boldsymbol{\alpha}^0)\tilde{\mathbf{X}} \leq -\mathbf{g}^0(\boldsymbol{\alpha}^0). \tag{5.49}$$

Since the right-hand side parts of (5.49) are non-negative, we have $\tilde{\mathbf{X}}_k = \mathbf{O}_n$, and by (5.36) $\bar{\mathbf{X}}_{kT} = \tilde{\mathbf{M}}_T^{-1}(\boldsymbol{\alpha}_{kT})\tilde{\mathbf{X}}_{kT}$. From here and (5.43), (5.48), we obtain

$$p \lim_{T\to\infty} \bar{\mathbf{X}}_{kT} = \mathbf{O}_n.$$

Therefore (5.31) follows by (5.2) and Assumption 5.1.6.

Lemma 5.4. *Let the conditions of Lemma 5.3 be satisfied and assume that Assumption 2.4 holds true. Then the Lagrange multipliers of auxiliary problem (5.33) converge in probability to 0 as $T \to \infty$:*

$$p \lim_{T\to\infty} \bar{\lambda}_i(\boldsymbol{\alpha}_{kT}) = \bar{\lambda}_i^0 = 0, \; i \in I. \tag{5.50}$$

Proof. The necessary and sufficient conditions for the existence of the minimum in (5.33) are

$$\bar{\bar{\mathbf{R}}}_T(\boldsymbol{\alpha}_{kT})\bar{\mathbf{X}}_{kT} - \bar{\mathbf{q}}_T(\boldsymbol{\alpha}_{kT}) + \sum_{i\in I} \bar{\lambda}_i(\boldsymbol{\alpha}_{kT})\bar{\mathbf{G}}'_{Ti}(\boldsymbol{\alpha}_{kT}) = \mathbf{O}_n, \tag{5.51}$$

$$\bar{\lambda}_i(\boldsymbol{\alpha}_{kT})(\bar{\mathbf{G}}'_{Ti}(\boldsymbol{\alpha}_{kT})\bar{\mathbf{X}}_{kT} + \bar{g}_i(\boldsymbol{\alpha}_{kT})) = 0, \quad \bar{\lambda}_i(\boldsymbol{\alpha}_{kT}) \geq 0, i \in I, \tag{5.52}$$

where $\bar{\mathbf{G}}'_{Ti}(\boldsymbol{\alpha}_{kT})$ is the ith ($i \in I$) line of the matrix $\bar{\mathbf{G}}_T(\boldsymbol{\alpha}_{kT})$ defined in (5.33).

From (5.52) and (5.45) we get

$$\bar{\lambda}_i(\boldsymbol{\alpha}_{kT})\Big((\sqrt{T}E_{Ti}^{-1})\bar{\mathbf{G}}'_{Ti}(\boldsymbol{\alpha}_{kT})\bar{\mathbf{X}}_{kT} + g_i(\boldsymbol{\alpha}^0)$$
$$+\nabla g'_i(\boldsymbol{\xi}_{ikT})\mathbf{E}_T^{-1}\mathbf{U}_{kT}\Big) = 0, \bar{\lambda}_i(\boldsymbol{\alpha}_{kT}) \geq 0, i \in I. \tag{5.53}$$

Taking into account Assumptions 5.1 and 5.2, conditions of the lemma and convergence of $\bar{\mathbf{X}}_{kT}$ to zero (see (5.53)) we get $p\lim_{T\to\infty} \bar{\lambda}_i(\boldsymbol{\alpha}_{kT}) = 0,\ i \in I$. Hence we obtain

$$p\lim_{T\to\infty} \bar{\lambda}_i(\boldsymbol{\alpha}_{kT}) = 0, \quad i \in I_2^0. \tag{5.54}$$

From (5.51) we have system of equations for $m_1 < n$ unknown parameters is

$$\sum_{i\in I_1^0} \bar{\lambda}_i(\boldsymbol{\alpha}_{kT})\bar{\mathbf{G}}'_{Ti}(\boldsymbol{\alpha}_{kT}) = \bar{\mathbf{q}}_T(\boldsymbol{\alpha}_{kT}) - \bar{\bar{\mathbf{R}}}_T(\boldsymbol{\alpha}_{kT})\bar{\mathbf{X}}_{kT} - \sum_{i\in I_2^0} \bar{\lambda}_i(\boldsymbol{\alpha}_{kT})\bar{\mathbf{G}}'_{Ti}(\boldsymbol{\alpha}_{kT}). \tag{5.55}$$

By (5.19), (5.31), (5.41), (5.42) and (5.54), the right-hand side of (5.55) converges in probability to 0 as $T \to \infty$. According to (5.42) and Assumption 2.4, the matrix in the left-hand side of (5.55) converges in probability to a matrix with linearly independent columns. From here and Lemma 2.5 we obtain $p\lim_{T\to\infty} \bar{\lambda}_i(\boldsymbol{\alpha}_{kT}) = 0$, $i \in I_1^0$, which implies together with (5.54) the limit (5.50). □

Put $v_{iT}(\boldsymbol{\alpha}_{kT}) = \sqrt{T}(\bar{\lambda}_i(\boldsymbol{\alpha}_{kT}) - \bar{\lambda}_i^0) = \sqrt{T}\bar{\lambda}_i(\boldsymbol{\alpha}_{kT}),\ i = \overline{1,m}$, since according to Lemma 5.4 $\bar{\lambda}_i^0 = 0$. Denote by $\mathbf{V}_{1k,T}$ the vector with components $v_i(\boldsymbol{\alpha}_{kT}),\ i \in I_1^0$, and by $\mathbf{V}_{2k,T}$ the vector with components $v_i(\boldsymbol{\alpha}_{kT}),\ i \in I_2^0$. Below we prove two lemmas for $\mathbf{V}_{ik,T},\ i = 1, 2$, using the arguments similar to those in the proofs of Lemmas 2.8 and 2.9.

Lemma 5.5. *Assume that the conditions of Lemma 5.3 are satisfied.*

$$p\lim_{T\to\infty} \mathbf{V}_{2k,T} = \mathbf{O}_{m_1}.$$

Proof. From (5.53) we have

$$v_{iT}\left[\left(\sqrt{T}E_{Ti}^{-1}\right)\bar{\mathbf{G}}'_{Ti}(\boldsymbol{\alpha}_{kT})\bar{\mathbf{X}}_{kT} + g_i(\boldsymbol{\alpha}^0) + \nabla g'_i(\boldsymbol{\xi}_{ikT})\mathbf{E}_T^{-1}\mathbf{U}_T\right] = 0, \quad i \in I. \tag{5.56}$$

Using (5.31) and that $g_i(\alpha^0) < 0$ for $i \in I_2^0$, we see that for $i \in I_2^0$ the random variables in square brackets converge in probability to some non-zero constants, which proves the lemma. □

We need some more notations. Denote

$$\mathbf{G}(\boldsymbol{\xi}_{kT}) = \begin{bmatrix} \nabla g_i'(\boldsymbol{\xi}_{1kT}) \\ \nabla g_i'(\boldsymbol{\xi}_{2kT}) \\ \vdots \\ \nabla g_i'(\boldsymbol{\xi}_{mkT}) \end{bmatrix},$$

where the entries $\boldsymbol{\xi}_{kT} = [\boldsymbol{\xi}'_{1kT} \quad \boldsymbol{\xi}'_{2kT} \quad \dots \boldsymbol{\xi}'_{mkT}]' \in \Re^{nm}$ and such that

$$p \lim_{T\to\infty} \boldsymbol{\xi}_{ikT} = \alpha^0, \quad i \in I. \tag{5.57}$$

By (5.57) we have

$$p \lim_{T\to\infty} \mathbf{G}(\boldsymbol{\xi}_{kT}) = \mathbf{G}(\mathbf{1}_n \otimes \alpha^0) = \mathbf{G}(\alpha^0), \tag{5.58}$$

where $\mathbf{1}_n$ is the n-dimensional vector with components equal to 1.

Further, denote

$$\bar{\mathbf{G}}_T(\boldsymbol{\xi}_{kT}) = \begin{bmatrix} \bar{\mathbf{G}}'_{T1}(\boldsymbol{\xi}_{1T}) \\ \bar{\mathbf{G}}'_{T2}(\boldsymbol{\xi}_{2kT}) \\ \vdots \\ \bar{\mathbf{G}}'_{Tm}(\boldsymbol{\xi}_{mkT}) \end{bmatrix} = \bar{\mathbf{E}}_T \mathbf{G}(\boldsymbol{\xi}_{kT}) \mathbf{E}_T^{-1}. \tag{5.59}$$

By (5.46), (5.57), (5.58) we get under Condition 5.2

$$p \lim_{T\to\infty} \bar{\mathbf{G}}_T(\boldsymbol{\xi}_{kT}) = \bar{\mathbf{G}}(\alpha^0). \tag{5.60}$$

It follows from (5.60)

$$p \lim_{T\to\infty} \bar{\mathbf{G}}_T^{(j)}(\boldsymbol{\xi}_{kT}) = \bar{\mathbf{G}}^{(j)}(\alpha^0), \quad j = 1, 2, \tag{5.61}$$

where $\bar{\mathbf{G}}_{\underline{T}}^{(j)}(\boldsymbol{\xi}_{kT})$ (respectively, $\bar{\mathbf{G}}^{(j)}(\alpha^0)$) is the matrix composed from lines of matrices $\bar{\mathbf{G}}_T(\boldsymbol{\xi}_{kT})$ (respectively, $\bar{\mathbf{G}}(\alpha^0)$) with indexes $i \in I_j^0$, $j = 1, 2$.

Finally, let

$$\bar{\mathbf{G}}_T^{(j)}(\alpha), \quad j = 1, 2, \tag{5.62}$$

be the matrix, composed from lines of $\bar{\mathbf{G}}'_{Ti}(\alpha)$, $i \in I_j^0$, of the matrix $\bar{\mathbf{G}}_T(\alpha)$ defined in (5.34).

Lemma 5.6. *Let the conditions of Lemma 5.4 be satisfied. Then for given* $\delta_k > 0$ *there exist* $e_k > 0$ *and* $\tau_k > 0$ *such that*

$$P\{||\mathbf{V}_{1k,T}|| \geq e_k\} < \delta_k, \quad T > \tau_k. \tag{5.63}$$

Proof. Since $g_i(\boldsymbol{\alpha}^0) = 0$, $i \in I_1^0$, we have by (5.53)

$$\mathbf{V}'_{1k,T}\left(\bar{\mathbf{G}}_T^{(1)}(\boldsymbol{\alpha}_{kT})\bar{\bar{\mathbf{X}}}_{kT} + \bar{\mathbf{G}}_T^{(1)}(\boldsymbol{\xi}_{kT})\mathbf{U}_{kT}\right) = 0, \tag{5.64}$$

where $\bar{\mathbf{G}}_T^{(1)}(\boldsymbol{\alpha}_{kT})$ is the matrix defined in (5.62), and $\bar{\bar{\mathbf{X}}}_{kT} = \sqrt{T}\bar{\mathbf{X}}_{kT}$.

Since $\bar{\bar{\mathbf{R}}}_T(\boldsymbol{\alpha}_{kT})$ is positive definite, we get by (5.51), (5.14)

$$\begin{aligned}\bar{\bar{\mathbf{X}}}_{kT} = \bar{\bar{\mathbf{R}}}_T^{-1}(\boldsymbol{\alpha}_{kT})\boldsymbol{\rho}_T(\boldsymbol{\alpha}_{kT}) - \bar{\bar{\mathbf{R}}}_T^{-1}(\boldsymbol{\alpha}_{kT})\left(\bar{\mathbf{G}}_T^{(1)}(\boldsymbol{\alpha}_{kT})\right)'\mathbf{V}_{1k,T}\\ -\bar{\bar{\mathbf{R}}}_T^{-1}(\boldsymbol{\alpha}_{kT})(\bar{\mathbf{G}}_T^{(2)}(\boldsymbol{\alpha}_{kT}))'\mathbf{V}_{2k,T},\end{aligned} \tag{5.65}$$

where, according to (5.4) and (5.40),

$$\boldsymbol{\rho}_T(\boldsymbol{\alpha}_{kT}) = \sqrt{T}\bar{\mathbf{q}}_T(\boldsymbol{\alpha}_{kT}) = \mathbf{Q}_T - \bar{\boldsymbol{\Phi}}_T\mathbf{U}_{kT}. \tag{5.66}$$

Using (5.64), we obtain from (5.65) after necessary transformations

$$\mathbf{V}'_{1k,T}\mathbf{A}_T\mathbf{V}_{1k,T} - \mathbf{V}'_{1k,T}\mathbf{h}_T = 0, \tag{5.67}$$

where

$$\mathbf{A}_T = \bar{\mathbf{G}}_T^{(1)}(\boldsymbol{\alpha}_{kT})\bar{\bar{\mathbf{R}}}_T^{-1}(\boldsymbol{\alpha}_{kT})(\bar{\mathbf{G}}_T^{(1)}(\boldsymbol{\alpha}_{kT}))', \tag{5.68}$$

and

$$\begin{aligned}\mathbf{h}_T = \bar{\mathbf{G}}_T^{(1)}(\boldsymbol{\alpha}_{kT})\bar{\bar{\mathbf{R}}}_T^{-1}(\boldsymbol{\alpha}_{kT})\mathbf{Q}_T(\boldsymbol{\alpha}_{kT}) + [\bar{\mathbf{G}}_T^{(1)}(\boldsymbol{\xi}_T) - \bar{\mathbf{G}}_T^{(1)}(\boldsymbol{\alpha}_{kT})\\ \times\bar{\bar{\mathbf{R}}}_T^{-1}(\boldsymbol{\alpha}_{kT})\bar{\boldsymbol{\Phi}}_T]\mathbf{U}_{kT} - \bar{\mathbf{G}}_T^{(1)}(\boldsymbol{\alpha}_{kT})\bar{\bar{\mathbf{R}}}_T^{-1}(\boldsymbol{\alpha}_{kT})(\bar{\mathbf{G}}_T^{(2)}(\boldsymbol{\alpha}_{kT}))'\mathbf{V}_{2k,T}.\end{aligned} \tag{5.69}$$

For the matrix $\mathbf{A}_T$ we have $\mathbf{C}'_T\mathbf{A}_T\mathbf{C}_T = \mathbf{N}_T$, where $\mathbf{C}_T$ is the orthogonal matrix, $\mathbf{N}_T = \mathrm{diag}(\nu_{1T}, \ldots, \nu_{nT})$ and ν_{iT} is the ith eigenvalue of $\mathbf{A}_T$.

By continuity of the elements of $\mathbf{G}(\boldsymbol{\alpha})$ in $\boldsymbol{\alpha}$ and Assumption 5.2, we see that for the matrices $\bar{\mathbf{G}}_T^{(j)}(\boldsymbol{\alpha}_{kT})$, $j = 1, 2$,

$$p\lim_{T\to\infty}\bar{\mathbf{G}}_T^{(j)}(\boldsymbol{\alpha}_{kT}) = \bar{\mathbf{G}}^{(j)}(\boldsymbol{\alpha}^0), \quad j = 1, 2, \tag{5.70}$$

where the matrix $\bar{\mathbf{G}}_T^{(j)}(\boldsymbol{\alpha}_{kT})$ is defined in (5.62).

According to (5.19) and Assumption 2.2, we obtain

$$p \lim_{T\to\infty} \mathbf{A}_T = \mathbf{A} = \bar{\mathbf{G}}^{(1)}(\boldsymbol{\alpha}^0)\bar{\mathbf{R}}^{-1}(\boldsymbol{\alpha}^0)(\bar{\mathbf{G}}^{(1)}(\boldsymbol{\alpha}^0))'.$$

By Assumptions 2.4, 5.1.3, the matrix $\mathbf{A}$ is positive definite, which implies by Lemma 2.4 that $p\lim_{T\to\infty}\mathbf{N}_T = \mathbf{N}$, $p\lim_{T\to\infty}\mathbf{C}_T = \mathbf{C}$, where $\mathbf{N} = \mathrm{diag}(\nu_1,\ldots,\nu_n)$, ν_i is the ith eigenvalue of $\mathbf{A}$, $\mathbf{C}$ is the orthogonal matrix such that $\mathbf{C}'\mathbf{A}\mathbf{C} = \mathbf{N}$.

Put $\tilde{\nu}_{iT} = \nu_{iT}$, if $\nu_{iT} > 0$, and $\tilde{\nu}_{iT} = 1$ otherwise. Denote $\tilde{\mathbf{N}}_T = \mathrm{diag}(\tilde{\nu}_{1T},\ldots,\tilde{\nu}_{nT})$. In the proof of Lemma 2.9 we obtained that

$$P\{\mathbf{N}_T = \tilde{\mathbf{N}}_T\} > 1 - \frac{\delta}{3}, \quad T > T_1. \tag{5.71}$$

Using the transformation $\mathbf{Y}_T = \tilde{\mathbf{N}}_T^{1/2}\mathbf{C}_T^{-1}\mathbf{V}_{1T}$, we obtain from (5.67) (and using (5.71)) the estimate

$$P\{\mathbf{Y}'_T\mathbf{Y}_T + 2\mathbf{Y}'_T\mathbf{K}_T = 0\} \geq P\{\mathbf{N}_T = \tilde{\mathbf{N}}_T\} > 1 - \frac{\delta}{3}, \quad T > T_1, \tag{5.72}$$

where $\mathbf{K}_T = (1/2)\tilde{\mathbf{N}}_T^{-1/2}\mathbf{C}'_T\mathbf{h}_T$.

From (5.15), (5.17), (5.19), (5.56), (5.61), (5.70) and Lemma 5.5 we get

$$\mathbf{h}_T \overset{p}{\Rightarrow} \bar{\mathbf{G}}_1(\boldsymbol{\alpha}^0)\bar{\mathbf{R}}^{-1}(\boldsymbol{\alpha}^0)\mathbf{Q}.$$

Then the limit distribution of $\mathbf{K}_T$ coincides with the distribution of the m_1-dimensional random variable

$$\mathbf{K} = -\frac{1}{2}\mathbf{N}^{-1/2}\mathbf{C}'\bar{\mathbf{G}}^{(1)}(\boldsymbol{\alpha}^0)\bar{\mathbf{R}}^{-1}(\boldsymbol{\alpha}^0)\mathbf{Q} \quad N(\mathbf{O}_n, \sigma^2\mathbf{J}_n).$$

Note that the expression (5.72) completely coincides with (2.56). Therefore, repeating the arguments from the proof of Lemma 2.9, we obtain (5.63). □

Lemma 5.7. *Under the conditions of Lemma 5.4, for given $\Delta_{k+1} > 0$, there exist $\omega_{k+1} > 0$ and $T_{k+1} > 0$ such that*

$$P\{||\mathbf{U}_{k+1,T}|| \geq \omega_{k+1}\} < \Delta_{k+1}, \quad T > T_{k+1},$$

where $\mathbf{U}_{k+1,T} = \mathbf{E}_T(\boldsymbol{\alpha}_{k+1,T} - \boldsymbol{\alpha}^0)$.

Proof. Put $\varepsilon = \varepsilon_1 + \varepsilon_2 + \varepsilon_3 + \varepsilon_4$, with $\varepsilon_i > 0$, $i = \overline{1,4}$. According to (5.65) and (5.66)

$$\begin{aligned} P\{||\bar{\bar{\mathbf{X}}}_{kT}|| \geq \varepsilon\} &\leq P\{||\bar{\bar{\mathbf{R}}}_T^{-1}(\boldsymbol{\alpha}_{kT})(\bar{\mathbf{G}}_T^{(1)}(\boldsymbol{\alpha}_{kT}))'|| \cdot ||\mathbf{V}_{1k,T}|| \geq \varepsilon_1\} \\ &+ P\{||\bar{\bar{\mathbf{R}}}_T^{-1}(\boldsymbol{\alpha}_{kT})\mathbf{Q}_T|| \geq \varepsilon_2\} + P\{||\bar{\bar{\mathbf{R}}}_T^{-1}(\boldsymbol{\alpha}_{kT})(\bar{\mathbf{G}}_T^{(2)}(\boldsymbol{\alpha}_{kT}))'\mathbf{V}_{2k,T}|| \geq \varepsilon_3\} \\ &+ P\{||\bar{\bar{\mathbf{R}}}_T^{-1}(\boldsymbol{\alpha}_{kT})\bar{\Phi}_T\mathbf{U}_{kT}|| \geq \varepsilon_4\}. \end{aligned} \tag{5.73}$$

The estimate for the first term can be obtained using Lemma 5.6 and following the proof of (2.67). We only need to replace $\mathbf{V}_{1T}$ by $\mathbf{V}_{1k,T}$, and $\tilde{\bar{\mathbf{\Phi}}}_T$ by $\bar{\bar{\mathbf{R}}}^{-1}{}_T(\boldsymbol{\alpha}_{kT})$. Then, for given $\delta_1 > 0$ there exist $\varepsilon_1 > 0$ and $\tau_1 > 0$, such that

$$P\left\{||\bar{\bar{\mathbf{R}}}_T^{-1}(\boldsymbol{\alpha}_{kT})(\bar{\mathbf{G}}_T^{(1)}(\boldsymbol{\alpha}_{kT}))'|| \cdot ||\mathbf{V}_{1k,T}|| \geq \varepsilon_1\right\} < \delta_1, \quad T > \tau_1. \tag{5.74}$$

Since $\bar{\bar{\mathbf{R}}}_T^{-1}(\boldsymbol{\alpha}_{kT})\mathbf{Q}_T$ converges in distribution to $\bar{\mathbf{R}}^{-1}(\boldsymbol{\alpha}^0)\mathbf{Q}$, for given $\delta_2 > 0$ there exist $\varepsilon_2 > 0$ and $\tau_2 > 0$, such that

$$P\{||\bar{\bar{\mathbf{R}}}_T^{-1}(\boldsymbol{\alpha}_{kT})\mathbf{Q}_T|| \geq \varepsilon_2\} < \delta_2, \quad T > \tau_2. \tag{5.75}$$

Taking into account Lemma 5.5, for given $\varepsilon_3 > 0$ and $\delta_3 > 0$ there exists $\tau_3 > 0$ such that

$$P\{||\bar{\bar{\mathbf{R}}}_T^{-1}(\boldsymbol{\alpha}_{kT})(\bar{\mathbf{G}}_T^{(2)}(\boldsymbol{\alpha}_{kT}))'\mathbf{V}_{2k,T}|| \geq \varepsilon_3\} < \delta_3, \quad T > \tau_3. \tag{5.76}$$

For $\mathbf{C}_T = \bar{\bar{\mathbf{R}}}_T^{-1}(\boldsymbol{\alpha}_{kT})\bar{\mathbf{\Phi}}_T$, we have

$$||\mathbf{c}_T\mathbf{U}_{kT}|| \leq ||\mathbf{c}_T - \mathbf{J}_n|| \cdot ||\mathbf{U}_{kT}|| + ||\mathbf{U}_{kT}||.$$

According to (5.17), (5.19) and (5.30), $p\lim_{T\to\infty}\mathbf{c}_T = \mathbf{J}_n$. Applying to the above inequality (5.9)–(5.11), we get for given δ_4,

$$P\{||\bar{\bar{\mathbf{R}}}_T^{-1}(\boldsymbol{\alpha}_{kT})\bar{\mathbf{\Phi}}_T\mathbf{U}_{kT}|| \geq \varepsilon_4\} < \delta_4, \quad T > \tau_4, \tag{5.77}$$

where ε_4 and τ_4 depend on δ_4.

Combining inequalities (5.74)–(5.77) with (5.73), we obtain that for any $\delta = \sum_{i=1}^4 \delta_i$ there exist $\varepsilon = \sum_{i=1}^4 \varepsilon_i$ and $\tau_0 = \max_{1\leq i\leq 4}\{\tau_i\}$, for which

$$P\{||\bar{\bar{\mathbf{X}}}_{kT}|| \geq \varepsilon\} < \delta, \quad T > \tau_0.$$

By (5.1) we have $||\mathbf{U}_{k+1,T}|| \leq ||\mathbf{U}_{kT}|| + \gamma_{kT}||\bar{\bar{\mathbf{X}}}_{kT}|| \leq ||\mathbf{U}_{kT}|| + ||\bar{\bar{\mathbf{X}}}_{kT}||$. Applying to it the formulas (5.9)–(5.11), by (5.30) and the previous inequality we derive the statement of the lemma. □

Now we can formulate the main result of this section.

Theorem 5.3. *Assume that:*

1. *Estimates of regression parameters are calculated by means of Algorithms A5.1 and A5.2.*
2. *Condition 5.1, Assumptions 2.2A, 2.8, 5.1, 5.2 (in case of using Algorithm A5.2, the matrix* $\mathbf{D}_T(\boldsymbol{\alpha})$ *should be of full rank in any point of a compact set) are satisfied.*

3. *The initial value* $\boldsymbol{\alpha}_{0T}$ *of the iterative process is the consistent estimate of* $\boldsymbol{\alpha}^0 \in \Theta$, *where* Θ *is a compact set, and the value* $\mathbf{U}_{0T} = \mathbf{E}_T(\boldsymbol{\alpha}_{0T} - \boldsymbol{\alpha}^0)$ *has a limit distribution.*

Then $\boldsymbol{\alpha}_{kT}$, $k = 1, 2 \ldots$ *are consistent estimates of* $\boldsymbol{\alpha}^0$.

Proof. Let $||\mathbf{U}_{kT}||$satisfy (5.30), and suppose that the estimate $\boldsymbol{\alpha}_{kT}$ is consistent. Then we derive from (5.1) and Lemma 5.3 (see (5.31)) that $\boldsymbol{\alpha}_{k+1,T}$ is also consistent. Since for $||\mathbf{U}_{k+1,T}||$ the conditions of Lemma 5.7 hold true, we get from Lemma 5.3 $p\lim_{T\to\infty} \mathbf{X}_{k+1,T} = \mathbf{O}_n$. Taking into account that the estimate $\boldsymbol{\alpha}_{0T}$ is consistent and k is random, we obtain the statement of the theorem. □

Corollary 5.2. *Under conditions of Theorem 5.3, (5.30) hold for* $k = 0, 1, 2, \ldots$.

Proof. For $k = 0$ the statement is true due to conditions of Theorem 5.3. For $k = 1, 2, \ldots$, the statement follows directly by Theorem 5.3 and Lemma 5.7. □

Theorem 5.4. *Suppose that conditions of Theorem 5.3 are satisfied, Assumption 2.4 is fulfilled, and* $\mathbf{U}_{0T} = \mathbf{E}_T(\boldsymbol{\alpha}_{0T} - \boldsymbol{\alpha}^0)$ *has a limiting distribution. Then the values* $\tilde{\mathbf{U}}_{kT} = \mathbf{E}_T(\tilde{\boldsymbol{\alpha}}_{kT} - \boldsymbol{\alpha}^0)$, $k = 1, 2, \ldots$, *where* $\tilde{\boldsymbol{\alpha}}_{kT} = \boldsymbol{\alpha}_{k-1,T} + \mathbf{X}_{k-1,T}$, *converge (in distribution) to the random variable* $\boldsymbol{U}$ *which is the solution to the quadratic programming problem*

$$\frac{1}{2}\mathbf{Z}'\bar{\mathbf{R}}(\boldsymbol{\alpha}^0)\mathbf{Z} - \mathbf{Q}'\mathbf{Z} \to \min, \quad \beta_i(\mathbf{Z}) = \bar{\mathbf{G}}_i'(\boldsymbol{\alpha}^0)\mathbf{Z} \le 0, \quad i \in I_1^0, \tag{5.78}$$

where $\bar{\mathbf{G}}_i'(\boldsymbol{\alpha}^0)$ *is the ith line of the matrix* $\bar{\mathbf{G}}(\boldsymbol{\alpha}^0)$ *defined in Assumption 5.2.*

Proof. Put

$$\bar{\mathbf{X}} = \left(\sqrt{T}\right)^{-1}(\mathbf{Z} - \mathbf{U}_{kT}), \quad \mathbf{U}_{kT} = \mathbf{E}_T(\boldsymbol{\alpha}_{kT} - \boldsymbol{\alpha}^0). \tag{5.79}$$

Substituting $\bar{\mathbf{X}}$ from (5.79) in (5.33), and taking into account expressions (5.45) and (5.66), we obtain after some transformations (we drop all terms which do not depend on $\mathbf{Z}$)

$$\frac{1}{2}\mathbf{Z}'\bar{\bar{\mathbf{R}}}_T(\boldsymbol{\alpha}_{kT})\mathbf{Z} - \mathbf{d}'_T(\boldsymbol{\alpha}_{kT})\mathbf{Z} \to \min, \quad \bar{\mathbf{G}}_{iT}'(\boldsymbol{\alpha}_{kT})\mathbf{Z} \le l_i(\boldsymbol{\alpha}_{kT}), \quad i \in I. \tag{5.80}$$

Here $\bar{\bar{\mathbf{R}}}_T(\boldsymbol{\alpha}_{kT})$ is defined in (5.3),

$$d_T(\boldsymbol{\alpha}_{kT}) = \left(\bar{\bar{R}}_T(\boldsymbol{\alpha}_{kT}) - \bar{\boldsymbol{\Phi}}_T(\boldsymbol{\xi}_{1T})\right)\mathbf{U}_{kT} + \mathbf{Q}_T, \tag{5.81}$$

$$l_i(\boldsymbol{\alpha}_{kT}) = \left(\bar{\mathbf{G}}_{iT}'(\boldsymbol{\alpha}_{kT}) - \bar{\mathbf{G}}_{Ti}'(\boldsymbol{\xi}_{iT})\right)\mathbf{U}_{kT}, \quad i \in I_1^0, \tag{5.82}$$

$$l_i(\boldsymbol{\alpha}_{kT}) = \left(\bar{\mathbf{G}}_{iT}'(\boldsymbol{\alpha}_{kT}) - \bar{\mathbf{G}}_{Ti}'(\boldsymbol{\xi}_{ikT})\right)\mathbf{U}_{kT} - \bar{E}_{Ti}g_{Ti}(\boldsymbol{\alpha}^0), \quad i \in I_2^0. \tag{5.83}$$

The solutions to the problems (5.33) and (5.80) are related by

$$\sqrt{T}\bar{\mathbf{X}}_{kT} = \mathbf{Z}_{k+1,T} - \mathbf{U}_{kT},$$

see (5.79), where $\mathbf{Z}_{k+1,T}$ is the solution to (5.80). From last formula and (5.32) get

$$\mathbf{Z}_{k+1,T} = \mathbf{E}_T(\boldsymbol{\alpha}_{kT} + \mathbf{X}_{kT} - \boldsymbol{\alpha}^0) = \mathbf{E}_T(\tilde{\boldsymbol{\alpha}}_{k+1,T} - \boldsymbol{\alpha}^0) = \tilde{\mathbf{U}}_{k+1,T}.$$

To investigate the statistical properties of $\tilde{\mathbf{U}}_{k+1,T}$, consider the quadratic programming problem

$$\frac{1}{2}\mathbf{Z}'\bar{\bar{\mathbf{R}}}_T(\boldsymbol{\alpha}_{kT})\mathbf{Z} - \mathbf{d}'_T(\boldsymbol{\alpha}_{kT})\mathbf{Z} \to \min, \quad \bar{\mathbf{G}}'_{iT}(\boldsymbol{\alpha}_{kT})\mathbf{Z} \le l_i(\boldsymbol{\alpha}_{kT}), \quad i \in I_1^0. \quad (5.84)$$

Denote its solution by $\mathbf{U}^*_{k+1,T}$. We show that

$$\mathbf{U}^*_{k+1,T} \overset{p}{\Rightarrow} \mathbf{U}, \quad T \to \infty, \quad (5.85)$$

where $\mathbf{U}$ is the solution to (5.78). Let

$$\mathbf{y} = \tilde{\mathbf{M}}_T(\boldsymbol{\alpha}_{kT})\mathbf{Z}, \quad (5.86)$$

where the matrix $\tilde{\mathbf{M}}_T(\boldsymbol{\alpha}_{kT})$ is defined in (5.35). Using (5.86), we obtain from (5.84)

$$\left.\begin{array}{l} \frac{1}{2}\mathbf{y}'\mathbf{y} - \mathbf{y}'\tilde{\mathbf{d}}_T(\boldsymbol{\alpha}_{kT}) \to \min, \\ \bar{\bar{\mathbf{G}}}_T^{(1)}(\boldsymbol{\alpha}_{kT})\mathbf{y} \le \mathbf{L}_1(\boldsymbol{\alpha}_{kT}). \end{array}\right\} \quad (5.87)$$

Here $\tilde{\mathbf{d}}_T(\boldsymbol{\alpha}_{kT}) = (\tilde{\mathbf{M}}_T^{-1}(\boldsymbol{\alpha}_{kT}))'\mathbf{d}_T(\boldsymbol{\alpha}_{kT})$, $\bar{\bar{\mathbf{G}}}_T^{(1)}(\boldsymbol{\alpha}_{kT}) = \bar{\mathbf{G}}_T^{(1)}(\boldsymbol{\alpha}_{kT})\tilde{\mathbf{M}}_T^{-1}(\boldsymbol{\alpha}_{kT})$, where the matrix $\bar{\mathbf{G}}_T^{(1)}(\boldsymbol{\alpha})$ is defined in (5.62), and $\mathbf{L}_1(\boldsymbol{\alpha})$ is a vector with components $l_1(\boldsymbol{\alpha})$, $i \in I_1^0$. Put $\mathbf{y} = \mathbf{M}(\boldsymbol{\alpha}^0)\mathbf{Z}$, with the matrix $\mathbf{M}(\boldsymbol{\alpha}^0)$ defined in (5.43). In this notation, (5.78) can be rewritten as

$$\frac{1}{2}\mathbf{y}'\mathbf{y} - \mathbf{y}'\tilde{\mathbf{Q}} \to \min, \quad \bar{\bar{\mathbf{G}}}^{(1)}(\boldsymbol{\alpha}^0)\mathbf{y} \le \mathbf{L}_1(\boldsymbol{\alpha}^0) = \mathbf{O}_{m_1}, \quad (5.88)$$

where $\bar{\bar{\mathbf{G}}}^{(1)}(\boldsymbol{\alpha}^0) = \bar{\mathbf{G}}^{(1)}(\boldsymbol{\alpha}^0)\tilde{\mathbf{M}}^{-1}(\boldsymbol{\alpha}^0)$, $\tilde{\mathbf{Q}} = (\mathbf{M}^{-1}(\boldsymbol{\alpha}^0))'\mathbf{Q}$.

The solutions to (5.87) and (5.88), are, respectively,

$$\mathbf{u}^*_{k+1,T} = F(\tilde{\mathbf{d}}_T(\boldsymbol{\alpha}_{kT}), \bar{\bar{\mathbf{G}}}_T^{(1)}(\boldsymbol{\alpha}_{kT}), \mathbf{L}_1(\boldsymbol{\alpha}_{kT})) \quad (5.89)$$

and

$$\mathbf{u} = \mathbf{F}\left(\tilde{\mathbf{Q}}, \bar{\bar{\mathbf{G}}}^{(1)}(\boldsymbol{\alpha}^0), \mathbf{L}_1(\boldsymbol{\alpha}^0)\right). \quad (5.90)$$

According to Lemmas 2.6 and 4.8, the function $F(\tilde{\mathbf{d}}_T, \bar{\bar{\mathbf{G}}}^{(1)}, \mathbf{L}_1)$ is continuous in $\tilde{\mathbf{d}}_T = \tilde{\mathbf{d}}_T(\boldsymbol{\alpha}_{kT})$, $\bar{\bar{\mathbf{G}}}^{(1)} = \bar{\bar{\mathbf{G}}}^{(1)}(\boldsymbol{\alpha}_{kT})$, and $\mathbf{L}_1 = \mathbf{L}_1(\boldsymbol{\alpha}_{kT})$, which in turn are continuous in $\boldsymbol{\alpha}_{kT}$. Under consistency of $\boldsymbol{\alpha}_{kT}$ (Theorem 5.3) we get, using (5.15), (5.17), (5.19), (5.43), (5.34), (5.81),

$$p\lim_{T\to\infty} \bar{\bar{\mathbf{G}}}^{(1)}(\boldsymbol{\alpha}_{kT}) = \bar{\bar{\mathbf{G}}}^{(1)}(\boldsymbol{\alpha}^0), \quad \tilde{\mathbf{d}}_T(\boldsymbol{\alpha}_{kT}) \overset{p}{\Rightarrow} \tilde{\mathbf{Q}}, \quad T \to \infty. \tag{5.91}$$

In the limit expression for $\tilde{\mathbf{d}}_T$, we use (5.81) and (5.30), which holds true for any $k \geq 1$ (cf. Lemma 5.7), and the fact that $\mathbf{U}_{0T}$ has a limit distribution. Since $\boldsymbol{\alpha}_{kT}$ is consistent we get by (5.82)

$$p\lim_{T\to\infty} \mathbf{L}_1(\boldsymbol{\alpha}_{kT}) = \mathbf{L}_1(\boldsymbol{\alpha}^0) = \mathbf{O}_{m_1}.$$

By (5.89), (5.90) and the last three limit expressions we derive $\mathbf{u}^*_{k+1,T} \overset{p}{\Rightarrow} \mathbf{u}, T \to \infty$. But $\mathbf{U}^*_{k+1,T} = \tilde{\mathbf{M}}_T^{-1}(\boldsymbol{\alpha}_{kT})\mathbf{u}^*_{k+1,T}$ and $\mathbf{U} = \mathbf{M}^{-1}(\boldsymbol{\alpha}^0)\mathbf{u}$, which implies (5.85).

Now we obtain the limit distribution of $\tilde{\mathbf{U}}_{k+1,T}$. Since $\bar{\bar{\mathbf{R}}}_T(\boldsymbol{\alpha}_{kT})$ is positive definite, $\tilde{\mathbf{U}}_{k+1,T} = \mathbf{U}^*_{k+1,T}$ provided that

$$\mathbf{U}^*_{k+1,T} \in \{\mathbf{Z} : \bar{\mathbf{G}}'_{iT}(\boldsymbol{\alpha}_{kT})\mathbf{Z} \leq l_i(\boldsymbol{\alpha}_{kT}), \quad i \in I\}.$$

The relation above takes place when $\bar{\mathbf{G}}'_{iT}(\boldsymbol{\alpha}_{kT})\mathbf{U}^*_{k+1,T} \leq l_i(\boldsymbol{\alpha}_{kT}),\ i \in I_2^0$, since $\mathbf{U}^*_{k+1,T}$ satisfies inequality (5.84). Thus,

$$P\{||\tilde{\mathbf{U}}_{k+1,T} - \mathbf{U}^*_{k+1,T}|| = 0\} = P\{\bar{\mathbf{G}}_T^{(2)}(\boldsymbol{\alpha}_{kT})\mathbf{U}^*_{k+1,T} \leq L_2(\boldsymbol{\alpha}_{kT})\}, \tag{5.92}$$

where the matrix $\bar{\mathbf{G}}_T^{(2)}(\boldsymbol{\alpha})$ is defined in (5.62), $\mathbf{L}_2(\boldsymbol{\alpha}_{kT})$ is a vector with components $l_i(\boldsymbol{\alpha}_{kT}),\ i \in I_2^0$.

Put

$$\boldsymbol{\Gamma}_T = -[\bar{\mathbf{G}}^{(2)}(\boldsymbol{\alpha}^0)\mathbf{U} - \bar{\mathbf{G}}_T^{(2)}(\boldsymbol{\alpha}_{kT})\mathbf{U}^*_{k+1,T} + (\bar{\mathbf{G}}_T^{(2)}(\boldsymbol{\alpha}_{kT}) - \bar{\mathbf{G}}_T^{(2)}(\boldsymbol{\xi}_{kT}))\mathbf{U}_{kT}],$$

where the matrix $\bar{\mathbf{G}}^{(2)}(\boldsymbol{\alpha}^0)$ is defined in (5.61), and components of $\boldsymbol{\xi}_{kT}$ satisfy the condition $||\boldsymbol{\xi}_{ikT} - \boldsymbol{\alpha}^0|| \leq ||\boldsymbol{\alpha}_{kT} - \boldsymbol{\alpha}^0||,\ i \in I$. Since $\boldsymbol{\alpha}_{kT}$ is consistent, this implies (5.57), cf. Theorem 5.3.

After simple transformations, the inequality $\bar{\mathbf{G}}_T^{(2)}(\boldsymbol{\alpha}_{kT})\mathbf{U}^*_{k+1,T} \leq \mathbf{L}_2(\boldsymbol{\alpha}_{kT})$ can be rewritten as $\bar{\mathbf{G}}^{(2)}(\boldsymbol{\alpha}^0)\mathbf{U} + \boldsymbol{\Gamma}_T \leq -\mathbf{E}_T\bar{\mathbf{g}}_T^{(2)}(\boldsymbol{\alpha}^0)$, where $\bar{\mathbf{g}}_T^{(2)}(\boldsymbol{\alpha}^0)$ is a vector with components $\bar{g}_{iT}(\boldsymbol{\alpha}^0) < 0,\ i \in I_2^0$, of $\bar{\mathbf{g}}_T(\boldsymbol{\alpha}^0)$. From here we obtain

$$\begin{aligned}
&P\{\bar{\mathbf{G}}^{(2)}(\boldsymbol{\alpha}^0)\mathbf{U} + \boldsymbol{\Gamma}_T \leq -\mathbf{E}_T\bar{\mathbf{g}}_T^{(2)}(\boldsymbol{\alpha}^0)\} \geq P\{\bar{\mathbf{G}}^{(2)}(\boldsymbol{\alpha}^0)\mathbf{U} \\
&\quad \leq -\mathbf{E}_T\bar{\mathbf{g}}_T^{(2)}(\boldsymbol{\alpha}^0), \Gamma_T \leq -\mathbf{E}_T\bar{\mathbf{g}}_T^{(2)}(\boldsymbol{\alpha}^0)\} \geq \sum_{i\in I_2^0} P\{(\bar{\mathbf{G}}^{(2)}(\boldsymbol{\alpha}^0)\mathbf{U})_i \\
&\quad \leq -E_{Ti}\bar{g}_{Ti}(\boldsymbol{\alpha}^0)\} - \sum_{i\in I_2^0} P\{|\Gamma_{Ti}| > -E_{Ti}\bar{g}_{Ti}(\boldsymbol{\alpha}^0)\} - m_2,
\end{aligned} \tag{5.93}$$

where $(\bar{\mathbf{G}}^{(2)}(\boldsymbol{\alpha}^0)\mathbf{U})_i$ and Γ_{Ti} are ith components of vectors $\bar{\mathbf{G}}^{(2)}(\boldsymbol{\alpha}^0)\mathbf{U}$ and Γ_T, and m_2 is the number of elements in the set I_2^0.

Because $\boldsymbol{\alpha}_{kT}$ and $\boldsymbol{\xi}_T$ are consistent, by Corollary 5.2 and (5.85), we get

$$p \lim_{T\to\infty} \Gamma_T = \mathbf{O}_{mn}. \tag{5.94}$$

From (5.34) we conclude, according to Condition 5.1.B.1, Assumption 5.2 and formula (2.14), that

$$-E_{Ti}\bar{g}_{Ti}(\boldsymbol{\alpha}^0) = -\frac{E_{Ti}\bar{E}_{Ti}}{\sqrt{T}} g_{Ti}(\boldsymbol{\alpha}^0) \to \infty, \quad T \to \infty, i \in I_2^0, \tag{5.95}$$

see (5.34). By (5.94) and (5.95), for any $\eta_{1i} > 0$ there exists $\theta_{1i} > 0$ such that

$$P\{|\Gamma_{Ti}| > -E_{Ti}\bar{g}_{Ti}(\boldsymbol{\alpha}^0)\} \le \eta_{1i}, \quad T > \theta_{1i}, i \in I_2^0. \tag{5.96}$$

Further, by (5.95), for any $\eta_{2i} > 0$ there exists $\theta_{2i} > 0$ for which

$$P\{(\bar{\mathbf{G}}^{(2)}(\boldsymbol{\alpha}^0)\mathbf{U})_i \le -E_{Ti}\bar{g}_{Ti}(\boldsymbol{\alpha}^0)\} \ge 1 - \eta_{2i}, \quad T > \theta_{2i} i \in I_2^0.$$

By (5.93) and (5.96) we obtain for $T > \theta = \max_{i\in I_2^0}(\theta_{1i}, \theta_{2i})$

$$P\{\bar{\mathbf{G}}^{(2)}(\boldsymbol{\alpha}^0)\mathbf{U} + \boldsymbol{\Gamma}_T \le -\mathbf{E}_T\bar{\mathbf{g}}_T^{(2)}(\boldsymbol{\alpha}^0)\} \ge 1 - \eta_1 - \eta_2, \tag{5.97}$$

where $\eta_1 = \sum_{i\in I_2^0} \eta_{1i}$, $\eta_2 = \sum_{i\in I_2^0} \eta_{2i}$. By (5.92) and (5.97), for any $\eta_1 > 0$, $\eta_2 > 0$

$$P\{||\tilde{\mathbf{U}}_{k+1,T} - \mathbf{U}^*_{k+1,T}|| = 0\} \ge 1 - \eta_1 - \eta_2, \quad T > \theta,$$

which together with (5.85) implies the statement of the theorem. □

Changing the algorithm similarly as in Corollary 5.1 (if there are no constraints) allows to derive the asymptotic distribution of $\mathbf{U}_{kT}$, which is the same as for $\tilde{\mathbf{U}}_{kT}$ for all k.

Theorem 5.5. *Assume that conditions of Theorem 5.4 hold true. Let k_0 be the iteration number, and $\gamma_{k_0T} = 1$, $\forall T$ $(k_0 \ge 0)$. Assume that starting from the point $\boldsymbol{\alpha}_{k_0T}$ Algorithms A5.1 and A5.2 converge. Then for $k > k_0$ $\mathbf{U}_{kT} = \mathbf{E}_T(\boldsymbol{\alpha}_{kT} - \boldsymbol{\alpha}^0)$ converges in distribution as $T \to \infty$ to a random variable $\mathbf{U}$ which is the solution to problem (5.78).*

Proof. Consider the problem

$$\frac{1}{2}\mathbf{x}'\bar{\mathbf{R}}(\boldsymbol{\alpha}^0)\mathbf{x} - (Q - \bar{\mathbf{R}}(\boldsymbol{\alpha}^0)\mathbf{U})'\mathbf{x} \to \min, \quad \bar{\mathbf{G}}^{(1)}(\boldsymbol{\alpha}^0)\mathbf{x} \le -\bar{\mathbf{G}}^{(1)}(\boldsymbol{\alpha}^0)\mathbf{U}, \tag{5.98}$$

where the matrix $\bar{\mathbf{G}}^{(1)}(\boldsymbol{\alpha}^0)$ consists of the lines of the matrix $\bar{\mathbf{G}}(\boldsymbol{\alpha}^0)$ with indexes $i \in I_1^0$ (the matrix $\bar{\mathbf{G}}(\boldsymbol{\alpha}^0)$ is defined in Assumption 5.2).

Let us find the solution $\mathbf{X}^0$ to problem (5.98). Write the dual problem:

$$\frac{1}{2}\mathbf{Z}'\bar{\mathbf{G}}^{(1)}(\boldsymbol{\alpha}^0)\bar{\mathbf{R}}^{-1}(\boldsymbol{\alpha}^0)(\bar{\mathbf{G}}^{(1)}(\boldsymbol{\alpha}^0))'\mathbf{Z}$$
$$-\mathbf{Q}'\bar{\mathbf{R}}^{-1}(\boldsymbol{\alpha}^0)\bar{\mathbf{G}}^{(1)}(\boldsymbol{\alpha}^0)\mathbf{Z} \to \min, \quad \mathbf{Z} \geq \mathbf{O}_{m_1}. \tag{5.99}$$

Denote by $\bar{\mathbf{V}}_1$ the solution to (5.99). It is easy to see that dual problem for (5.78) coincides with (5.99). Hence, $\bar{\mathbf{V}}_1 = \mathbf{V}_1$, where $\mathbf{V}_1 \in \Re^{m_1}$ is the dual variable for problem (5.78).

The necessary and sufficient conditions of the existence of a minimum in (5.98) are

$$\begin{cases} \bar{\mathbf{R}}(\boldsymbol{\alpha}^0)\mathbf{X}^0 + \bar{\mathbf{R}}(\boldsymbol{\alpha}^0)\mathbf{U} - \mathbf{Q} + (\bar{\mathbf{G}}^{(1)}(\boldsymbol{\alpha}^0))'\bar{\mathbf{V}}_1 = \mathbf{O}_n, \\ \bar{v}_i(\bar{\mathbf{G}}_i^{(1)}(\boldsymbol{\alpha}^0))'(\mathbf{X}^0 + \mathbf{U}) = 0, \quad i \in I_1^0, \end{cases} \tag{5.100}$$

where $\bar{v}_i$ is the Lagrange multiplier, $\bar{\mathbf{V}}$ is the vector with components $\bar{v}_i$, $i \in I_1^0$.

Further, the necessary and sufficient conditions for the existence of a minimum in (5.78) are

$$\begin{cases} \bar{\mathbf{R}}(\boldsymbol{\alpha}^0)\mathbf{U} - \mathbf{Q} + (\bar{\mathbf{G}}^{(1)}(\boldsymbol{\alpha}^0))'\mathbf{V}_1 = \mathbf{O}_n, \\ v_i(\bar{\mathbf{G}}_i^{(1)}(\boldsymbol{\alpha}^0))'\mathbf{U} = 0, \quad i \in I_1^0, \end{cases} \tag{5.101}$$

where v_i is the Lagrange multiplier, and $\bar{\mathbf{V}}$ is the vector with components $\bar{v}_i$, $i \in I_1^0$.

Since $\bar{\mathbf{V}}_1 = \mathbf{V}_1$, we see from the first equations in (5.100) and (5.101) that $\bar{\mathbf{R}}(\boldsymbol{\alpha}^0)\mathbf{X} = \mathbf{O}_n$. The matrix $\bar{\mathbf{R}}(\boldsymbol{\alpha}^0)$ is non-degenerate (see Assumption 5.1.B.3), which implies $\mathbf{X}^0 = \mathbf{O}_n$. One can show that $\mathbf{X}^0$ satisfies also the second condition in (5.100). Thus, the solution to (5.98) is

$$\mathbf{X}^0 = \mathbf{O}_n. \tag{5.102}$$

Taking into account (5.40) and (5.45), we can rewrite auxiliary problem (5.33) as:

$$\begin{cases} \frac{1}{2}\mathbf{x}'\bar{\bar{\mathbf{R}}}_T(\boldsymbol{\alpha}_{kT})\mathbf{x} - (\mathbf{Q}_T - \bar{\boldsymbol{\Phi}}_T\mathbf{U}_{kT})'\mathbf{x} \to \min, \\ \bar{\mathbf{G}}_T^{(1)}(\boldsymbol{\alpha}_{kT})\mathbf{x} \leq -\bar{\mathbf{G}}_T^{(1)}(\boldsymbol{\xi}_T)\mathbf{U}_{kT}, \\ \bar{\mathbf{G}}_T^{(2)}(\boldsymbol{\alpha}_{kT})\mathbf{x} \leq -\bar{\mathbf{G}}_T^{(2)}(\boldsymbol{\xi}_{kT})\mathbf{U}_{kT} - \bar{\mathbf{E}}_T\mathbf{g}^{(2)}(\boldsymbol{\alpha}^0), \end{cases} \tag{5.103}$$

where $\mathbf{x} = \sqrt{T}\mathbf{X}$, and $\mathbf{g}^{(2)}(\boldsymbol{\alpha}^0)$ is a vector with components $\mathbf{g}_i(\boldsymbol{\alpha}^0)$, $i \in I_2^0$.

Consider another problem:

$$\begin{cases} \frac{1}{2}\mathbf{x}'\bar{\bar{\mathbf{R}}}_T(\boldsymbol{\alpha}_{kT})\mathbf{x} - (\mathbf{Q}_T - \bar{\Phi}_T\mathbf{U}_{kT})'\mathbf{x} \to \min, \\ \bar{\mathbf{G}}_T^{(1)}(\boldsymbol{\alpha}_{kT})\mathbf{x} \leq -\bar{\mathbf{G}}_T^{(1)}(\boldsymbol{\xi}_{kT})\mathbf{U}_{kT}, \end{cases} \tag{5.104}$$

which differs from (5.103) in the absence of the second group of constraints.

Denote the solutions to (5.103) and (5.104), respectively, by $\bar{\bar{\mathbf{X}}}_{kT} = \sqrt{T}\bar{\mathbf{X}}_{kT}$ and $\mathbf{X}^*_{kT}$. Put $\mathbf{y} = \tilde{\mathbf{M}}_T(\boldsymbol{\alpha}_{kT})\mathbf{x}$, where the matrix $\tilde{\mathbf{M}}_T(\boldsymbol{\alpha}_{kT})$ is defined in (5.35). From (5.104) we get

$$\begin{cases} \frac{1}{2}\mathbf{y}'\mathbf{y} - \mathbf{y}'[(\tilde{\mathbf{M}}_T^{-1}(\boldsymbol{\alpha}_{kT}))'(\mathbf{Q}_T - \bar{\boldsymbol{\Phi}}_T\mathbf{U}_{kT})] \to \min, \\ \bar{\mathbf{G}}_T^{(1)}(\boldsymbol{\alpha}_{kT})\mathbf{y} \leq -\bar{\mathbf{G}}_T^{(1)}(\boldsymbol{\xi}_{kT})\mathbf{U}_{kT}, \end{cases} \tag{5.105}$$

where the matrix $\bar{\bar{\mathbf{G}}}_T^{(1)}(\boldsymbol{\alpha}_{kT})$ is defined in (5.87).

Now put $\mathbf{y} = \mathbf{M}(\boldsymbol{\alpha}^0)\mathbf{x}$, where $\mathbf{M}(\boldsymbol{\alpha}^0)$ is defined in (5.43). By (5.98) we arrive at

$$\begin{cases} \frac{1}{2}\mathbf{y}'\mathbf{y} - \mathbf{y}'[(\mathbf{M}^{-1}(\boldsymbol{\alpha}^0))'(\mathbf{Q} - \bar{\mathbf{R}}(\boldsymbol{\alpha}^0)\mathbf{U})] \to \min, \\ \bar{\bar{\mathbf{G}}}^{(1)}(\boldsymbol{\alpha}^0)y \leq -\bar{\mathbf{G}}^{(1)}(\boldsymbol{\alpha}^0)U. \end{cases} \tag{5.106}$$

Denote the solutions to (5.105) and (5.106), respectively, by $\mathbf{x}^*_{kT}$ and $\mathbf{x}^0$. These solutions are related to the solutions $\mathbf{X}^*_{kT}$ (i.e. the solution to (5.104)) and $\mathbf{X}^0$ (i.e. the solution to (5.98)) as follows:

$$\mathbf{X}^*_{kT} = \tilde{\mathbf{M}}_T^{-1}(\boldsymbol{\alpha}_{kT})\mathbf{x}^*_{kT}, \quad \mathbf{X}^0 = \mathbf{M}^{-1}(\boldsymbol{\alpha}^0)\mathbf{x}^0. \tag{5.107}$$

Assume that for any $k > k_0$

$$\mathbf{U}_{kT} \overset{p}{\Rightarrow} \mathbf{U}, \quad T \to \infty. \tag{5.108}$$

Taking into account (5.15), (5.17) and the limit (5.91), we get according to Lemmas 2.6 and 4.8, that if (5.108) holds true, then $\mathbf{x}^*_{kT} \overset{p}{\Rightarrow} \mathbf{x}^0 = \mathbf{M}(\boldsymbol{\alpha}^0)\mathbf{X}^0 = \mathbf{O}_n$, $T \to \infty$ (cf. (5.102)). Then, taking into account (5.43), we obtain from (5.107)

$$p \lim_{T\to\infty} \mathbf{X}^*_{kT} = \mathbf{O}_n. \tag{5.109}$$

Let us find the limit distribution of the solution $\bar{\mathbf{X}}_{kT}$ to (5.103) as $T \to \infty$. Since $\bar{\bar{\mathbf{R}}}_T(\boldsymbol{\alpha}_{kT})$ is positive definite we have $\bar{\mathbf{X}}_{kT} = \mathbf{X}^*_{kT}$, provided that $\mathbf{X}^*_{kT}$ satisfies the second system of restrictions in (5.103). Thus, we get the expression similar to (5.92):

$$P\left\{||\bar{\bar{\mathbf{X}}}_{kT} - \mathbf{X}^*_{kT}|| = 0\right\} = P\left\{\bar{\mathbf{G}}_T^{(2)}(\boldsymbol{\alpha}_{kT})\mathbf{X}^*_{kT} + \bar{\mathbf{G}}_T^{(2)}(\boldsymbol{\xi}_{kT})\mathbf{U}_{kT} \leq -\mathbf{E}_T\mathbf{g}^{(2)}(\boldsymbol{\alpha}^0)\right\}. \tag{5.110}$$

Let

$$\Gamma_T = -\left[\bar{\mathbf{G}}^{(2)}(\boldsymbol{\alpha}^0)\mathbf{U} - \bar{\mathbf{G}}_T^{(2)}(\boldsymbol{\alpha}_{kT})\mathbf{X}^*_{kT} - \bar{\mathbf{G}}_T^{(2)}(\boldsymbol{\xi}_{kT})\mathbf{U}_{kT}\right].$$

From (5.110) we have

$$P\left\{||\bar{\bar{\mathbf{X}}}_{kT} - \mathbf{X}^*_{kT}|| = 0\right\} = P\left\{\bar{\mathbf{G}}^{(2)}(\boldsymbol{\alpha}^0)\mathbf{U} + \Gamma_T \leq -\bar{\mathbf{E}}_T\bar{\mathbf{g}}^{(2)}(\boldsymbol{\alpha}^0)\right\}. \tag{5.111}$$

Suppose that (5.108) holds true; then it implies (5.109). From the existence of these limits, consistency of $\boldsymbol{\alpha}_{kT}$ and $\boldsymbol{\xi}_{kT}$, and the existence of limits (5.70) and (5.61), we have $p \lim_{T\to\infty} \Gamma_T = \mathbf{O}_n$. Then (5.93) implies inequality (5.97). From (5.111), taking into account (5.97), we obtain

$$P\left\{||\bar{\bar{\mathbf{X}}}_{kT} - \mathbf{X}^*_{kT}|| = 0\right\} \geq 1 - \eta_1 - \eta_2, \quad T > \theta,$$

where η_1, η_2 are some positive numbers, such that $\eta_1 + \eta_2 < 1$. From here and (5.109), we derive the existence of the limit for the solution to (5.103), provided that (5.108) holds true:

$$p \lim_{T\to\infty} \bar{\bar{\mathbf{X}}}_{kT} = \mathbf{O}_n. \tag{5.112}$$

We have $\bar{\mathbf{X}}_{kT} = \left(\sqrt{T}\right)^{-1} \bar{\bar{\mathbf{X}}}_{kT}$. Then by (5.32) we see that the solution to the auxiliary problem satisfies

$$\bar{\mathbf{X}}_{kT} = \mathbf{E}_T^{-1} \bar{\bar{\mathbf{X}}}_{kT}.$$

From here and (5.1) we derive

$$\mathbf{U}_{k+1,T} = \mathbf{U}_{kT} + \gamma_{kT} \bar{\bar{\mathbf{X}}}_{kT}.$$

From the above equality and (5.112) we obtain (note that γ_{kT} is bounded)

$$\mathbf{U}_{k+1,T} \overset{p}{\Rightarrow} \mathbf{U} \quad \text{if } \mathbf{U}_{kT} \overset{p}{\Rightarrow} \mathbf{U}, k > k_0. \tag{5.113}$$

According to Theorem 5.4, the asymptotic distribution of $\tilde{\mathbf{U}}_{k_0+1,T}$ coincides with the distribution of $\mathbf{U}$. But $\mathbf{U}_{k_0+1,T} = \tilde{\mathbf{U}}_{k_0+1,T}$. Thus using (5.113) we obtain the statement of the theorem. □

Consider now the matrix of m.s.e. of parameter estimates and their biases at each iteration. We generalize the concept of active and passive constraints for the estimate $\boldsymbol{\alpha}_T$, introduced in Sect. 4.2, to its kth iteration $\boldsymbol{\alpha}_{kT}$. Then according to (4.8) and (4.9), the ith constraint on kth iteration is active with accuracy up to $\xi > 0$, if $-\xi \leq g_i(\boldsymbol{\alpha}_{kT}) \leq 0$, and is inactive, if $g_i(\boldsymbol{\alpha}_{kT}) < -\xi$. The value ξ is defined as in Sect. 4.2, i.e. it satisfies the condition (4.10).

Denote by $p_{lT}^{(k)}$, $l = \overline{1, L}$ (where L is the number of possible combinations of active restrictions) the probability that on kth iteration $\boldsymbol{\alpha}_{kT}$ corresponds to lth combination of active restrictions. By Theorem 5.3 on the consistency of $\boldsymbol{\alpha}_{kT}$, Theorem 4.1 (Sect. 4.2) holds true for $p_{lT}^{(k)}$, $l = \overline{1, L}$, $k = 1, 2, \ldots$.

As it was done in Sect. 4.2, put $\gamma_{lT}^{(k)} = 1$ if on kth iteration we obtain lth combination of active and inactive restrictions, otherwise put $\gamma_{lT}^{(k)} = 0$.

Let

$$\eta_{iT}^{(k)} = \begin{cases} 1 & \text{if } -\xi \leq g_i(\boldsymbol{\alpha}_{kT}) \leq 0, \\ 0 & \text{if } \quad g_i(\boldsymbol{\alpha}_{kT}) < -\xi, \end{cases} \qquad k = 1, 2, \ldots.$$

Since for $p_{lT}^{(k)}$, $l = \overline{1, L}$, $k = 1, 2, \dots$ Theorem 4.1 holds true, one can apply Lemma 4.1 for $\gamma_{lT}^{(k)}$, $l = \overline{1, L}$, and $\eta_{iT}^{(k)}$, $i \in I$, $k = 1, 2, \dots$. This implies, in particular, that

$$p \lim_{T\to\infty} \eta_{iT}^{(k)} = 1, \quad i \in I_1^0, \qquad p \lim_{T\to\infty} \eta_{iT}^{(k)} = 0, \quad i \in I_2^0. \tag{5.114}$$

Now we determine the estimate σ_{kT}^2 of σ^2 obtained on kth iteration. Similarly to (4.18), we have

$$\sigma_{kT}^2 = \left(T - n + \sum_{i\in I} \eta_{iT}^{(k)}\right)^{-1} \sum_{t=1}^{T} (y_t - f_t(\boldsymbol{\alpha}_{kT}))^2. \tag{5.115}$$

Theorem 5.6. *Assume that Assumptions 2.8, 5.1 and conditions of Theorem 5.5 are satisfied,* $\boldsymbol{\alpha}_{kT}$ *is the consistent estimate of* $\boldsymbol{\alpha}^0$ *, and that* $\mathbf{U}_{kT} = \mathbf{E}_T(\boldsymbol{\alpha}_{kT} - \boldsymbol{\alpha}^0)$ *has a limit distribution. Then* $p \lim_{T\to\infty} \sigma_{kT}^2 = \sigma^2$.

Proof. We have

$$\sigma_{kT}^2 = \left(\frac{T}{T - n + \sum_{i\in I} \eta_{iT}^{(k)}}\right) S_T(\boldsymbol{\alpha}_{kT}), \quad S_T(\boldsymbol{\alpha}) = T^{-1} \sum_{t=1}^{T} (y_t - f_t(\boldsymbol{\alpha}))^2. \tag{5.116}$$

Denote the first multiplier in (5.116) by $\sigma_{kT}^2(1)$. From (5.114),

$$p \lim_{T\to\infty} \sigma_{kT}^2(1) = 1. \tag{5.117}$$

By Taylor expansion,

$$\begin{aligned} S_T(\boldsymbol{\alpha}_{kT}) &= S_T(\boldsymbol{\alpha}^0) + \nabla S_T'(\boldsymbol{\alpha}^0)(\boldsymbol{\alpha}_{kT} - \boldsymbol{\alpha}^0) \\ &\quad + \frac{1}{2}(\boldsymbol{\alpha}_{kT} - \boldsymbol{\alpha}^0)' \boldsymbol{\Phi}_T (\boldsymbol{\alpha}_{kT} - \boldsymbol{\alpha}^0), \end{aligned} \tag{5.118}$$

where $\boldsymbol{\Phi}_T$ is the matrix of second derivatives of the function $S_T(\boldsymbol{\alpha})$, calculated in $\boldsymbol{\alpha} = \boldsymbol{\xi}_{1T}$. The (k, l)-elements of the matrix $\boldsymbol{\Phi}_T$ are given in (2.27). Moreover, $||\boldsymbol{\xi}_{1T} - \boldsymbol{\alpha}^0|| \leq ||\boldsymbol{\alpha}_{kT} - \boldsymbol{\alpha}^0||$, $\boldsymbol{\xi}_{1T} \in \Re^n$. Then, since the estimate $\boldsymbol{\alpha}^0$ is consistent, we get

$$p \lim_{T\to\infty} \boldsymbol{\xi}_{1T} = \boldsymbol{\alpha}^0. \tag{5.119}$$

Since $S_T(\boldsymbol{\alpha}^0) = T^{-1} \sum_{t=1}^{T} \varepsilon_t^2$, we have by Assumption 2.8

$$p \lim_{T\to\infty} S_T(\boldsymbol{\alpha}^0) = \sigma^2. \tag{5.120}$$

According to Sect. 2.2, see (2.31), we have $p\lim_{T\to\infty} \nabla S_T(\boldsymbol{\alpha}^0) = \mathbf{O}_n$. Taking into account the consistency of $\boldsymbol{\alpha}_{kT}$, we obtain that the second summand in (5.118) converges in probability to zero as $T \to \infty$.

Consider the third term in (5.118). Denote by $\bar{\boldsymbol{\Phi}}_T = T\mathbf{E}_T^{-1}\boldsymbol{\Phi}_T\mathbf{E}_T^{-1}$ the matrix with (k,l)th elements given in (5.16). In this notation, the third term in (5.118) is $(1/2T)\mathbf{U}'_{kT}\bar{\boldsymbol{\Phi}}_T\mathbf{U}_{kT}$. By (5.119), there exists the limit (5.17), and, moreover, $\mathbf{U}_{kT}$ has a limit distribution (by assumptions of the theorem). Therefore, the third term in (5.118) converges in probability to zero, as well as the second term. To complete the proof, we use (5.120) and (5.117). □

Following the proofs from Sect. 4.2 and taking into account Theorem 5.6, one can conclude that all results of Sects. 4.2 and 4.3 remain true for consistent truncated estimates of the matrix of m.s.e. estimates of regression parameters $\mathbf{K}_T^{(k)} = \sum_{l=1}^{L} \mathbf{K}_{lT}^{(k)}\gamma_{lT}^{(k)}$ and their bias $\boldsymbol{\Psi}_T^{(k)} = \sum_{l=1}^{L} \boldsymbol{\Psi}_{lT}^{(k)}\gamma_{lT}^{(k)}$ on kth iteration (for which conditions of Theorem 5.5 also hold true). Here $\mathbf{K}_{lT}^{(k)}$ is the estimate of $M\{\mathbf{U}_l\mathbf{U}'_l\}$, and $\boldsymbol{\Psi}_{lT}^{(k)}$ is the estimate of $M\{\mathbf{U}_l\}$. In our notation $\mathbf{U}_l$ is the solution to the quadratic programming problem (4.15).

Consider in detail the truncated sample estimate $\hat{\mathbf{K}}_{kT}$ of the matrix of m.s.e. of the regression parameter $\boldsymbol{\alpha}_{kT}$, performed on kth iteration. Similarly to (4.53) we have

$$\hat{\mathbf{K}}_{kT} = \mathbf{K}_{lT}^{(k)}(\mathbf{R}_T(\hat{\boldsymbol{\alpha}}_{kT}), \mathbf{G}_l(\hat{\boldsymbol{\alpha}}_{kT}), \hat{\sigma}_{kT}),$$

where $l = l(\hat{\boldsymbol{\alpha}}_{kT})$ is the number of combinations of active restrictions with accuracy ξ, which correspond to the estimate $\hat{\boldsymbol{\alpha}}_{kT}$ for $\boldsymbol{\alpha}^0$, calculated on the kth iteration.

According to the expression (5.115) of the sample estimate for σ^2 we obtain

$$\hat{\sigma}_{kT}^2 = \left(T - n + \sum_{i\in I}\hat{\eta}_{iT}^{(k)}\right)^{-1} \sum_{t=1}^{T}(y_t - f_t(\hat{\boldsymbol{\alpha}}_{kT}))^2,$$

where

$$\hat{\eta}_{iT} = \begin{cases} 1 & \text{if} - \xi \le \mathbf{g}_i(\hat{\boldsymbol{\alpha}}_{kT}) \le 0, \\ 0 & \text{if}\ \mathbf{g}_i(\hat{\boldsymbol{\alpha}}_{kT}) < -\xi. \end{cases}$$

By definition of $\mathbf{K}_l^{(k)}$,

$$\hat{\mathbf{K}}_{kT} = M\{\hat{\mathbf{U}}_{kT}\hat{\mathbf{U}}'_{kT}\}, \tag{5.121}$$

where $\hat{\mathbf{U}}_{kT}$ is the solution to the problem

$$\frac{1}{2}\mathbf{Z}'\bar{\mathbf{R}}_T(\hat{\boldsymbol{\alpha}}_{kT})\mathbf{Z} - \hat{\mathbf{Q}}'_T\mathbf{Z} \to \min, \quad \bar{\mathbf{G}}'_i(\hat{\boldsymbol{\alpha}}_{kT})\mathbf{Z} \le \mathbf{O}_{m_T}, \quad i \in I(\hat{\boldsymbol{\alpha}}_{kT}).$$

Here $I(\hat{\boldsymbol{\alpha}}_{kT}) = \{i : -\xi \le g_i(\hat{\boldsymbol{\alpha}}_{kT}) \le 0, i \in I\}$, $m_T = |I(\hat{\boldsymbol{\alpha}}_{kT})|$ is the number of elements in $I(\hat{\boldsymbol{\alpha}}_{kT})$, and $\hat{\mathbf{Q}}_T = \mathbf{E}_T^{-1}\mathbf{D}'_T(\boldsymbol{\alpha}_{kT})\boldsymbol{\varepsilon}_T$. For large T according to Assumption 2.8 (Sect. 2.5.1) on the noise in the regression $\boldsymbol{\varepsilon}_T$, we may assume that $\hat{\mathbf{Q}} \sim N(\mathbf{O}_T, \hat{\sigma}_{kT}^2\bar{\mathbf{R}}_T(\hat{\boldsymbol{\alpha}}_{kT}))$.

This fact also takes place if T is small, and $\boldsymbol{\varepsilon}_T \sim N(\mathbf{O}_T, \sigma^2 J_T)$, i.e. components of $\boldsymbol{\varepsilon}_T$ are independent and normally distributed.

For calculation of $\hat{\mathbf{K}}_{kT}$ by (5.121) one can use the same methods as for the matrix $\hat{\mathbf{K}}_T$, defined in (4.56).

In case of no constraints we have according to Theorems 5.2 and 5.6 $\mathbf{K}_{kT} = \sigma_{kT}^2 \bar{\mathbf{R}}_T^{-1}(\boldsymbol{\alpha}_{kT})$ and $\boldsymbol{\Psi}_{kT} = \mathbf{O}_n$. The sample estimate $\hat{\mathbf{K}}_{kT}$ is obtained from the expression for $\mathbf{K}_{kT}$ when $\boldsymbol{\alpha}_{kT} = \hat{\boldsymbol{\alpha}}_{kT}$.

It is natural to take as the initial approximation of $\boldsymbol{\alpha}_{0T}$ the estimate of the regression parameters under no constraints. Obtained results can also be generalized to the case of inequality constraints, see the problems (1.74) and (1.75).

Chapter 6
Prediction of Linear Regression Evaluated Subject to Inequality Constraints on Parameters

In this chapter we investigate statistical properties of the prediction in a regression with constraints. This problem is very complicated, since even in the case of linear regression and linear constraints the estimation of parameters is a nonlinear problem. Especially, the problem of interval prediction, i.e., of finding the confidence interval for the estimated value, is highly non-trivial. In Sect. 6.1 the interval prediction is constructed based on the distribution function of the prediction error, whose parameters are the true regression parameter $\boldsymbol{\alpha}^0$ and the variance σ^2 of the noise. Section 6.2 is devoted to the interval prediction based on the conditional distribution function of the prediction error.

6.1 Dispersion of the Regression Prediction with Inequality Constraints: Interval Prediction Under Known Distribution Function of Errors

In this section we investigate statistical properties of the prediction of dependent variable in the linear regression $y_t = \mathbf{x}'_t\boldsymbol{\alpha} + \varepsilon_t$ with constraints $\mathbf{g}'_i\boldsymbol{\alpha} - b_i \leq 0$, $i = \overline{1,m}$. We construct the prediction for the moment $t = \theta > T$, where T is the number of observations, i.e., $\hat{y}_\theta = \mathbf{x}'_\theta\boldsymbol{\alpha}_T$. The true value of the prediction is $y_\theta = \mathbf{x}'_\theta\boldsymbol{\alpha}^0 + \varepsilon_\theta$, and the error is of the form

$$\xi_\theta = \hat{y}_\theta - y_\theta = \mathbf{x}'_\theta(\boldsymbol{\alpha}_T - \boldsymbol{\alpha}^0) - \varepsilon_\theta. \tag{6.1}$$

Here $\boldsymbol{\alpha}_T$ is the solution to the problem (4.58), see Sect. 4.4.

Considering that the regressor has a trend and satisfies Assumptions 2.9 and 2.11, and that constraints satisfy Assumptions 2.10 and 2.12, we can transform this problem. Using the notation introduced in Assumption 2.12, we can write the problem of finding $\boldsymbol{\alpha}_T$ as follows:

$$\frac{1}{2}\mathbf{Y}'\mathbf{R}_T\mathbf{Y} - \mathbf{Y}'\mathbf{Q}_T \to \min, \quad \tilde{\mathbf{G}}_T\mathbf{Y} \leq \overline{\mathbf{E}}_T(\mathbf{b} - \mathbf{G}\boldsymbol{\alpha}^0), \tag{6.2}$$

P.S. Knopov and A.S. Korkhin, *Regression Analysis Under A Priori Parameter Restrictions,* Springer Optimization and Its Applications 54,
DOI 10.1007/978-1-4614-0574-0_6,

where

$$\mathbf{Y} = \mathbf{E}_T(\boldsymbol{\alpha} - \boldsymbol{\alpha}^0), \quad \mathbf{Q}_T = \mathbf{E}_T^{-1}\mathbf{X}'_T\boldsymbol{\varepsilon}_T, \tag{6.3}$$

and $\mathbf{X}_T = [x_1 \quad x_2 \quad \ldots \quad x_T]'$, $\boldsymbol{\varepsilon}_T = [\varepsilon_1 \quad \varepsilon_2 \quad \ldots \quad \varepsilon_T]'$.

Put $\mathbf{Y} = \mathbf{L}_T\mathbf{Z}$, where the transformation matrix is

$$\mathbf{L}_T = \mathbf{C}_T\mathbf{M}_T^{-1/2}\mathbf{H}'_T\boldsymbol{\Pi}_T. \tag{6.4}$$

Then we obtain from (6.2) the minimization problem

$$\frac{1}{2}\mathbf{Z}'\mathbf{Z} - \mathbf{Z}'\mathbf{P}^* \to \min, \quad \mathbf{N}_T\mathbf{Z} \leq \overline{\mathbf{E}}_T(\mathbf{b} - \mathbf{G}\boldsymbol{\alpha}^0), \quad \mathbf{P}^* \sim N(\mathbf{O}_n, \sigma^2\mathbf{J}_n), \tag{6.5}$$

where $\mathbf{N}_T = [\mathbf{B}_T \vdots \mathbf{O}_{m,n-m}]$, and the triangular matrix $\mathbf{B}_T$ was determined in Sect. 4.6.1 (note that in Sect. 4.6.1 index T in $\mathbf{B}_T$ was omitted). Taking into account (6.3), we have

$$\mathbf{P}^* = \mathbf{L}'_T\mathbf{Q}_T = \mathbf{L}'_T\mathbf{E}_T^{-1}\mathbf{X}'_T\boldsymbol{\varepsilon}_T. \tag{6.6}$$

Matrices $\mathbf{C}_T, \mathbf{M}_T, \mathbf{H}_T$ used in (6.4) were introduced in Sect. 4.6.1 for transforming the problem (4.149), and $\boldsymbol{\Pi}_T = \begin{bmatrix} \tilde{\Pi}_T & \mathbf{O}_{m,n-m} \\ \mathbf{O}_{n-m,m} & \mathbf{J}_{n-m} \end{bmatrix}$, see Sect. 4.6.1 for the definition of the diagonal matrix $\tilde{\Pi}_T$.

By Assumption 2.8 we derive that the prediction dispersion is given by

$$\delta_\theta^2 = E\{(\tilde{y}_\theta - y_\theta)^2\} = \mathbf{x}'_\theta(E\{(\boldsymbol{\alpha}_T - \boldsymbol{\alpha}^0)(\boldsymbol{\alpha}_T - \boldsymbol{\alpha}^0)'\})\mathbf{x}_\theta + \sigma^2, \tag{6.7}$$

where $E\{(\boldsymbol{\alpha}_T - \boldsymbol{\alpha}^0)(\boldsymbol{\alpha}_T - \boldsymbol{\alpha}^0)'\} = \mathbf{E}_T^{-1}\mathbf{K}_T^0\mathbf{E}_T^{-1}$. Here $\mathbf{K}_T^0 = E\{\mathbf{U}_T\mathbf{U}'_T\}$ and $\mathbf{U}_T = \mathbf{E}_T(\boldsymbol{\alpha}_T - \boldsymbol{\alpha}^0)$ are the normalized matrix of m.s.e. of the regression parameter estimate, introduced in (4.111); the matrix $\mathbf{E}_T$ is defined in Assumption 2.8.

According to Sect. 4.1 and Theorem 4.3, the expectation of the prediction is $E\{\hat{y}_\theta\} = \mathbf{x}'_\theta E\{\boldsymbol{\alpha}_T\} \neq \mathbf{x}'_\theta\boldsymbol{\alpha}^0 = y_\theta$, as $E\{\boldsymbol{\alpha}_T\} \neq E\{\boldsymbol{\alpha}^0\}$. Thus, in contrast to the prediction in the regression without constraints, in general prediction is biased.

Define the confidential interval for the true value of y_θ. For this we find the distribution of the prediction error ξ_θ under the assumption that the regression errors are normally distributed, i.e., Assumption 4.1 holds true. From (6.1) we have

$$\xi_\theta = \sigma\mathbf{w}'_\theta\mathbf{V}_T - \varepsilon_\theta, \tag{6.8}$$

where $\mathbf{w}_\theta = \mathbf{L}'_T\mathbf{E}_T^{-1}\mathbf{x}_\theta$, the matrix $\mathbf{L}_T$ is defined in (6.4); $\mathbf{V}_T = \mathbf{L}_T^{-1}\mathbf{E}_T(\boldsymbol{\alpha}_T - \boldsymbol{\alpha}^0)/\sigma$ is the solution to the problem

$$\frac{1}{2}\mathbf{Z}'\mathbf{Z} - \mathbf{U}'\mathbf{Z} \to \min, \quad \mathbf{N}_T\mathbf{Z} \leq \sigma^{-1}\overline{\mathbf{E}}_T(\mathbf{b} - \mathbf{G}\boldsymbol{\alpha}^0), \quad \mathbf{U} \sim N(\mathbf{O}_n, \mathbf{J}_n), \tag{6.9}$$

which follows from (6.5) by division of the cost function by σ^2 and by division of both parts of the constraints by σ. In (6.9) we have $\mathbf{V}_T = \mathbf{L}_T^{-1}\mathbf{E}_T(\boldsymbol{\alpha} - \boldsymbol{\alpha}^0)/\sigma$.

According to Sect. 4.6.1, we have in (6.9) $\mathbf{U} = \begin{bmatrix} \mathbf{u} \\ \overline{\mathbf{U}} \end{bmatrix}$, where $\mathbf{u} \in \Re^m$, $\overline{\mathbf{U}} \in \Re^{n-m}$. Let $\mathbf{w}_\theta = \begin{bmatrix} \mathbf{w}_{1\theta} \\ \mathbf{w}_{2\theta} \end{bmatrix}$, $\mathbf{V}_T = \begin{bmatrix} \mathbf{V}_{1T} \\ \mathbf{V}_{2T} \end{bmatrix}$, where $\mathbf{w}_{1\theta} \in \Re^m$, $\mathbf{w}_{2\theta} \in \Re^{n-m}$; $\mathbf{V}_{1T} \in \Re^m$, $\mathbf{V}_{2T} \in \Re^{n-m}$, and $\mathbf{V}_{2T} = \overline{\mathbf{U}}$. $\mathbf{V}_{1T}$ is the solution to the problem

$$\frac{1}{2}\mathbf{z}'\mathbf{z} - \mathbf{u}'\mathbf{z} \to \min, \quad \mathbf{u} \sim N(\mathbf{O}_m, \mathbf{J}_m), \quad \mathbf{B}_T\mathbf{z} \le \sigma^{-1}\overline{\mathbf{E}}_T(\mathbf{b} - \mathbf{G}\boldsymbol{\alpha}^0). \quad (6.10)$$

Using this notation we obtain from (6.8)

$$\xi_\theta = \sigma \mathbf{w}'_\theta \mathbf{V}_T - \varepsilon_\theta = \xi_{1\theta} + \xi_{2\theta}, \quad \xi_{1\theta} = \sigma \mathbf{w}'_{1\theta}\mathbf{V}_{1T}, \quad \xi_{2\theta} = \sigma \mathbf{w}'_{2\theta}\mathbf{V}_{2T} - \varepsilon_\theta. \quad (6.11)$$

Independence of $\mathbf{V}_{1T}$ and $\mathbf{V}_{2T} = \overline{\mathbf{U}}$ follows from the independence of $\mathbf{u}$ and $\overline{\mathbf{U}}$, since $\mathbf{V}_{1T}$ is the function of $\mathbf{u}$. This property of $\mathbf{V}_{1T}$ and $\mathbf{V}_{2T}$ implies independence of $\xi_{1\theta}$ and $\xi_{2\theta}$ in (6.8). Therefore the distribution function of ξ_θ is the convolution of distribution functions of $\xi_{1\theta}$ and $\xi_{2\theta}$.

Denote the distribution functions of ξ_θ, $\xi_{1\theta}$, and $\xi_{2\theta}$, respectively, by $F_\theta(x) = P\{\xi_\theta \le x\}$, $F_{1\theta}(x) = P\{\xi_{1\theta} \le x\}$, $F_{2\theta}(x) = P\{\xi_{2\theta} \le x\}$.

Theorem 6.1. *Under Assumptions 2.9, 4.1, and 4.2, the distribution function of the prediction error $\xi_\theta = \tilde{y}_\theta - y_\theta$ is a convolution of functions $F_{1\theta}(x)$ and $F_{2\theta}(x)$*

$$F_\theta(t) = \int_{-\infty}^{\infty} F_{1\theta}(t - x)\mathrm{d}F_{2\theta}(x). \quad (6.12)$$

Let $\nu = P\{l_1 \le \xi_\theta \le l_2\}$, where l_1 and l_2 are determined as the solutions to the minimization problem

$$l_2 - l_1 \to \min, \quad F_\theta(l_2) - F_\theta(l_1) = \nu, \quad l_2 \ge l_1,$$

where $F_\theta(x)$ is given by (6.12).

Then the confidential interval for y_θ can be found from the inequality $l_1 \le \xi_\theta \le l_2$, where $\xi_\theta = \tilde{y}_\theta - y_\theta : \tilde{y}_\theta + l_2 \le y_\theta \le \tilde{y}_\theta + l_1$.

Let us compare the accuracy of prediction by regression with (respectively, without) the constraints on parameters. When the regression parameter is estimated by the least squares estimates method (LS) without constraints, the dispersion of the estimate is given by

$$\tilde{\delta}_\theta^2 = \mathbf{x}'_\theta(E\{(\boldsymbol{\alpha}_T^* - \boldsymbol{\alpha}^0)(\boldsymbol{\alpha}_T^* - \boldsymbol{\alpha}^0)'\})\mathbf{x}_\theta + \sigma^2, \quad (6.13)$$

where $E\{(\boldsymbol{\alpha}_T^* - \boldsymbol{\alpha}^0)(\boldsymbol{\alpha}_T^* - \boldsymbol{\alpha}^0)'\} = \mathbf{E}_T^{-1}\tilde{\mathbf{K}}_T^0\mathbf{E}_T^{-1}$, $\boldsymbol{\alpha}_T^*$ is the LS estimate of the regression parameter. Here $\tilde{\mathbf{K}}_T^0 = \sigma^2\mathbf{R}_T$ is the normalized matrix of m.s.e of the estimate $\boldsymbol{\alpha}_T^*$.

By (6.7) and (6.13) we have $\tilde{\delta}_\theta^2 - \delta_\theta^2 = \mathbf{x}'_\theta\mathbf{E}_T^{-1}(\tilde{\mathbf{K}}_T^0 - \mathbf{K}_T^0)\mathbf{E}_T^{-1}\mathbf{x}_\theta$. From this expression we see that the difference of the dispersions of prediction is completely determined by the difference of matrixes of m.s.e. of the estimate of regression parameters $\tilde{\mathbf{K}}_T^0 - \mathbf{K}_T^0$. In case of one constraint this difference $\tilde{\mathbf{K}}_T - \mathbf{K}_T$ was investigated in Theorem 4.7, under the assumption that in $\tilde{\mathbf{K}}_T$ and $\mathbf{K}_T$ the same estimate for σ^2 is used. According to Sect. 4.7, Theorem 4.7 will remain in true if we replace $\tilde{\mathbf{K}}_T$ and $\mathbf{K}_T$, respectively, with $\mathbf{K}_T^0$ and $\tilde{\mathbf{K}}_T^0$. Thus we obtain the theorem formulated below.

Theorem 6.2. *If* $m = 1$ *and Assumptions 2.9, 4.1, and 4.2 hold true, then*

(a) $\tilde{\delta}_\theta^2 - \delta_\theta^2 \geq 0$ for $n > 1$
(b) $\tilde{\delta}_\theta^2 - \delta_\theta^2 > 0$ for $n = 1$
(c) $\tilde{\delta}_\theta^2 - \delta_\theta^2 > 0$ for $n > 1$, provided that the first component of the n-dimensional vector $\mathbf{\Psi}_\theta = \mathbf{\Pi}_T \mathbf{H}_T \mathbf{M}_T^{-1/2} \mathbf{C}'_T \mathbf{E}_T^{-1} \mathbf{x}_\theta$ is not equal to zero

Denote estimates of prediction dispersions given in (6.7) and (6.13), respectively, by d_θ^2 and $\tilde{d}_\theta^2$. Comparing these estimates we assume, as in Sects. 4.6 and 4.7, that they are calculated with the same estimate of the dispersion of the noise σ_T^2. Then we obtain $d_\theta^2 = \mathbf{x}'_\theta \mathbf{E}_T^{-1} \mathbf{K}_T \mathbf{E}_T^{-1} \mathbf{x}_\theta + \sigma_T^2$, $\tilde{d}_\theta^2 = \mathbf{x}'_\theta \mathbf{E}_T^{-1} \tilde{\mathbf{K}}_T \mathbf{E}_T^{-1} \mathbf{x}_\theta + \sigma_T^2$, implying that $\tilde{d}_\theta^2 - d_\theta^2 = \mathbf{x}'_\theta \mathbf{E}_T^{-1} (\tilde{\mathbf{K}}_T - \mathbf{K}_T) \mathbf{E}_T^{-1} \mathbf{x}_\theta$. The result below follows from the above expression and Theorem 4.7.

Theorem 6.3. *Let* $m = 1$, *and assume that Assumptions 2.9, 4.1, and 4.2 hold true. Then*

(a) $\tilde{a}_\theta^2 - d_\theta^2 \geq 0$ for $n > 1$
(b) $\tilde{b}_\theta^2 - d_\theta^2 > 0$ for $n = 1$
(c) $\tilde{c}_\theta^2 - d_\theta^2 > 0$ for $n > 1$, if the first component of the n-dimensional vector $\mathbf{\Psi}_\theta = \mathbf{\Pi}_T \mathbf{H}_T \mathbf{M}_T^{-1/2} \mathbf{C}'_T \mathbf{E}_T^{-1} \mathbf{x}_\theta$ is not equal to zero

If there are two constraints, then one can conclude about the accuracy of the prediction from the theorem below.

Theorem 6.4. *If* $m = 2$, *and Assumptions 2.9, 4.1, and 4.2 hold true, then* $\tilde{d}_\theta^2 - d_\theta^2 \geq 0$, *provided that the dispersion is calculated as* $d_\theta^2 = \mathbf{x}'_\theta \mathbf{E}_T^{-1} \mathbf{K}_T^{\mathrm{tr}} \mathbf{E}_T^{-1} \mathbf{x}_\theta + \sigma_T^2$.

Proof. The statement of the theorem follows from $\tilde{d}_\theta^2 - d_\theta^2 = \mathbf{x}'_\theta \mathbf{E}_T^{-1} (\tilde{\mathbf{K}}_T - \mathbf{K}_T^{\mathrm{tr}}) \mathbf{E}_T^{-1} \mathbf{x}_\theta$ and Theorem 4.9. □

Let us replace in (6.7) and (6.13) σ^2 with its estimate: when the estimation is performed under constraints, we replace σ^2 with σ_T^2, and when the estimation is performed without constraints, we replace σ^2 with $(\sigma_T^*)^2$. We obtain the following expressions for dispersions of the prediction:

with restrictions: $d_\theta^2 = \mathbf{x}'_\theta \mathbf{E}_T^{-1} \mathbf{K}_T \mathbf{E}_T^{-1} \mathbf{x}_\theta + \sigma_T^2$
without restrictions: $\tilde{d}_\theta^2 = \mathbf{x}'_\theta \mathbf{E}_T^{-1} \tilde{\mathbf{K}}_T \mathbf{E}_T^{-1} \mathbf{x}_\theta + (\sigma_T^*)^2$

Theorem 6.5. *If Assumptions 2.9, 2.11, 4.1, and 4.2 hold true, then* $p\lim_{T\to\infty}(\tilde{d}_\theta^2 - d_\theta^2) \to 0$.

Proof. From two formulas above we derive

$$\tilde{d}_\theta^2 - d_\theta^2 = \mathbf{x}'_\theta \mathbf{E}_T^{-1}(\tilde{\mathbf{K}}_T - \mathbf{K}_T)\mathbf{E}_T^{-1}\mathbf{x}_\theta + (\sigma_T^*)^2 - \sigma_T^2.$$

Note that $\tilde{\mathbf{K}}_T$ is the consistent LS estimate of the covariance. According to Theorem 4.6, the matrix $\mathbf{K}_T$ is the consistent estimate of the m.s.e. of the parameter estimate (under constraints). Therefore, the difference $\tilde{\mathbf{K}}_T - \mathbf{K}_T$ has a limit in probability as $T \to \infty$. Under Assumption 2.11, $\mathbf{E}_T^{-1}\mathbf{x}_\theta \to \mathbf{O}_n$ as $T \to \infty$, which implies that the first term in the expression for $\tilde{d}_\theta^2 - d_\theta^2$ converges in probability to zero. Taking into account that both $(\sigma_T^*)^2$ and σ_T^2 are the consistent estimates of σ^2 (see Lemma 4.10 for σ_T^2), we arrive at the statement of the Theorem. □

According to Theorem 6.5, when the sample size increases, the difference in accuracy of the prediction with and without constraints decreases.

Except the interval prediction, the results obtained in this section can be applied for samples with arbitrary size, provided that the noise is normally distributed. It is also possible to apply these results for large samples if the noise is not normally distributed, but satisfies Assumption 2.8, because in this case the random variable $\mathbf{U}$ in (6.9) is normal.

Described method of determining the confidence interval for the prediction is based on the calculation of the error function under the condition that the right-hand side in (6.10) is known, i.e., it is assumed that dispersions of σ^2 and $\boldsymbol{\alpha}^0$ are known. If these values are unknown, one should replace them with the corresponding estimates. In this case the right-hand side of inequality (6.10) depends on the estimates $\boldsymbol{\alpha}^0$ and σ^2. We consider this problem in the next section.

6.2 Interval Prediction Under Unknown Variance of the Noise

6.2.1 *Computation of the Conditional Distribution Function of the Prediction Error*

To determine the confidence interval for the dependent variable $\hat{y}_{T+\tau} = \mathbf{x}'_{T+\tau}\boldsymbol{\alpha}_T$, we find the distribution function of $\Theta_{T+\tau} = \xi_{T+\tau}/\sigma_T^*$, where $T + \tau = \theta$, $\tau > 0$ is an integer, and $(\sigma_T^*)^2$ is the estimate of the dispersion σ^2 of the noise by LS, i.e.,

$$(\sigma_T^*)^2 = \frac{\sum_{t=1}^{T}(y_t - \mathbf{x}'_t\boldsymbol{\alpha}_T^*)^2}{T-n} = \frac{\boldsymbol{\varepsilon}'_T \mathbf{a}_T \boldsymbol{\varepsilon}_T}{T-n}, \tag{6.14}$$

where $\boldsymbol{\alpha}_T^*$ is the estimate of $\boldsymbol{\alpha}^0$ by LS (without constraints); $\boldsymbol{\varepsilon}_T = [\varepsilon_1, \varepsilon_2, \ldots, \varepsilon_T]'$; $\mathbf{a}_T = \mathbf{J}_T - \mathbf{X}_T(\mathbf{X}'_T\mathbf{X}_T)^{-1}\mathbf{X}'_T$ is the idempotent matrix, whose trace is equal to $T - n$.

From (6.1) we have

$$\Theta_{T+\tau} = \frac{\xi_{T+\tau}}{\sigma_T^*} = \frac{\mathbf{x}'_{T+\tau}(\boldsymbol{\alpha}_T - \boldsymbol{\alpha}^0)}{\sigma_T^*} - \frac{\varepsilon_{T+\tau}}{\sigma_T^*}. \tag{6.15}$$

Set

$$\overline{\mathbf{P}}_{T+\tau} = \begin{bmatrix} \overline{\mathbf{P}}_1 \\ \overline{\mathbf{P}}_{2,T+\tau} \end{bmatrix} = \begin{bmatrix} \mathbf{P}^* \\ \varepsilon_{T+\tau} \end{bmatrix}, \quad \mathbf{P}^* = \begin{bmatrix} \mathbf{P}_1^* \\ \mathbf{P}_2^* \end{bmatrix}, \quad \overline{\mathbf{P}}_{2,T+\tau} = \begin{bmatrix} \mathbf{P}_2^* \\ \varepsilon_{T+\tau} \end{bmatrix}, \tag{6.16}$$

where $\overline{\mathbf{P}}_{T+\tau} \in \Re^{n+1}$, the vector $\mathbf{P}^*$ is defined in (6.5), and $\overline{\mathbf{P}}_1, \mathbf{P}_1^* \in \Re^m$, $\overline{\mathbf{P}}_1 = \mathbf{P}_1^*$, $\overline{\mathbf{P}}_{2,T+\tau} \in \Re^{n+1-m}$, $\mathbf{P}_2^* \in \Re^{n-m}$.

Denote

$$\mathbf{P}_{T+\tau} = \frac{1}{\sigma_T^*}\overline{\mathbf{P}}_{T+\tau}, \quad \mathbf{p}_T = \mathbf{P}_{1T} = \frac{1}{\sigma_T^*}\overline{\mathbf{P}}_1, \quad \mathbf{P}_{2,T+\tau} = \frac{1}{\sigma_T^*}\overline{\mathbf{P}}_{2,T+\tau}. \tag{6.17}$$

Divide the objective function of the problem (6.5) by $(\sigma_T^*)^2$, and both parts of its constraints by σ_T^*. Further, in the left-hand side of the constraint we replace $\boldsymbol{\alpha}^0$ by $\boldsymbol{\alpha}_T$. After such transformations, we come from (6.5) to the problem

$$\frac{1}{2}\mathbf{Z}'\mathbf{Z} - \mathbf{Z}''\mathbf{P}^*(\sigma_T^*)^{-1} \to \min, \quad \mathbf{N}_T\mathbf{Z} \leq (\sigma_T^*)^{-1}\overline{\mathbf{E}}_T(\mathbf{b} - \mathbf{G}\boldsymbol{\alpha}_T), \tag{6.18}$$

where $\mathbf{Z} = \mathbf{L}_T^{-1}\mathbf{E}_T(\boldsymbol{\alpha} - \boldsymbol{\alpha}^0)/\sigma_T^*$.

In order to solve (6.18) we fix for a concrete sample the right side of its constraint. Replacing in the constraints the random variables σ_T^* and $\boldsymbol{\alpha}_T$ with their estimates, we can rewrite the problem as

$$\frac{1}{2}\mathbf{Z}'\mathbf{Z} - \mathbf{Z}'\mathbf{P}^*(\sigma_T^*)^{-1} \to \min, \quad \mathbf{N}_T\mathbf{Z} \leq \hat{\mathbf{b}}_T, \quad \hat{\mathbf{b}}_T = (\hat{\sigma}_T^*)^{-1}\overline{\mathbf{E}}_T(\mathbf{b} - \mathbf{G}\boldsymbol{\alpha}_T). \tag{6.19}$$

As it was done in Sect. 6.1, we represent the solution $\mathbf{V}_T$ to the problem (6.19) in the form of two subvectors $\mathbf{V}_T = \begin{bmatrix} \mathbf{V}_{1T} \\ \mathbf{V}_{2T} \end{bmatrix}$, where $\mathbf{V}_{1T} \in \Re^m$, $\mathbf{V}_{2T} \in \Re^{n-m}$. From (6.19) and the structure of the matrix $\mathbf{N}_T$ (see the comments to (6.5)), we see that $\mathbf{V}_{2T} = (1/\sigma_T^*)\mathbf{P}_2^* \in \Re^{n-m}$, $\mathbf{V}_{1T}$ is the solution to the problem

$$\frac{1}{2}\mathbf{z}'\mathbf{z} - \mathbf{p}'_T\mathbf{z} \to \min, \quad \mathbf{B}_T\mathbf{z} \leq \hat{\mathbf{b}}_T, \tag{6.20}$$

where $\mathbf{z} \in \Re^m$, and the vector $\mathbf{p}_T \in \Re^m$ is defined in (6.17).

Further we regard the distribution $\mathbf{V}_{1T}$ as conditional, assuming that $\hat{\mathbf{b}}_T$ is fixed.

Proposition 6.1. *Suppose that Assumption 4.1 holds true. Then the vector* $\mathbf{P}_{T+\tau}$, *determined in (6.17), has* $n+1$ *dimensional Student distribution with* $q = T - n$ *degrees of freedom.*

Proof. From (6.6) and (6.16) we have

$$\overline{\mathbf{P}}_{T+\tau} = \begin{bmatrix} \mathbf{P}^* \\ \varepsilon_{T+\tau} \end{bmatrix} = \begin{bmatrix} \mathbf{L}'_T \mathbf{E}_T^{-1} \mathbf{X}'_T & \mathbf{O}_{n1} \\ \mathbf{O}_{1T} & -1 \end{bmatrix} \overline{\boldsymbol{\varepsilon}}_T, \quad \text{where } \overline{\boldsymbol{\varepsilon}}_T = \begin{bmatrix} \boldsymbol{\varepsilon}_T \\ \varepsilon_{T+\tau} \end{bmatrix}.$$

Let us show that $\overline{\mathbf{P}}_{T+\tau}$ does not depend on σ_T^2. From (6.14) we obtain $(\sigma_T^*)^2 = \overline{\boldsymbol{\varepsilon}'}_T \overline{\mathbf{a}}_T \overline{\boldsymbol{\varepsilon}}_T / T - n$, where $\overline{\mathbf{a}}_T = \begin{bmatrix} \mathbf{a}_T & \mathbf{O}_{T1} \\ \mathbf{O}_{1T} & 0 \end{bmatrix}$. Taking into account this expression for $\overline{\mathbf{a}}_T$ one can show that the product of matrices $\overline{\mathbf{L}}_T \overline{\mathbf{a}}_T = \begin{bmatrix} \mathbf{L}'_T \mathbf{E}_T^{-1} \mathbf{X}'_T & \mathbf{O}_{n1} \\ \mathbf{O}_{1T} & -1 \end{bmatrix} \overline{\mathbf{a}}_T$ is a zero matrix $\mathbf{O}_{n+1,T+1}$. Then according to Demidenko (1981, Appendix P.5), $\overline{\mathbf{P}}_{T+\tau}$ and $(T-n)(\sigma_T^*)^2$ are independent.

On the other hand, according to Assumption 4.1, components $\overline{\boldsymbol{\varepsilon}}_T$ are independent and normally distributed: $\overline{\boldsymbol{\varepsilon}}_T \sim N(\mathbf{O}_{n+1}, \sigma^2 \mathbf{J}_{n+1})$, and the trace of the matrix $\overline{\mathbf{a}}_T$ is $\operatorname{tr} \overline{\mathbf{a}}_T = T - n$. Thus, $E\{\overline{\boldsymbol{\varepsilon}'}_T \overline{\mathbf{a}}_T \overline{\boldsymbol{\varepsilon}}_T\} = (T-n)\sigma^2$. Therefore $E\{((\sigma_T^*)^2)/\sigma^2\} = 1$, which implies that $(\sigma_T^*)^2$ has χ^2 distribution with $q = T - n$ degrees of freedom. By (6.17) we have $\mathbf{P}_{T+\tau} = \overline{\mathbf{P}}_{T+\tau}/\sigma_T^*$, which implies the statement of the proposition. □

Thus, $\mathbf{P}_{T+\tau} \sim T(\mathbf{O}_{n+1}, (q/q-2)\mathbf{J}_{n+1})$, where $T(\mathbf{M}_1, \mathbf{M}_2)$ denotes multi-dimensional Student distribution with expectation $\mathbf{M}_1$ and covariance matrix $\mathbf{M}_2$. By (6.16) and (6.17), $\mathbf{P}_{T+\tau} \in \Re^{n+1}$ can be written in the form

$$\mathbf{P}_{T+\tau} = \begin{bmatrix} \mathbf{p}_T \\ p_{m+1,T} \\ p_{m+2,T} \\ \vdots \\ p_{nT} \\ p_{n+1,T+\tau} \end{bmatrix} = \begin{bmatrix} \mathbf{P}_{1T} \\ \mathbf{P}_{2,T+\tau} \end{bmatrix}, \tag{6.21}$$

where $\mathbf{p}_T = \mathbf{P}_{1T} \in \Re^m$, $p_{n+1,T+\tau} = \varepsilon_{T+\tau}/\sigma_T^*$. Moreover,

$$\mathbf{P}_{2,T+\tau} = \begin{bmatrix} p_{m+1,T} \\ \vdots \\ p_{nT} \\ \cdots \\ p_{n+1,T+\tau} \end{bmatrix} = \begin{bmatrix} \mathbf{P}_2^*/\sigma_T^* \\ \cdots \\ p_{n+1,T+\tau} \end{bmatrix}.$$

According to Zellner (1971, Chapter 12), $\mathbf{P}_{2,T+\tau}$, which is a marginal of the vector $\mathbf{P}_{T+\tau}$, has the multi-dimensional Student distribution with q degrees of freedom:

$$\mathbf{P}_{2,T+\tau} \sim T\left(\mathbf{O}_{n-1}, \frac{q}{q-2}\mathbf{J}_{n-1}\right). \tag{6.22}$$

Denote

$$\mathbf{A}_{T+\tau} = \begin{bmatrix} \mathbf{L}'_T \mathbf{E}_T^{-1} \mathbf{x}_{T+\tau} \\ -1 \end{bmatrix} = \begin{bmatrix} \mathbf{A}_{1,T+\tau} \\ \mathbf{A}_{2,T+\tau} \end{bmatrix}, \tag{6.23}$$

where $\mathbf{A}_{1,T+\tau} \in \Re^m$, $\mathbf{A}_{2,T+\tau} \in \Re^{n+1-m}$.

From (6.15) and (6.23) we have

$$\Theta_{T+\tau} = \mathbf{A}'_{1,T+\tau}\mathbf{V}_{1T} + \mathbf{A}'_{2,T+\tau}\mathbf{P}_{2,T+\tau}. \tag{6.24}$$

Set

$$\vartheta_{T+\tau} = \frac{\mathbf{A}'_{2,T+\tau}\mathbf{P}_{2,T+\tau}}{||\mathbf{A}_{2,T+\tau}||}. \tag{6.25}$$

According to (6.16), $\overline{\mathbf{P}}_{2,T+\tau} = \sigma_T^* \mathbf{P}_{2,T+\tau} \sim N(\mathbf{O}_{n+1-m}, \sigma^2 \mathbf{J}_{n+1-m})$, which implies that

$$e_{T+\tau} = \frac{\mathbf{A}'_{2,T+\tau}\overline{\mathbf{P}}_{2,T+\tau}}{||\mathbf{A}_{2,T+\tau}||} \sim N(0, \sigma^2). \tag{6.26}$$

After some transformations we arrive at

$$\vartheta_{T+\tau} = \frac{e_{T+\tau}/\sigma}{\sigma_T^*/\sigma}. \tag{6.27}$$

But $e_{T+\tau}/\sigma \sim N(0,1)$, and the random variable $(\sigma_T^*)^2/\sigma^2$ has the χ^2 distribution with $q = T - n$ degrees of freedom. Therefore we derive from (6.27) that $\vartheta_{T+\tau}$ has one-dimensional Student distribution with q degrees of freedom. From (6.24) to (6.27) we obtain

$$\Theta_{T+\tau} = \mathbf{A}'_{1,T+\tau}\mathbf{V}_{1T} + ||\mathbf{A}_{2,T+\tau}||\vartheta_{T+\tau}. \tag{6.28}$$

Suppose that we have no constraints. In this case, the solution to (6.20) is $\mathbf{V}_{1T} = \mathbf{p}_T$. According to Zellner (1971, Chap. 12), the distribution of the m-dimensional vector $\mathbf{p}_T$ is the marginal distribution of the vector $\mathbf{P}_{T+\tau} \sim T(\mathbf{O}_{n+1}, (q/q-2)\mathbf{J}_{n+1})$. Thus, $\mathbf{p}_T \sim T(\mathbf{O}_m, (q/q-2)\mathbf{J}_m)$. By (6.21), (6.25) and Assumption 6.1, ϑ is not correlated with $\mathbf{p}_T$. Thus, in this case (see Zellner 1971, Chap. 12), $\Theta_{T+\tau} = ||\mathbf{A}_{T+\tau}||\eta_{T+\tau}$, where $\eta_{T+\tau} \sim T(0, q/q-2)$. Using (6.23) we obtain $||\mathbf{A}_{T+\tau}||^2 = 1 + \mathbf{x}'_{T+\tau}\mathbf{E}_T^{-1}\mathbf{L}_T\mathbf{L}'_T\mathbf{E}_T^{-1}\mathbf{x}_{T+\tau} = 1 + \mathbf{x}'_{T+\tau}\mathbf{R}_T^{-1}\mathbf{x}_{T+\tau}$. Finally, by (6.15), the standardized prediction error $\zeta_{T+\tau}/\sigma_T^*(1 + \mathbf{x}'_{T+\tau}\mathbf{R}_T^{-1}\mathbf{x}_{T+\tau})^{1/2}$ has one-dimensional Student distribution with q degrees of freedom. We note that the denominator in this fraction is the estimate of the m.s.e. of the prediction.

Thus, in case of no constraints we obtained from (6.28) the known result used for finding a confidence interval for prediction of the dependent variable in classical regression analysis (Demidenko 1981, § 2.2):

$$\hat{y}_{T+\tau} - d_\theta t_p(q) \leq y_{T+\tau} \leq \hat{y}_{T+\tau} + d_\theta t_p(q), \tag{6.29}$$

where $d_{\theta}^2 = (\sigma_T^*)^2(1 + \mathbf{x}'_{T+\tau}\mathbf{R}_T^{-1}\mathbf{x}_{T+\tau})$ is the estimate of the prediction dispersion, $t_p(q)$ is the p-quantile of Student distribution with q degrees of freedom, defined by $P\{t > t_p(q)\} = p$.

Set

$$\mathbf{U}_{T+\tau} = -\frac{\mathbf{A}'_{1,T+\tau}}{||\mathbf{A}_{2,T+\tau}||}. \tag{6.30}$$

By (6.28), the distribution function of $\Theta_{T+\tau}$ is equal to

$$F_{T+\tau}(h) = P\{\Theta_{T+\tau} \leq h\} = P\{-\mathbf{U}'_{T+\tau}\mathbf{V}_{1T} + \vartheta_{T+\tau} \leq H\}, \tag{6.31}$$

where $H = H(h) = h/||\mathbf{A}_{2,T+\tau}||$.

Introduce the sets

$$\Phi_0 = \{\mathbf{z} : \mathbf{B}_T\mathbf{z} \leq \mathbf{b}_T, \mathbf{z} \in \Re^m\},$$
$$\Phi = \Phi(h) = \{\mathbf{Z} : z_{m+1} \leq U'_{T+\tau}\mathbf{z} + H(h) = L_{m+1,1}^{(m)}(\mathbf{z}, h), \mathbf{z} \in \Re^m\},$$
$$H(h) = \frac{h}{||\mathbf{A}_{2,T+\tau}||}, \tag{6.32}$$

and

$$\omega_0 = \{\mathbf{Z} : \mathbf{B}_T\mathbf{z} \leq \mathbf{b}_T, \mathbf{z} \in \Re^m, \mathbf{Z} \in \Re^{m+1}\},$$
$$\omega_i = \{Z : \mathbf{B}_T\mathbf{z} \leq \mathbf{b}_T, \bar{z} \in \Re^m, Z \in \Re^{m+1}\}, \quad i = \overline{1, m},$$
$$\omega_{ij} = \omega_i \cap \omega_j, \quad (i, j) \in \Psi_2, \quad |\Psi_2| = C_m^2,$$
$$\omega_{ijk} = \omega_i \cap \omega_j \cap \omega_k, \quad (i, j, k) \in \Psi_3, \quad |\Psi_3| = C_m^3, \ldots,$$
$$\omega_{12\ldots m} = \{\mathbf{Z} : \mathbf{B}_T\mathbf{z} \leq \mathbf{b}, \mathbf{z} \in \Re^m, \mathbf{Z} \in \Re^{m+1}\}, \tag{6.33}$$

where $\mathbf{z} = [z_1 \ \ldots \ z_m]'$, $\mathbf{Z} = [\mathbf{z} \vdots z_{m+1}]'$, $|\Psi_i|$, $i=2, 3, \ldots$, is the number of elements in Ψ_i, and C_m^i, $i = 2, 3, \ldots$ is the number of combinations from m elements in i.

In (6.32) and further the indexes of the linear function $L_{ij}(\cdot)$ denote the following: i is the number of the variable, which is determined by this function, $i \leq m + 1$; j is the number of the function (the numeration is done inside the variable), the upper index m is determined by the number of constraints in the estimation problem.

Set

$$\overline{\omega}_i = \omega_i \cap \Phi \cap \Phi_0, i = \overline{1, m}; \overline{\omega}_{ij} = \omega_{ij} \cap \Phi \cap \Phi_0, (i, j) \in \Psi_2; \overline{\omega}_{ijk}$$
$$= \omega_{ijk} \cap \Phi \cap \Phi_0, (i, j, k) \in \Psi_3; \ldots,$$

According to (6.31), the distribution function of $\Theta_{T+\tau}$ can be written in this notation as follows:

$$F_{T+\tau}(h) = \sum_{i=0}^{m} \int_{z \in \overline{\omega}_i} \phi_i(z) dz + \sum_{(i,j) \in \Psi_2} \int_{z \in \overline{\omega}_{ij}} \phi_{ij}(z) dz + \sum_{(i,j,k) \in \Psi_3} \int_{z \in \overline{\omega}_{ijk}} \phi_{ijk}(z) dz + \cdots + \int_{z \in \overline{\omega}_{12\ldots m}} \phi_{12\ldots m}(z) dz, \quad (6.34)$$

where $\phi_i(z), i = \overline{0,m}; \phi_{ij}(z), (i,j) \in \Psi_2; \phi_{ijk}(z), (i,j,k) \in \Psi_3; \ldots, \phi_{12\ldots m}(z)$ are, respectively, the densities of probability distributions supported in the sets $\omega_i, i = \overline{0,m}; \overline{\omega}_{ij}, (i,j) \in \Psi_2; \overline{\omega}_{ijk}, (i,j,k) \in \Psi_3; \ldots; \overline{\omega}_{12\ldots m}$.

Making orthogonal transformations of the space of regression parameters, described in Sect. 4.6.2, one can calculate $F_{T+\tau}(h)$ using (6.34) for $m \leq 3$ and arbitrary dimension n of the regression parameter.

6.2.2 Calculation of Confidence Intervals for Prediction

Suppose that ν is the (assigned) probability with which the confidence interval covers $y_{T+\tau}$, i.e., $\nu = P\{l_\Theta(\nu_1) \leq \Theta_{T+\tau} \leq l_\Theta(\nu_2)\}$, where $\Theta_{T+\tau}$ is determined in (6.31), $l_\Theta(\nu_1)$ and $l_\Theta(\nu_2)$ are the quantiles of orders, respectively, ν_1 and ν_2, related to $\Theta_{T+\tau}$, and let $\nu = \nu_1 - \nu_2$. Assume that the probabilities ν_1 and ν_2 (and, consequently, the quantiles $l_\Theta(\nu_1)$ and $l_\Theta(\nu_2)$) are unknown.

Taking into account (6.15), the confidence interval for $y_{T+\tau}$ is of the form

$$y^*_{T+\tau} + \sigma^*_T l_\Theta(\nu_1) \leq y_{T+\tau} \leq y^*_{T+\tau} + \sigma^*_T l_\Theta(\nu_2). \quad (6.35)$$

In the particular case when there are no constraints, we obtain from (6.35) and (6.29)

$$l_\Theta(\nu_1) = -t_p(q)(1+\mathbf{x}'_{T+\tau}\mathbf{R}_T^{-1}\mathbf{x}_{T+\tau})^{1/2}, \quad l_\Theta(\nu_2) = t_p(q)(1+\mathbf{x}'_{T+\tau}\mathbf{R}_T^{-1}\mathbf{x}_{T+\tau})^{1/2}.$$

Moreover, we get $\nu_1 = p/2, \nu_2 = 1 - (p/2), \nu = 1 - p$.

In the general case one can find the values $l_{1\Theta}(\nu_1)$ and $l_{2\Theta}(\nu_2)$ in (6.35) by solving the problem

$$\begin{cases} l_\Theta(\nu_2) - l_\Theta(\nu_1) \to \min, \\ F_{T+\tau}(l_\Theta(\nu_2)) - F_{T+\tau}(l_\Theta(\nu_1)) = \nu, \\ l_\Theta(\nu_2) \geq l_\Theta(\nu_1), \end{cases}$$

where the distribution function $F_{T+\tau}(h)$ of $\Theta_{T+\tau}$ is given by (6.31).

The above algorithms are implemented (using Mathcad) when the number of constraints m is 1 or 2. Calculations for the real data and computer simulations show that in case of inequality constraints the prediction error is smaller and the confidence interval is more accurate (see Korkhin and Minakova 2009).

Bibliographic Remarks

Chapter 1

In this chapter, we describe methods of solving optimization problems in frames of regression analysis. In Sect. 1.1 we use the results from Lawson and Hanson (1974). Section 1.2 is based on the papers Korkhin (1978, 1999). The method described in this section is based on the method of linearization (Pshenichnyi 1983).

Chapter 2

The results of Sect. 2.1 concerning general approaches of finding consistent estimates in nonlinear non-parametric regression models are presented in the papers (Dorogovtsev and Knopov 1976, 1977; Knopov 1997a–c, 1998; Knopov and Kasitskaya 1995, 1999). These papers are based on the fundamental results of the theory of asymptotical estimation proved in Le Cam (1953), Huber (1867), Jennrich (1969), Pfanzagl (1969), and Dorogovtsev (1982). The results stated in Sects. 2.2–2.4 on the limit distribution of a multi-dimension regression parameter with nonlinear constrains which define a convex admissible domain, are described in Korkhin (1985). In Sect. 2.3, this method of obtaining limit distributions was adopted to the case of non-convex constrains, and in Sect. 2.4, to the case of constraint-inequalities. Section 2.5 is based on the results proved in Korkhin (2009). Limit distribution of the estimate of the non-linear regression parameter under another assumptions on the regression function and on constrains was obtained in Wang (1996).

P.S. Knopov and A.S. Korkhin, *Regression Analysis Under A Priori Parameter Restrictions*, Springer Optimization and Its Applications 54,
DOI 10.1007/978-1-4614-0574-0,

Chapter 3

Section 3.1 is based on the results proved in Knopov (1997a–c, 1998) and Knopov and Kasitskaya (1995, 1999, 2002). The general approach for investigation of consistency of estimates is presented in Hampel et al. (1986), Huber (1967, 1981), Chiang Tse-pei (1959), Dorogovtsev (1975, 1976, 1982, 1992), Jennrich (1969), Pfanzagl (1969), Dzhaparidze and Sieders (1987), Ibramhalilov and Skorokhod (1980), Ivanov and Leonenko (1989), Knopov (1980), Grenander (1950), Heble (1961), Ibragimov and Has'minskii (1981), Ibragimov and Linnik (1971), Wald (1949), Walker (1973), Prakasa Rao (1987), Yadrenko (1980), Zwanzig (1997) and others. Section 3.2 is devoted to consistency conditions for regression models for long memory systems. This section is based on the results proved in Ivanov and Leonenko (2001, 2002) and Moldavskaya (2002). In Sect. 3.3, we present some results concerning the method of empirical means in stochastic optimization problems. We quote some results from Dupacova and Wets (1986) and Norkin (1989, 1992). Section 3.4 is devoted to asymptotic distributions of estimates obtained by the method of empirical means for systems with weak or strong dependence. This section is based on Ivanov and Leonenko (2001, 2002), Knopov (1976a–c, 1981, 1997a), Knopov and Kasitskaya (1989a, b, 1999, 2002) and Moldavskaya (2007). In Sect. 3.5, we present some results on large deviations for estimates obtained by the method of empirical means for, in general, independent variables. Presented results are taken from Knopov and Kasitskaya (2004, 2005, 2010).

Chapter 4

Sections 4.2–4.3 are devoted to the construction of the truncated estimate of the matrix of m.s.e. of the estimate of multi-dimensional regression parameter. In such a construction inactive constraints are not taken into account. Another approach (which takes into account all constraints) is considered in Sects. 4.4–4.7. This chapter is based on the results obtained in Korkhin (1994, 1998, 2002a, 2005, 2006a, b, 2009).

Chapter 5

Asymptotic properties of iterative estimates of regression parameters in nonlinear regression with and without constraints are investigated, in Sects. 5.1 and 5.2, respectively. The results of this chapter are based on Korkhin (2002b, c).

Chapter 6

Section 6.1 is devoted to interval prediction under known distribution function of the prediction error. In Sect. 6.2 interval prediction is constructed in the case when the noise dispersion is unknown. The results of this chapter rely on Korkhin (2002a–d).

References

Ajvazyan, S.A., Rozanov, Yu.A. Asymptotically effective estimates of regression coefficients in the case of dependent errors. Theory Probab. Appl., 8, 216 (1963).
Anderson, T.W. The Statistical Analysis of Time Series, Wiley, New York (1971).
Attouch, H. Variational Convergence for Functions and Operators. Research Notes in Mathematics, Pitman, London (1984).
Bard, Y. Nonlinear Parameter Estimation. Academic Press, New York (1974).
Chiang Tse-pei. On the estimation of regression coefficients of a continuous parameter time series wits a stationary residual. Theory Prob. Appl., 4, 405–423 (1959).
Danford, J.-D, Schwartz, J. Linear Operators, PART 1. General Theory, Interscience Publishers, New York (1957).
Demidenko, E.Z. Linear and nonlinear regression. Financy and statistica, Nauka, Moscow (1981) (in Russian).
Demidenko, E.Z. Optimization and Regression. Nauka: Moscow (1989) (in Russian).
Deutschel, J.D., Strook, D.W. Large Deviations. Academic Press, Boston (1989).
Dobrushin, R.L., Mayor, P. Non-central theorems for non-linear functionals of gaussian fields. Z.Wahrsch.verw.Gebiete, 50, 52–72 (1972).
Donoghue, J.F. Distributions and Fourier Transforms. Academic Press, New York (1969).
Dorogovtsev, A.Ya. On Limit Behavior of One Estimate of the Regression Function. Teor. Veroyatnost. Matem. Statist., 13, 38–46 (1975) (in Russian).
Dorogovtsev, A.Ya. On One Statement Useful for estimates Consistency Proving. Teor. Veroyatnost. Matem. Statist 14, 34–41 (1976) (in Russian).
Dorogovtsev, A.Ya. The Theory of Estimation of Random Processes Parameters. Vishcha Shkola. Kiev (1982) (in Russian).
Dorogovtsev, A.Ya. Consistency of Least Squares Estimator of an Infinite Dimensional Parameter. Siber Math. J., 33, 65–69 (1992) (in Russian).
Dorogovtsev, A.Ya., Knopov, P.S. On Consistency of Estimates of the Continuous in the Domain Function by Observations of Its Values in a Finite Set of Points with Random Errors. Dokl. Akad. Nauk Ukrain. S.S.R.A, 12, 1065–1069 (1976) (in Russian).
Dorogovtsev, A.Ya., Knopov, P.S. Asymptotic Properties of One Nonparametric Function of Two Variables. Teor. Sluch. Processov, 5, 27–35 (1977) (in Russian).
Draper, N., Smith, H. Applied Regression Analysis. 3rd Edition, Wiley, New York (1998).
Dupacova, J. Experience in stochastic programmings models. Servey Math. Programming Proc. of the 9th Intern. Math. Programming Symp.. Budapest: Akademia Kiado, 99–105 (1979).
Dupacova, J., Wets, R. Asymptotic behavior of statistical estimators and optimal solutions for stochastic optimization problems. WP-86–41, IIASA, Laxenburg (1986).

P.S. Knopov and A.S. Korkhin, *Regression Analysis Under A Priori Parameter Restrictions,* Springer Optimization and Its Applications 54,
DOI 10.1007/978-1-4614-0574-0, © Springer Science+Business Media, LLC 2012

Dzhaparidze, K., Sieders, A. A Large Deviation result for Parameter Estimators and Its Applications to Non-linear Regression Analysis. The Ann. Statist., 15, 1031–1049 (1987).
Ermoliev, Yu. M., Knopov, P.S. Method of empirical means in stochastic programming problems. Cybern. Syst. Anal. 6, 773–785 (2006).
Ermoliev, Yu. M., Norkin, V.I. Methods for solution of nonconvex nonsmooth stochastic optimization problems. Cybern. Syst. Anal., 5, 60–81 (2003).
Ermoliev, Yu. M., Wets, R, (eds.) Techniques for Stochastic Optimization. Springer, Berlin (1988).
Ermol'eva, T. Yu., Knopov, P.S. Asymptotic properties of the estimate of an average stverage loss as foundation of certain actuarial tariffings. Publ. Inst. Stat. Univ. Paris 31 (2–3), 67–71 (1986).
Girko, V.L. Theory of Random Determinants, Vishcha Shkola, Kiev (1980) (in Russian).
Grenander, U. Stochastic Processes and Statistical Inference. Almqvist & Wiksells Boktryckeri AB. Stockholm, (1950).
Gross, J. Linear Regressions. Springer-Verlag, New York (2003).
Hampel, F.R., Ronchetti, E.M., Rousseeuw, P.J., Stahel, W.J.: Robust Statistics: Approach Based on Influence Functions. Wiley, New York (1986).
Hannan, E.J. Multiple Time Series, Wiley, New York (1976).
Heble, M.P. A Regression Problem Concerning Stationary Processes. Trans. Amer. Math. Soc., 99(2), 350–371 (1961).
Holevo, A.S. On Asymptotic Normality of Estimates of Regression Coefficients. Theory Prob. Appl., 16(4), 724–728 (1971).
Huber, P.J. The Behavior of Maximum Likelihood Estimates under Nonstandard Conditions. Proc. 5th Berkeley Symp. on Mathematical Statistics and Probability. I, University of California Press: Berkeley, 221–234 (1967).
Huber, P.J. Robust Statistics, Wiley, New York (1981).
Ibragimov, I.A., Has'minskii, R.Z. Statistical Estimation. Asymptotic Theory. Springer New York (1981).
Ibragimov, I.A., Linnik, Yu.V. Independent and Stationary Sequences of Random Variables. Nauka, Moscow, Edited by JFC, Wolters-Noordhoff Series of Monographs and Textbooks on Pure and Applied Mathematics (1971).
Ibramhalilov, I.S., Skorokhod, A.V. Consistency Estimators of Random Processes Parameters. Naukova Dumka: Kiev (1980) (in Russian).
Ivanov, A.V. On Consistency and Asymptotic Normality of Least Moduli Estimator. Ukrain. Math. J., 36, 267–272 (1984a) (in Russian).
Ivanov, A.V. Two theorems on consistency of a least squares estimator. Prob Theory Math. Statist., 28, 25–34 (1984b).
Ivanov, A.V. Asymptotic Theory of Nonlinear Regression. Kluwer Academic Publishers, Dordrecht (1997).
Ivanov, A.V. and Leonenko N.N. Statistical Analysis of Random Fields. Kluwer Academic Publishers: Dordrecht (1989).
Ivanov, A.V., Leonenko, N.N. Asymptotic inferences for a nonlinear Regression with a long-range dependence. Prob Theory Math. Statist., 63, 65–85 (2001).
Ivanov, A.V., Leonenko, N.N. Asymptotic behavior of M-estimators in continuous-time non-linear regression with long-range dependent errors. Random Oper. and Stoch. Equ., 10, (3), 201–222 (2002).
Jennrich, R.I. Asymptotic Properties of Non-Linear Least Squares Estimators. Ann. Math. Statist., 40, 633–643 (1969).
Johnston, J. Econometric Methods. McGraw Hill, New York (1963).
Kaniovskii, Yu. M., King, A., Wets, R.J.-B. Probabilistic bounds (via large deviations) for the solutions of stochastic programming problems. Annals of Oper. res. 56, 189–208 (1995).
Kankova, V. Optimum solution of a stochastic optimization problem with unknown parameters. Trans. 7th Prague Conf. Prague: Akademia, 239–244 (1979).
Karmanov, V.G. Mathematical Programming. Nauka, Moscow (1975) (in Russian).

Kasitskaya, E.J., Knopov, P.S. Asymptotic Behavior of Empirical Estimates in Stochastic Programming Problems. Dokl. Akad. Nauk S.S.S.R., 315(2), 279–281 (1990) (in Russian).

Kasitskaya, E.J., Knopov, P.S. On Convergence of Empirical Estimates in Stochastic Optimization Problems. Cybernetics 27(2), 297–303 (1991a).

Kasitskaya, E.J., Knopov, P.S. About One Approach to the Nonlinear Estimating Problem. Proc. 6th USSR-Japan Symposium on Probability Theory and Mathematical Statistics, Kiev, World Scientific: Singapore, 151–157 (1991b).

Khenkin, M.Z., Volinsky, E.I., The searching algorithm of solution of general problem of the mathematical programming. Journal of calculus mathematics and mathematical physics, 1, 61–71 (1976) (in Russian).

King, A. Asymptotic Behavior of Solutions in Stochastic Optimization: Nonsmooth Analysis and the Derivation of Non-Normal Limit Distributions. Dissertation, University of Washington (1986).

King, A. Asymptotic Distribution for Solutions in Stochastic Optimization and Generalized M-Estimator. WP-88, IIASA: Laxenburg, Austria (1988).

King, A., Wets R.J.-B. Epi-consistency of convex stochastic programs. WP-88–57, IIASA, Laxenburg, Austria (1988).

Knopov, P.S. On Some Estimates of Nonlinear Parameters for the Stochastic Field Regression. Prob Theory Math. Stat., 14, 67–74 (1976a).

Knopov, P.S. On Asymptotic Properties of Some Nonlinear Regression Estimates. Prob Theory Math. Stat., 15, 73–82 (1976b).

Knopov, P.S. On Asymptotic Behavior of Periodogram Estimates of Parameters in Nonlinear Regression Models. Dokl. Akad. Nauk Ukrain. S.S.R.A, 11, 942–945 (1979c) (in Russian).

Knopov, P.S. On Some Problems of Nonparametric Estimation of Stochastic Fields. Dokl. Akad. Nauk Ukrain. S.S.R.A, 9, 79–82 (1980) (in Russian).

Knopov, P.S. Optimal Estimators of Parameters of Stochastic Systems. Naukova Dumka: Kiev (1981) (in Russian).

Knopov, P.S. Asymptotic Properties of One Class of M-Estimates. Cybern. Syst. Anal., 33, (4), 468–481 (1997a).

Knopov, P.S. On a Nonstationary Model of M - Estimators with Discrete Time. Prob Theory Math. Stat., 57, 60–66 (1997b).

Knopov, P.S. On some classes of M-estimates for non-stationary regression models. Theory of Stochastic Processes. 222–227 (1997c).

Knopov, P.S. Estimates of Parameters for Nonidentically Distributed Random Variables, Prob Theory Math. Stat., 58, 38–44 (1998).

Knopov, P.S., Kasitskaya, E.J. On Asymptotic Behavior of Nonlinear Parameters of Random Functions. Proc. 4th International Conference on Probability Theory and Mathematical Statistics, Vilnus, 65–70 (1989a).

Knopov, P.S., Kasitskaya, E.J. Asymptotic Properties of Least Squares Estimates of Gaussian Regression for Random Fields. Cybernetics 25(5), 641–648 (1989b).

Knopov, P.S., Kasitskaya, E.J. Properties of Empirical Estimates in Stochastic Optimization and Identification Problems. Ann. Oper. Res., 56, 225–239 (1995).

Knopov, P.S., Kasitskaya, E.J. Consistency of Least Squares Estimates for Parameters of the Gaussian Regression Model. Cybern. Syst. Anal. 35(1), 19–25 (1999).

Knopov, P.S., Kasitskaya, E.J. Empirical Estimates in stochastic optimization and identification. Kluwer Academic Publishers, New York (2002).

Knopov, P.S., Kasitskaya, E.I. Large deviations of empirical estimates in stochastic programming problems. Cybern. Syst. Anal. 40 (4), 510–516 (2004).

Knopov, P.S., Kasitskaya, E.I. On the convergence of empirical estimates in problems of stochastic programming for processes with discrete time. Cybern. Syst. Anal. 41(1), 144–147 (2005).

Knopov, P.S., Kasitskaya, E.I. On large deviations of empirical estimates in a stochastic programming problem with time-dependent observations. Cybern. Syst. Anal. 46 (5), 724–728 (2010).

Korkhin, A.S.: An algorithm of non-linear regression analysis with a priori constraints on regression coefficients. Èkon. Mat. Metody 14 (2), 345–356 (1978) (in Russian).

Korkhin, A.S. Certain properties of the estimates of the regression parameters under a priori constraint-inequalities. Cybernetics 21(6), 858–870 (1985).

Korkhin, A.S. Calculation of statistic properties of regression parameter estimates with a priori constraints-inequalities. Ehkon. Mat. Metody 30(2), 146–163 (1994) (in Russian).

Korkhin, A.S. Parameter estimation accuracy for nonlinear regression with nonlinear constraints. Cybern. Syst. Anal. 34(5), 663–672 (1998).

Korkhin, A.S. Solution of Problems of the Nonlinear Least-Squares Method with Nonlinear Constraints. Journal of Automatization and Informatic Sciences, 6, 110–120 (1999).

Korkhin, A.S. Estimation accuracy of linear regression parameters with regard to inequality constraints based on a truncated matrix of mean square errors of parameter estimates. Cybern. Syst. Anal. 38 (6), 900–903 (2002a).

Korkhin, A.S. Some asymptotic properties of estimations of nonlinear regression parameters, calculated by a iterative method. Mathematical modeling, 1, 44–48 (2002b) (in Russian).

Korkhin, A.S. About consistency of estimation a nonlinear regression parameter which is calculated by a recurrent method. Transactions of National Mining University, 5, 69–72 (2002c).

Korkhin, A.S. Point and interval forecasts on a regression model estimated with allowance for inequality constrains. Theory of Stochastic Processes, 1–2, 209–212 (2002d).

Korkhin, A.S. Determining sample characteristics and their asymptotic linear regression properties estimated using inequality constraints. Cybern. Syst. Anal. 41(3), 445–456 (2005).

Korkhin, A.S. Estimating the matrix of root-mean-square errors of estimates of linear regression parameters for an arbitrary number of regressors and three inequality constraints. Cybern. Syst. Anal., 42 (3), 342–356 (2006a).

Korkhin, A.S. Linear regression and two disparity-inequalities. Èkon. Mat. Metody, 42(1), 119–127 (2006b) (in Russian).

Korkhin, A.S. Linear regression with nonstationary variables and constraints on its parameters. Cybern. Syst. Anal., 45(3), 373–386 (2009).

Korkhin, A.S., Minakova, O.P. Computer statistics. Dnepropetrovsk, National Mining University, Part II (2009).

Koroljuk, V.S., Portenko, N.I., Skorokhod, A.V., Turbin, A.F. Handbook of Probability Theory and Mathematical Statistics. Nauka, Moscow (1985) (in Russian).

Kukush, A.G. Asymptotic Properties of the Estimator of a Nonlinear Regression Infinite Dimensional Parameter. Math. Today, 5, 84–105 (1989) (in Russian).

Lawson, C.L., Hanson, R.J: Solving Least Squares Problems. Prentice-Hall, Inc., Englewood Cliffs (1974).

Le Cam, L. On some asymptotic properties of maximum likelihood estimates and related Bayes estimates. Univ. California, Publ. Statist, 1, 277–330 (1953).

Liese, F., Vajda, I. Consistence of M-Estimates in general Regression Models, Journal of Multivariate Analysis. 50, 93–114 (1994).

Liew, C.K. Inequality Constrained Least – Squares Estimation. Journal of the American Statistical Association, 71(355), 746 – 751 (1976).

Ljung, L. System Identification: Theory for the User. Prentice-Hall, Englewood Cliffs (1987).

Loeve, M. Probability Theory. Van Nostrand: Princeton University Press, Princeton, NJ (1963).

Lutkepohl, H. Non-linear least squares estimation under non-linear equality constraints. Economics Letters, 2–3., 191–196 (1983).

Malinvaud, E. Methods of Statistic in Econometric. Dunod, Paris (1969).

Major, P. Wiener-Ito integrals. Lecture Notes Math., 849. Springer-Verlag, Berlin (1981).

Maindonald, J.H., Statistical Computation. John Wiley & Sons, New York (1984).

McDonald, G.C. Constrained regression estimates of technology effects on fuel economy. Journal of Quality Technology, 31 (2), 235–245 (1999).

Moldavskaya, E.M. Theorem useful in proving of estimates consistency in long-memory regression models. Bulletin of the University of Kiev. Series: Physics, Mathematics. 1.58–65 (2002).

Moldavskaya, E.M. Asymptotic distributions of the least square estimator of linear regression coefficients with non-linear constraints and strong dependence. Prob Theory Math. Stat., 75, 121–137 (2007).

Nagaraj, N.K., Fuller, W.A. Estimation of the parameters of linear time series models subject to nonlinear restrictions. The Annals of Statistics, 19(3), 1143–1154 (1991).

Nemirovskii, A.S., Poljak, B.T., Tsybakov, A.V. Signal processing by the nonparametric maximum-likelihood method. Probl. Inf. Transm. 20(3), 177–192 (1984).

Norkin, V.I. Stability of stochastic optimization models and statistical methods of stochastic programming. Preprint 89–53, Glushkov Institute of Cybernetics, Kiev, 1989 (In Russian).

Norkin, V.I. Convergence of the empirical mean method in statistics and stochastic programming. Cybern. Syst. Anal. 28, No. 2; 253–264 (1992).

Pfanzagl, J. On the Measurability and Consistency of Minimum Contrast Estimates. Metrika, 14, 249–272, (1969).

Pfanzagl, J., Wefelmeyer, W. Asymptotic Expansions for General Statistical Models, Lecture Notes in Statistics, 31, Springer-Verlag, Berlin (1985).

Pflug, G. Optimization of Stochastic Models. Kluwer Academic Publishers, Dordrecht (1996).

Poljak, B. Introduction to Optimization, Nauka: Moscow. (1983) (in Russian).

Prakasa Rao, B.L.S. Asymptotic Theory of Statistical Inference. Wiley, New York (1987).

Pshenichnyi, B.N. Linearization Method. Nauka, Moscow (1983) (in Russian)

Rao, C.R. Linear Statistical Inference and Its Applications. Wiley, New York (1965).

Rezk, G. Inequality Restrictions in Regression Analysis. Journal of Development Economics. 18, 110–129 (1996).

Salinetti, G., Wets, R.J-B. On the Convergence in Distribution of Measurable Multifunctions (Random Sets). Normal Integrands, Stochastic Processes and Stochastic Infima. Math. Oper. Res., 11, 385–419, (1986).

Schmetterer, L. Introduction to Mathematical Statistics, Springer.Verlag: Berlin, (1974).

Shapiro, A. Asymptotic Properties of Statistical Estimators in Stochastic Programming. Ann. Statist., 17, 841–858, (1989).

Shapiro, A. Asymptotic Analysis of Stochastic Programs. Ann. Oper. Res., 30, 169–186 (1991).

Shapiro, A., Dentcheva, D., Ruszczyski, A. Lectures on Stochastic Programming: Modeling and Theory. SIAM, Philadelphia, (2009).

Taqqu, M.S. Convergence of iterated processes of arbitrary Hermite rank. Z. Wahrsch. Verw.Gebiete., 50, 53–58 (1979).

Thomson, M. Some results on the statistical properties of an inequality constraints least squares estimator in a linear model with two regressors. Journal of Econometrics. 19. 215–231 (1982).

Thomson, M, Schmidt P. A note on the comparison of the mean square error of inequality constrained least squares and other related estimators. Review of Economics and Statistics, 64, 174–176 (1982).

Tikhonov, A.N., Arsenin, V.Ja. Methods of solution of well-posed problem. Nauka, Moscow (1979) (in Russian).

Vapnik, V.N. Estimation of Dependence Based on Empirical Data. Springer, NY, (1982).

Vapnik, V.N. Weak convergence and empirical processes with application statistics. Springer Series in Statistics. New York, NY, Springer. xvi, 9(1996).

Van de Geer, S. The Method of sieves and minimum contrast estimators. Allerton Press, Inc.m, 4(1), 20–38 (1995).

Yadrenko, M.I. The Spectral Theory of Random Fields. Vishcha Shkola, Kiev (1980) (in Russian).

Yubi, E. Statistical investigation of stochastic programming problems and a method for their solution. Izv. Akad. Nauk Ehst. SSR, Fiz. Mat. 26, 369–375 (1977) (in Russian).

Wald, A. Note on the Consistency of the Maximum Likelihood Estimate. Ann. Math. Statist., 20(2), 595–601 (1949).

Walker, A.M. On the Estimation of a Harmonic Component in a Time Series with Stationary Dependent Residuals. Adv. Appl. Prob., 5, 217–241 (1973).

Wang, J. Asymptotics of least-square estimator for constrained nonlinear regression. The Annals of Statistics. 24 (3), 1316–1326 (1996).

Watson, G.N. A Treatise of Theory of Bessel Functions. University Press, Cambridge (1944).

Wets, R. A statistical approach to the solution of stochastic programs with (convex0 simple recourse. Working paper. Univ. of Kentuckey, Lexington (1979).

Wets, R.J-B. Stochastic Programming: Solution Techniques and Approximation Schemes Mathematical Programming. The State of the Art, Springer-Verlag, Berlin, 566–603 (1983).

Wilks, S.S. Mathematical Statistics. Wiley, New York (1962).

Zellner, A. An Introduction to Bayesian Inference in Econometrics, Wiley & Sons, New York (1971).

Zwanzig, S. On L1-norm estimators in nonlinear regression and in nonlinear error-in variables models, in "L1-Statistical Procedures and Related Topics". IMS Lecture Notes- Monographs Series, 31, 101–118 (1997).

Index

P.S. Knopov and A.S. Korkhin, *Regression Analysis Under A Priori Parameter Restrictions,* Springer Optimization and Its Applications 54,
DOI 10.1007/978-1-4614-0574-0, © Springer Science+Business Media, LLC 2012

L

M

N

O

P

Q

R

S

T

U

V

Printed by Books on Demand, Germany